LabVIEW
Applications
and Solutions

ISBN 0-13-096423-9

9 780130 964236

90000

 **NATIONAL INSTRUMENTS** | **VIRTUAL INSTRUMENTATION SERIES**

Lisa K. Wells • Jeffrey Travis
■ LabView For Everyone

Mahesh L. Chugani • Abhay R. Samant • Michael Cerna
■ LabVIEW Signal Processing

Barry E. Paton
■ Sensors, Transducers, & LabVIEW

Rahman Jamal • Herbert Pichlik
■ LabVIEW Applications and Solutions

LabVIEW™
Applications
and Solutions

▲ Rahman Jamal
▲ Herbert Pichlik

Prentice Hall PTR, Upper Saddle River, NJ 07458
http://www.phptr.com

Library of Congress Cataloging-in-Publication Data

Jamal, Rahman.
    LabVIEW applications and solutions/Rahman Jamal, Herbert Pichlik.
        p.   cm. — (National instruments virtual instrumentation series)
    Includes bibliographical references and index.
    LabVIEW.
    ISBN 0-13-096423-9
    1. Scientific apparatus and instruments — Computer simulation.
    2. Computer graphics. I. Pichlik, Herbert. II. Title. III. Series.
    Q185.J26 1998   98-34131
    006—dc21   CIP

Editorial/Production Supervision: Jane Bonnell   Manufacturing Manager: Alan Fischer
Acquisitions Editor: Bernard M. Goodwin   Cover Design: Talar Agasyan
Editorial Assistant: Diane Spina   Cover Design Director: Jerry Votta
Translator: Angelika Shafir   Series Design: Gail Cocker-Bogusz
Marketing Manager: Kaylie Smith

© 1999 by Markt&Technik Buch- und Software Verlag GmbH
85540 Haar bei München/Germany
Original title: LabVIEW: Programmiersprache der vierten Generation by R. Jamal & H. Pichlik

Published by Prentice Hall PTR
Prentice-Hall, Inc.
A Simon & Schuster Company
Upper Saddle River, NJ 07458

Prentice Hall books are widely used by corporations and government agencies for training,
marketing, and resale.
The publisher offers discounts on this book when ordered in bulk quantities. For more information, contact:
Corporate Sales Department, Phone: 800-382-3419; FAX: 201-236-7141;
E-mail: corpsales@prenhall.com
Or write: Prentice Hall PTR, Corporate Sales Dept., One Lake Street, Upper Saddle River, NJ 07458

Quotation on p. 31 reprinted from *Wholeness and the Implicate Order* by David Bohm by permission of
Routledge, a division of Routledge, Chapman & Hall, Inc. Quotation on p. 193 from *Neural Networks and
Fuzzy Systems* by Bart Kosko, © 1991; reprinted by permission of Prentice-Hall, Inc., Upper Saddle River, NJ.

Printed in the United States of America

10 9 8 7 6 5 4 3 2 1

ISBN 0-13-096423-9

Prentice-Hall International (UK) Limited, London
Prentice-Hall of Australia Pty. Limited, Sydney
Prentice-Hall Canada Inc., Toronto
Prentice-Hall Hispanoamericana, S.A., Mexico
Prentice-Hall of India Private Limited, New Delhi
Prentice-Hall of Japan, Inc., Tokyo
Simon & Schuster Asia Pte. Ltd., Singapore
Editora Prentice-Hall do Brasil, Ltda., Rio de Janeiro

To Mimi
    — Rahman

To Karin, Sabine, Melissa, Tobias, and Benjamin
    — Herbert

Contents

▼ **2**

LabVIEW Basics 15

▼ **3**

Data Acquisition and Instrument Control in LabVIEW 31

▼ **4**

Commercial Communication Applications in LabVIEW 119

▼5

LabVIEW and Automation Technology   169

▼6

LabVIEW and Fuzzy Logic   193

▼7

LabVIEW and Genetic Algorithms   211

▼8

Mathematics and Simulation in LabVIEW   221

▼ 9

Fourier Transformation in LabVIEW   229

▼ 10

Time-Frequency Analysis of Signals   285

▼ 11

Designing Digital Filters in LabVIEW   305

▼ 12

Image Processing in LabVIEW   315

▼ 13

Quality, Reliability, and Maintainability of LabVIEW Programs   325

▼ 14

Statistical Process Control in LabVIEW   337

▼ 15

LabVIEW and Quality Management   345

▼ 16

LabVIEW in Medical Applications   357

▼ 17

BioBench   373

▼ 18

Control and Simulation in G — An Integrated Environment for Dynamic Systems   379

▼ 19

Network-centric Test and Measurement System   405

Preface

Measurement and automation technology is experiencing a phase of radical change. After a brief Windows 95 euphoria, the trend is clearly pointing toward Windows NT, as far as operating systems are concerned. With regard to hardware, ISA/EISA bus systems are increasingly replaced by the superior PCI bus. Formerly stiff boundaries between measurement and automation technologies blur, clearly because the flexible possibilities of data acquisition, analysis, control, and visualization merge more and more to form a software environment entirely in the sense of a holistic approach. We no longer distinguish between classical measurement technology software and process visualization tools. As a global player in the area of measurement and automation technology, National Instruments drives this trend to "unification" under the slogan "the software is the instrument." The company introduced the concept of virtual instruments with its graphical LabVIEW (Laboratory Virtual Instrument Engineering Workbench) programming system in the eighties. This concept represented the first leap from conventional measuring devices to adaptable virtual measurement and automation systems.

LabVIEW, originally developed for the measurement and automation technology, has been advancing more and more as an alternative to conventional programming languages.

Along with the C/C++ programming languages, LabVIEW is among the most frequently used programming languages for technical and scientific applications today.

LabVIEW is a revolutionary paradigm that can be used to solve technical and commercial problems. The implicitly parallel and modular development system with its graphical G compiler is impressive in its offering of high runtime speeds and very short program development cycles. A number of extensive libraries built in LabVIEW allow rapid prototyping as well as elegant and powerful application development. In addition to measurement and control engineering, process visualization, laboratory automation, and image processing, the fields of application of this platform-independent development software extend to communications technology, statistics, mathematics, simulation, and commercial data processing.

This holistic approach has been widely adopted by users and manufacturers alike to build applications based on industry standards, standard computers, and standard operating systems. The number of organizations that discover and use the efficiency of scalable applications based on virtual instruments increases continually. Annual user symposia are organized worldwide under the motto "virtual instruments in practice," where users exchange hands-on experience, concepts, and results in applying virtual instruments. Today, many consider this approach to be the only way to achieve cost and time savings and, consequently, a short time to market. Particularly in the age of the global Internet, this new system generation also gains increasing importance with regard to distributed, modular applications. It opens new horizons in the sense of an open and communicative measurement and automation technology.

Objectives of This Book

This book provides an insight into the capabilities of LabVIEW, describing graphical flow programming with LabVIEW, illustrated by many examples and practical applications. It lets you dive into a totally new programming world using "virtual instruments." Information technology problems are reduced to a graphical formulation of solutions.

Coding and documenting happen in the background. Applications from almost all areas of graphical flow programming and relevant background information impressively demonstrate the range of potential applications and the power, quality, and reliability of LabVIEW applications.

This book represents crucial reading for instructors, scientists, students, hardware and software developers, and decision-makers in research, academia, and industry. Professional LabVIEW developers, novices to the field, and those curious look beyond conventional paradigms are provided with an extensive overview on supported hardware, fields of applications, and interfaces to other hardware and software systems. In addition to documenting the current state of the art in virtual instrumentation, this book suggests potential future LabVIEW uses.

Organization of This Book

This book is divided into three main parts: introduction; communication technologies and mechanisms; and analysis and evaluation methods and application examples. Each part is divided into chapters, and the overall structure reflects the following sequence of basic questions:

- What are virtual instruments?

- What is LabVIEW?

- What requirements does LabVIEW place on the underlying hardware?

- What can we do with LabVIEW today?

- What are the benefits of using LabVIEW?

- What is the outlook of LabVIEW?

The structure of the book follows a top-down approach. However, the text enters, at each step, into rather more detail than a strictly logical organization would require. We hope that the resulting redundancies are in practice beneficial to the overall understanding, providing explanations given in different contexts and from different perspectives.

Acknowledgments

Many people deserve credit for their contributions, their direct help, fruitful technical discussions, or their contributions in the form of basic material and application examples. In particular, we wish to thank the following contributors (in alphabetical order): Gerd Bauer, Peter Herrmann, Dr. Christian Nef, Martin Studtfeld, and Andreas Zimmer.

We are particularly grateful to Prof. Dr. Norbert Stockhausen, who enriched this book with valuable examples and suggestions in the field of digital signal processing, and Prof. Dipl.-Ing. Norbert Dahmen who, with his sound fuzzy knowledge, helped fuzzy logic attain its solid position in the world of graphical data flow programming, next to the classical control engineering. We also extend our gratitude to Dr. Lothar Wenzel for his valuable contribution in the field of control and simulation.

We thank all those LabVIEW users who have supplied new ideas for this project by asking technical questions and providing comments. We are grateful to the Technical Support Department of National Instruments Germany GmbH, particularly Heinrich Illig, Matthias Vogel, and Georg Sinkovic; and Michael Dams, Manager of NI Germany.

Our special thanks go to the LabVIEW developers in Austin, Texas, in particular to Brian Powell, Steve Rogers, Dean Luick, Greg Fowler, Greg McKaskle, David Beisner, Ray Almgren, Tamra Kerns, and Ravi Marawar.

Last but not least, we wish to warmly thank LabVIEW coinventor Jeff Kodosky for his invaluable encouragement and comments in numerous discussions.

Finally, this text would probably not have been finished without the continuing encouragement from our families: Karin, Sabine, Tobias, Benjamin, and Melissa Pichlik; and Farida, Munira, Hamida, and Sadrudin Jamal.

Munich, Germany, May 1998
Rahman Jamal and Herbert Pichlik

About the Authors

Following his electrical engineering studies at the University of Paderborn, Germany, Rahman Jamal joined National Instruments, Austin, Texas, in 1990, as application engineer. Six months later, he moved to National Instruments Germany GmbH, where he played an important role in establishing this subsidiary. In 1993, he assumed the position of Application Engineering Manager. Since 1997, he has been Technical Manager, responsible for applications, training, and strategic marketing at NI Germany. Rahman Jamal was born in 1965; he is enthusiastically dedicated to graphical data flow programming. In his spare time, he is primarily concerned with interdisciplinary themes that bridge the gap between science, music, art, and literature. He has written several books and is the author of over 100 national and international papers and articles. Rahman Jamal, who is well known from his many lectures as a true LabVIEW expert, deals mainly with the philosophic and cognitive aspects of optimum man-machine interfaces, in addition to the purely pragmatic aspects of graphical paradigms.

Herbert Pichlik was born in 1958; he studied electrical engineering at the Georg-Simon-Ohm University of Nuremberg. He started his professional career in 1985 when he joined Philips Kommunikations Industrie AG (PKI) as a software development engineer. Later, he moved to the quality management department at PKI. After a short period at LGA, he joined Quelle AG in 1990, where he has been in charge of measuring and test instrument management as well as test instrument development. Herbert Pichlik has written and coauthored several books and dozens of papers and articles. Since 1992, when he assumed responsibility for a large number of different projects, he has worked intensively with LabVIEW. Herbert Pichlik is an internationally awarded synergist, enthusiastic squash player, father of four children, and owner of several patents in the field of analog and digital integrated circuit technologies; he started lecturing in graphical data flow programming at the Nuremberg University as a sideline in 1997.

Virtual Instruments

The limits of my language mean the limits of my world.

— Ludwig Wittgenstein
Tractatus Logico-Philosophicus

The evolution of programming paradigms was marked by the transition from machine language to assembly languages, and from these to functional, procedural, and, eventually, object-oriented, high-level languages. During the past ten years, a totally new paradigm evolved along these lines, offering problem-solving approaches in a new dimension — graphical data flow programming in the form of so-called "virtual instruments." This chapter covers the evolution of textual and graphical programming, the basics of virtual instrumentation, and various virtual instruments embedded in LabVIEW.

1.1   Introduction

The pace of innovations in the development of hardware has remained on a constantly high level during the past centuries. The "law" of Intel collaborator Moore (doubling of the integration density of semiconductor components approximately every two years) appears to be valid beyond the turn of the

1

millennium. While dynamic memories had approximately 1K to 4K storage cells at the end of the seventies, development patterns with a storage capacity of one gigabit have already been introduced by several semiconductor manufacturers. It can be expected that series production of these highly complex components will start soon.

The increase of the integration density and clock frequency, the use of novel processor architectures (Superscalar Pipeline, Harvard, speculative execution units, etc.), the development of increasingly sophisticated memory and bus systems, as well as the massive use of paralleling structures led to increasingly powerful hardware systems. Software development does not keep up with this pace. The gap between the innovation speed in hardware and software developments is growing continually.

In addition to a lengthy engineering tradition in the hardware area, early introduction and widely proliferated standardization (technology, casing, etc.) have a positive effect on the innovation climate because new developments can greatly profit from existing approaches.

The history of software development has shown a large number of historically founded developments, partly relying on the organization of the underlying hardware and its performance capability and partly depending on the paradigms used. These historical developments do not build a suitable basis for a healthy software evolution. The focus of the programming paradigms must shift from concentrating on the syntax, operations, and primitive data types to the recognition that the general way of doing things is much more than the syntax and grammar of a language.

1.2   Programming Paradigms

Different programming paradigms emphasize different aspects of problem solving. Each programming paradigm has its own strengths and weaknesses. However, almost all current paradigms build on cryptic, text-oriented programming grammars, which are mastered only by experienced application developers. The following section provides a brief overview of the evolution of textual and graphical languages from a user's perspective. Subsequently, the term "virtual instrument" is introduced in a global context. Finally, the inherent significance of this term is explained in the LabVIEW context.

❏ 1.2.1   Textual Programming Languages

Textual programming languages evolved from attempts to represent algorithms in an exact and readable manner. Understandably, these first textual descriptions relied heavily on a combination of natural languages and abstract mathematical formalisms. As a consequence, they inherited a complex syntax, influenced not only by the simple character-oriented input and output capabilities of early hardware and execution environments available for running algorithms but also from their natural language ancestry.

These historically founded developments have led to far-reaching, unfortunate consequences. First of all, the possibilities these programs provide for the representation of algorithms refer more closely to the computing hardware than the perceptual or cognitive capabilities of the programmer. For instance, the components of early text-based languages resulted from attempts to describe the functionality of hardware elements, such as registers, accumulators, etc. Variables were employed to represent addresses, statements such as `if-then` or `goto` served to symbolize conditional and unconditional branches.

Second, because the means of representation or description of such programs is text, the resulting representations are inherently linear and sequential. These unfortunate properties unnecessarily force programmers to think and express their ideas in terms constrained by the machine and enforce restrictive structures on the programs and algorithms. These limitations may be acceptable for single-processor systems but become a major hindrance in multiprocessor systems, in particular with regard to distributed and parallel computing architectures.

Another extensively discussed shortcoming of textual descriptions is the syntactical complexity resulting partly from its natural language foundation. In contrast to their natural language counterparts, however, the complex command syntax and the overall cryptic grammar of textual languages force the programmer to deal with minor irrelevant syntactic details rather than to focus on his actual application. Moreover, the cryptic nature of the syntax makes it extremely difficult to interpret the semantic relationships among program entities. The lack of a real correspondence between the physical objects involved in an application and the abstract textual code lead to a high cognitive load of programming.

❑ 1.2.2   Graphical Programming Languages

The visually oriented human mind can capture and grasp complex inter-relations much more easily when they are depicted in a pictorial form. Thus, it comes as no surprise that the use of pictorials has been a firmly established medium of expression in the fields of science and education as well as in many other areas. In fact, visualization has traditionally played a key role in the development of programs for digital computers. For instance, diagrammatic methods, such as flow charts, structure diagrams, Petri nets, and Nassi-Schneidermann diagrams, have historically been em-ployed as a heuristic aid to specify algorithms, data structures, and other interdependencies.

In contrast to textual representations with their sequential, linear ordering principles (e.g., causality or historical process), visual descriptions enable us to perceive those very crucial connections that manifest them-selves as loop structures, dynamic interactions, communication networks, concurrent processes, and so on. Perhaps our inability to reflect in terms of visual networks is to some extent due to our limited descriptive system of textual language.

"The purpose of computing is insight, not numbers." This well-known statement made by Hamming in 1966 reflects the importance of visual representation of data. Large amounts of data are generally produced in many applications as results of data acquisition, numerical intensive computing, etc. A deeper insight can be gained using visualization elements or tools that reveal characteristics not necessarily visible in a purely numerical form.

With regard to such reflections and arguments, it becomes clear that visual representations or models help to make abstract concepts and algorithms more tangible, allowing users to express their ideas in a more comprehensive manner. A method of communicating with a machine that takes advantage of these principles would offer the ability to develop programs to those who lack the time, education, or skills.

In the quest for a visual language that can be used effectively by a broad range of people with different programming skill levels, Jeff Kodosky, cofounder and vice president of R&D of National Instruments, introduced at the beginning of the eighties a programming system based on a dataflow model extended with graphical control-flow structures: the LabVIEW (Laboratory Virtual Instrument Engineering Workbench) development en-vironment with its embedded G programming language. As the metaphor of virtual instruments used in LabVIEW suggests, the development of LabVIEW was motivated by the laboratory automation area, whereby LabVIEW programs are regarded as a hierarchy of instrument-like modules, called *virtual instruments*, composed of execution-time user interfaces (front panels) with design-time visual programming (block diagrams). This melding of development and execution can be considered as one of the major benefits of LabVIEW's graphical programming environment, where the edit/compile/link/run sequence of traditional programming is replaced by the draw/run cycle.

LabVIEW has been in use since October 1986 and is employed in a wide variety of industries, such as automated testing, industrial automation, laboratory automation, automotive engineering, personal instrumentation, etc., to build virtual instrumentation systems. LabVIEW has gained much popularity over the past years, primarily because many scientists and engineers have experienced improvements in programming efficiency due to the natural understandability of LabVIEW's graphical programming tools. Among the many well-known advantages of LabVIEW, such as ease of use, natural representation, rapid prototyping, and code reusability, we consider the possibility of graphically building large real-world applications, especially by nonexperts, as being one of the most important. The design engineer can rapidly prototype, implement, and modify a system using a single graphical language without the aid of traditional computer scientists. And because graphical programs are often similar to the standard flowcharts, to which many engineers are accustomed, LabVIEW has demonstrated real capability of reducing software development time.

1.3   Virtual Instrumentation

Virtual instruments have become the catchword in measurement technology. What does the "test engineer" understand by this term? When attempting to answer this question, almost everyone ties a different idea or concept to this catchword. For many, it represents a control instrument based on standard personal computers to store, evaluate, and represent test data. According to this notion, data is acquired through special measuring devices attached to a personal computer over a serial or parallel cable. Other people think that it means a computer equipped with application and driver software and a built-in transmitter as a sort of low-cost alternative to relatively expensive stand-alone measuring devices. Both ideas are correct, but only up to a certain point. They cover only part of this concept. Before discussing the exact definition, we will describe the principal types of computer-assisted test data acquisition.

Test data can be acquired in a computer in different ways. It is important to first understand the underlying architecture of a measuring device. A traditional measuring device always consists of three components, shown in Figure 1-1, which perform the following tasks:

- Acquire the measurement parameters (data acquisition)

- Adapt and process the measured signal (analysis)

- Output the measured value (presentation)

Such measuring devices generally have fixed functions defined by the manufacturer, and they are characterized by a manufacturer-specific architecture and an inflexible user interface. Consequently, they cannot be adapted easily to changing needs. Both the operation and the documentation are entirely manual. To add long measuring sequences that require constant changes of settings, a large amount of time is used just to set the measuring devices and to document the measured values.

☐ 1.3.1   PC-Assisted Measuring Technology with Stand-Alone Measuring Devices

The situation in measuring-device technology began to change with the advent of personal computers. Over the course of time, many different execution forms of PC-assisted measuring technology have emerged. The simplest way of using the personal computer in measurement applications consists in the control of measuring devices through a serial interface (RS-232) or a parallel interface, particularly developed for measurement technology (IEEE-488). With the IEEE-488 measuring device bus (also called HP-IB, IEC 625, or GPIB), a manufacturer-independent interface was created in the 1970s, which allows connection of measuring and control devices from different manufacturers and their remote control over a standard PC. This form of measurement data acquisition is undoubtedly the most popular, offering the

Figure 1-1

Structure of a measuring device

largest choice of devices. At the end of the 1980s, so-called VXI systems entered the powerful high-end market for measuring technological applications. VXI systems follow the "instrument on a card" concept. This means that measuring devices are plugged into a mainframe computer. These measuring devices no longer dispose of operating and display elements, so that a controlling computer is essential.

❏ 1.3.2   Computer-Assisted Measurement Technology with Multifunction Cards

In addition to the IEEE-488, VXI, and RS-232/485, which are standardized interfaces allowing communication between personal computers and stand-alone measuring devices, many measurement and control applications use data acquisition cards built directly into the computer. These cards are usually designed as so-called multifunction cards (*DAQ = Data AcQuisition*). That is, you can use them both to measure and output analog parameters and to accept or create digital signals. This form of measuring data acquisition is becoming increasingly popular. The reasons are obvious. In addition to the favorable price, they offer an unparalleled flexibility that enables you to handle a number of tasks in measurement and control technology — including environments where applications change — with one single piece of equipment, provided only that powerful software tools are used.

❑ 1.3.3   Computer-Assisted Measurement Technology with PC-Based Instruments

Today, there is basic consensus in the measurement technology industry that the classical laboratory measuring devices cannot be replaced by PC plug-in cards over the long term. Instead, these cards are increasingly used as an addition wherever this appears meaningful from the user's view. Despite some performance improvements achieved in conventional measurement technologies, users attach great importance to user-specific adaptation capabilities, particularly in view of open-standard personal computers. Modern oscilloscopes, multimeters, and functional generators have to perform many different analytical functions, they have to comply with the programming environments that can be used to create measuring protocols, and they have to be capable of connecting to the Internet. Using the advantages of a personal computer under Windows NT and the PCI bus and thus combining computer power with adaptability, we obtain a new world of measurement technology, now commonly described as PC-based instruments, which specifically meets these demands.

❑ 1.3.4   Field Buses in Measurement and Control Applications

As field bus systems proliferate, the desire to use these bus systems to also acquire test measurement data increases. Field buses connect field devices, sensors, actors, etc., to superior stations, such as process computers or programmable logic controllers (PLCs). A large number of different field bus systems are available from different vendors. However, notice that three field bus systems — CAN (Controller Area Network), Interbus-S, and Profibus-DP — currently hold the largest market share in Europe, and their popularity in measurement technology applications rises.

Another type of field bus, which is increasingly gaining popularity, is the Foundation Fieldbus — a key, enabling open system technology based on an international standard. Foundation Fieldbus provides a new paradigm for building scalable, open control systems. This new technology will provide true interoperability and digital integration of smart field and control devices.

☐ 1.3.5 Image Processing in Measurement and Control Applications

In practice, the acquisition of data in "number column form" is normally adequate in most applications to document problems or operating states in measurement and control applications. It is often necessary to acquire and process image data. Quality assurance and production are not the only sectors to demand a combination of optical measuring values and analog measuring parameters. The fields of use of image processing are manifold. The most important are the presentation and evaluation of thermograms and the preparation of visual data for pattern recognition.

☐ 1.3.6 Motion Control

Motion control and machine vision are often used in automation applications that require precise inspection and positioning. Other applications include part placement, alignment, microscopy, and inspection.

1.4 Definition — Virtual Instruments

Based on this background information, we are now able to define the term "virtual instrument" in a more accurate way. We speak about a virtual instrument when we create measuring systems composed of a standard personal computer, suitable software, and appropriate measuring hardware tailored to the measuring task, which is normally available only in specifically designed stand-alone measuring devices. Virtual instruments represent a visualization and centralization of complex measurement systems on a standard personal computer in the form of a virtual user interface. The user sees a uniform, comprehensive single system, i.e., a complete application, consisting of many individual measuring components. This fundamental concept is the quantum leap from the conventional measuring device over computer-assisted measurement technologies to adaptable virtual measuring systems. This represents a shift from manufacturer-defined measuring devices to user-defined measuring systems. The main benefits of this concept are:

- A virtual instrument can contain any combination of industry-standard hardware to acquire or output data: IEEE-488.2, RS-232 devices, VXI/MXI systems, field buses (CAN, Interbus-S, Profibus, Foundation Fieldbus, etc.), multifunction plug-in cards, DAQ instruments, image processing components, external black-box systems, or motion control. Figure 1-2 illustrates the hardware architecture.

- The capabilities to analyze and represent measured data reach far beyond the boundaries of conventional measurement technology.

- A powerful software development environment and a set of hardware components allow creation of a number of virtual instruments to cover a wide range of test functions and applications.

The spectrum of virtual instruments used in practice includes classical laboratory automation, process visualization and control, automotive and aviation industries, medical applications, manufacturing industry, and research and science.

1.5   Software Aspects

The previous sections discussed the hardware of measuring data acquisition applications, and this section concentrates on software for computer-assisted acquisition of measuring data. Software forms the key to powerful virtual instruments because the typical functions of measuring devices are defined only by the software. Currently, there are two different paradigms in the area of measurement and automation technology for the programming of a virtual instrument: The first is a text-based approach, where the control program is defined through a standard programming language. The second is a dataflow-oriented programming methodology, where the control program is

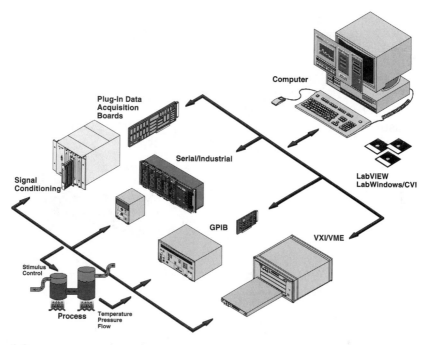

Figure 1-2

Hardware architecture of virtual instruments

implemented through a graphical block diagram. The difference between the two is not so much the implementation of the user interface, but rather the creation of the control program. Figure 1-3 illustrates the architecture.

❑ 1.5.1   Graphical Programming

Graphical programming languages follow the innovative approach of simplifying program development by using graphical elements so that experienced programmers and engineers as well as scientists with little experience in traditional programming can create complex applications on their own. In this context, the graphical representation of a control program is not to be understood as a mere illustration; instead, it represents the executable program. In 1986, National Instruments developed a fully featured graphical programming language, LabVIEW, on the basis of the data

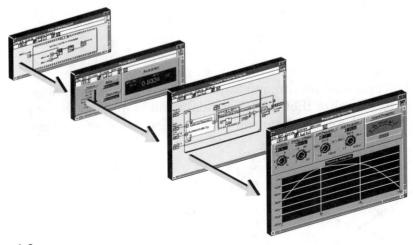

Figure 1-3

Software architecture of virtual instruments

flow concept. In a LabVIEW program, the structure of a hierarchy of virtual instruments is reflected in the form of software modules, the individual components of which are interactive surfaces (front panels), controlled by data flow diagrams (block diagrams). Today, LabVIEW is in its fifth generation; it is platform independent, and its use extends beyond the classical measurement and test technology as an universal graphical programming language. This popularity is based mainly on outstanding capabilities, such as quick implementation, testing, and maintenance of prototypes by means of graphical programming. These properties offer great advantages mainly to engineers and scientists who have little experience in conventional programming, but who have to or want to implement complex applications themselves.

❏ 1.5.2   Virtual Instruments and Standardization

VISA (Virtual Instrumentation Software Architecture) represents a milestone in the history of I/O driver software standardization. More than 60 measurement technology manufacturers participated in the standardization process. VISA is the first concept to allow individual system components from different manufacturers to interoperate. VISA represents a uniform, manufacturer-independent groundwork for the development and integration of software components, e.g., device drivers, front panels, and application software that is independent of measuring device types, buses, operating systems, programming languages, and network mechanisms. Although VISA was specified by the *VXI Plug&Play alliance*, the standard is not limited to the VXI bus. It is based on a holistic approach, and it also contains guidelines for future bus systems and communication technologies. At present, work is in progress to formalize IEEE 1226.5 standardization. In view of the properties of LabVIEW, these development environments were selected as standard tools for the VISA standard.

❑ 1.5.3   Virtual Instruments and the Internet

Networks are advancing at incredible speed. Catchwords like *connectivity*, *data highway*, and *worldwide networking* characterize this trend. New communication mechanisms, such as the Internet and the World Wide Web (WWW), have become an integral part of the engineer's working environment. These advanced technologies allow development of applications that had been almost unthinkable or impractical in the past. Particularly, the WWW allows the user to access any data in any place of the world by a mouse click, without knowing the details of the underlying network structure. Which new possibilities are offered by the Internet for virtual instruments? There are many application cases for which networked virtual instrumentation would be the ideal solution. We will use two examples to explain.

A classical application for test engineers is the nonlocal data monitoring of an experiment running in a test laboratory. Of course, this use requires a connection between the test computer and the central computer. A commonplace example would be the monitoring of a decentralized test bed or control equipment through an Internet browser. The test bed or the control equipment could, in turn, consist of many smaller nonlocal, autonomously working virtual instruments in the form of measuring and test stations.

1.6   Summary

The LabVIEW graphical programming language has been under continuous development for almost ten years. It has had a decisive impact on the concept of virtual instrumentation, a philosophy that has established itself well as a useful standard in many areas of measurement and control applications. Virtual instrumentation means far more than graphical user interfaces or a collection of icons promising to be user friendly. It is a philosophy that achieves a breakthrough with regard to productivity and efficiency and where problem solutions are not defined by the manufacturer, but by the user. Particularly in the age of the global Internet, this new system generation increasingly gains importance with regard to distributed, modular applications. It opens entirely new horizons.

2

LabVIEW Basics

There are indeed things that cannot be put into words. They make themselves manifest. They are what is mystical.

— Ludwig Wittgenstein
Tractatus Logico-Philosophicus

LabVIEW uses terminology, icons, and ideas familiar to scientists and engineers. It relies on graphical symbols rather than textual language to describe programming actions. The principle of data flow, in which functions execute only after receiving the necessary data, governs execution in a straightforward manner. You can use LabVIEW even if you have little or no programming experience, but you will find knowledge of programming fundamentals very helpful. This chapter covers the basics of LabVIEW.

2.1   Introduction

In LabVIEW, a program is called *VI* (*Virtual Instrument*) because it models the appearance and function of a physical instrument. This metaphor reflects the origins of LabVIEW in measurement applications, as mentioned earlier. This explains why the software is structured in *front panels* and *block diagrams*.

Analogous to the front panels of measuring instruments, which feature operating and display elements and serve as the actual interface between the user and the device, in LabVIEW the front panel is the user interface of a program and allows the user to interact with the program and to visualize the output from applications. The functionality of an instrument is determined by electronic and electric components, interconnected in a suitable manner by electric circuits. Similarly, a block diagram is composed of graphical symbols, connected through software-based data lines in the form of wires.

The block diagram forms the actual program code — a graphical source code that is "drawn" as a schematic diagram in the sense of a signal flow representation, in contrast to the conventional textual source code. The components of this block diagram can be simple mathematical operators or user-defined subVIs (similar to subroutines). These components point to the third component of a VI, the *icon/connector*, which serves as an interface of a subVI to the calling program.

To understand the components of a VI, consider the example in Figure 2-1.

The example shown in Figure 2-1 simulates a real transmission path, consisting of a sender/receiver. The block diagram shows that the transmit signal is exposed to noise during the transmission, so it has to be filtered at the receiving end to restore the pure transmit signal. This small virtual instrument is more than ten years old. In fact, it is one of the first LabVIEW example VIs, but it has lost nothing of its illustrative character and clarity.

2.2   Program Structures

Structured programming is an established programming methodology, and its benefits are well known. A program structure consists of a *body*, which separates the structure from the actual program, and a *program part* within the structure, which is subject to certain conditions (iterative or conditional execution). Particular attention is paid to the separation of the structure's body from the remaining program part. In a conventional language like C, the fundamental property of a structure is created by placing a "text part" within brackets. Access to the code inside the body is limited in order to linearly pass data through the structure's header. In a graphical environment the ob-

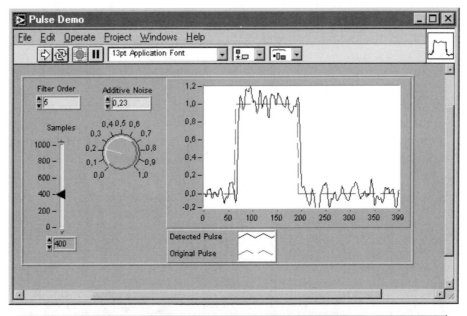

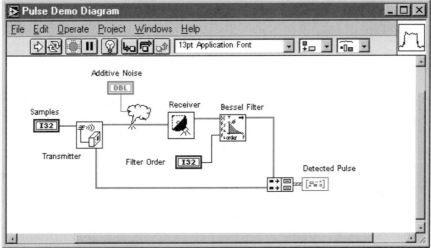

Figure 2-1
Example of a VI (front panel, block diagram, connector)

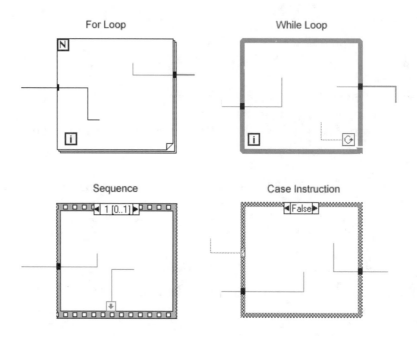

Figure 2-2
Diagram structures

vious method of separating body and program is a graphical one, that is, a kind of frame around parts of the block diagram. This true two-dimensional representation allows access to the structure (frame) from all sides, improving both the flexibility and interpretability of LabVIEW.

Delimiting a program part by a frame as the basis of a structure is one aspect, and defining the behavior of the delimited part is another, as shown in Figure 2-2. LabVIEW structures differ in their graphical characteristic in the block diagram. In addition to the well-known structures, while and for loops and case statements, the graphical representation contains a *sequence structure*, which forces program parts to be executed sequentially.

The while loop is a conditional loop; it iterates over the subprograms it contains until it hits a break condition. It contains two special *terminals*: *i* is a data source, which is incremented by the integral value 1 on each iteration, starting from zero. The second *terminal*, serving to control the loop, is a data sink for a Boolean value that, depending on its condition (true/false), decides whether the subprogram is to be executed once more. So-called *tunnels* ensure that data is supplied to or output from the loop body.

The for loop is similar to the while loop, except that it ends after the nth cycle. This means that, in this structure, the number of iterations is defined in advance by the structure's N terminal. This serves as data sink for integral values between 1 and n. Here, too, data is transmitted through tunnels.

The case structure consists of two or more subprograms. You can think of them as a type of stack of superimposed block diagrams. A selector terminal determines the card and block diagram to be executed. This selector is a data sink for Boolean values (two choices to select from) or integral positive values (n choices to select from).

The sequential structure is similar to a case structure, except it has no selector. The subdiagrams it contains are handled sequentially. To calculate data $x(i)$ from data $x(i - n)$ in repetitive structures (for and while loops), the *shift register* concept is introduced, where $i,n \in N+$. During an iteration, shift registers provide n data, which had been generated by one or more previous iterations. From a heuristic view, the shift register is linked directly to the body of the structure itself (see Figure 2-3). At this point, it is worth looking at the implementation of the classical Newtonian method of approximation implemented in LabVIEW. As is generally known, this recursive method is formulated as follows:

$$x_{i+1} = x_i - f(x_i)/f'(x_i)$$

The constant on the left puts the initial value on top of the shift register. Each iteration creates a better approximation, replacing the previous value in the shift register. The while loop iterates until the absolute value of the difference of consecutive approximations is bigger than the desired value. Once the loop ends, the shift register passes the last approximation to the display on the right of the loop.

2.3   Hierarchy and Modularity

One main characteristic of structured programming is the meaningful use of subprograms with their well-known advantages, such as good readability, modularity, easy maintenance, etc. It is necessary to have a well-defined abstraction mechanism with regard to passing data between modules to handle

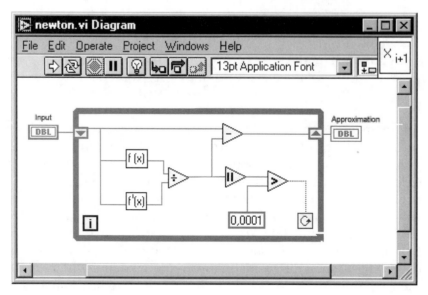

Figure 2-3
Newtonian approximation method

this type of software module. LabVIEW meets this requirement by providing a block diagram and an icon/connector. Such icons can be freely defined by the user. LabVIEW provides connections (*terminals*), which are allocated to the data sources and data sinks of the diagram. LabVIEW has always offered the possibility of dividing a program part into partial functions, i.e., subprograms (*subVIs*), since its first release at the beginning of the 1980s. This concept has been followed subsequently and expanded during the course of years. The `create subVI from selection` function allows you to abstract any part of a block diagram as a subVI, as shown in Figure 2-4. The subVI created in this way is connected automatically, and the panel elements created in this process are labeled at the same time.

A subVI compares well with an integrated circuit, which can also be switched through various inputs and outputs. You can access any such module from the icon of a higher program as often as needed.

LabVIEW does not limit this module and hierarchy creation in any way, so other subprograms can be used within subprograms (see Figure 2-5). LabVIEW does not impose any limit to the nesting depth. Its hierarchical windows provide an overview of the hierarchical VI structure. These windows are an ideal tool for finding your way around in complex program structures.

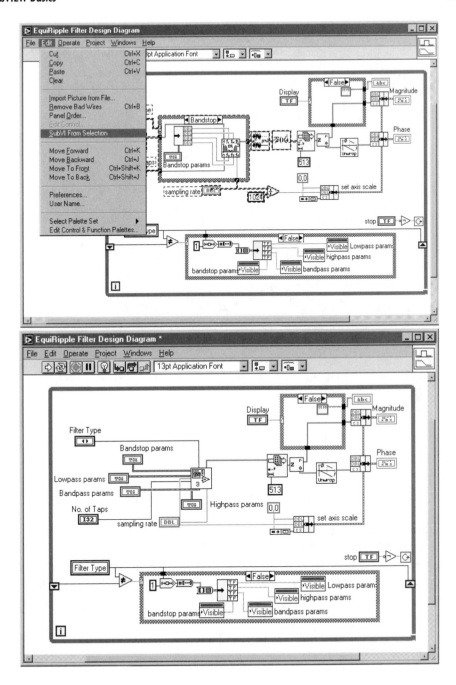

Figure 2-4
SubVI from selection

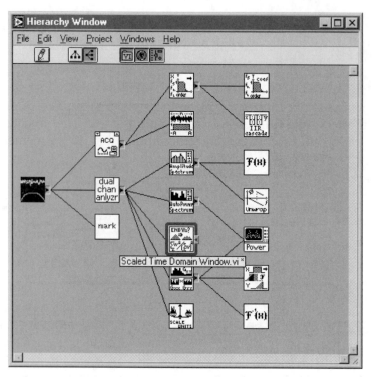

Figure 2-5
Hierarchy of LabVIEW programs

2.4   Multitasking, Multithreading, and Multiprocessing

The data flow engine embedded in LabVIEW provides parallelism and multi-tasking constructs. In contrast to conventional textual programming languages, which work sequentially and let you write parallel code only with complicated programming effort (e.g., by using semaphores), LabVIEW allows you to create concurring, individually running, and well-synchronized program parts. In the simplest case, one or several existing virtual instruments are placed into a diagram to allow concurrent execution of VIs in one application. The reason is that LabVIEW is driven purely by data flow, as opposed to textual paradigms, which build on control flow mechanisms. The data flow approach is much more elegant, and a natural way of formulating solutions to problems. The strengths of the data flow approach become evident particularly in today's massively parallel structures (within individual processors and by using multiprocessor systems) and powerful computer architectures. Finally, it is possible to bring hardware and software innovations to an equal level. This possibility offers invaluable benefits. Instead of having to take care of irrelevant parallel mechanisms specific to a given implementation, LabVIEW programmers can formulate problems and realize solutions in the full sense of rapid prototyping. The inventors of this graphical data flow paradigm laid the foundation to implement multitasking/multithreading/multiprocessing structures in the pioneering era of LabVIEW.

The terms multitasking, multithreading, and multiprocessing are different technologies often used interchangeably. However, a lot of confusion also exists about what these technologies offer and the advantages they provide. The ability of an operating system to switch between tasks is referred to as *multitasking*. This mechanism creates the impression that all the execution applications are progressing concurrently.

Multithreading extends the concept of multitasking, giving applications the ability to designate different tasks to individual threads executed in parallel. The operating system can divide the processing time on these threads, similar to the way it divides processing time among entire applications in a multitasking system.

The term *multiprocessing* refers to multiple processors in one single computing domain. Whereas a single processor can take some advantage of multithreading, multithreaded programs with parallel executing threads can take full advantage of any number of processors within a multiprocessing machine. Figures 2-6 and 2-7 depict the distinguishing factors of a single-threaded in contrast to a multithreaded environment.

In LabVIEW, multithreading is built into the environment, including management of all threads. In particular, LabVIEW's programming environment lends itself naturally to building subsystems essential to multithreaded applications. The syntax of LabVIEW agrees with the logic of multithreading and provides the flexibility to set priorities and choose custom configurations for DAQ and instrument control applications.

As computer technology constantly evolves, developers can not only take advantage of multithreading in LabVIEW to improve their system performance and reliability today, but also can lay the foundation to port to multiprocessor systems of tomorrow. In the DAQ section, we will see the concrete advantages of the different concepts introduced in this section in real-world DAQ applications.

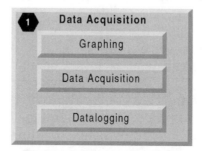

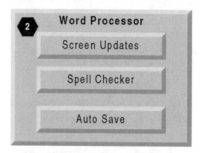

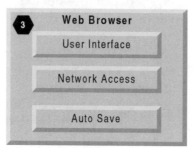

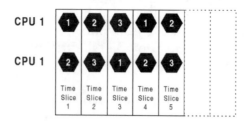

Figure 2-6
Parallelism with equal execution priority

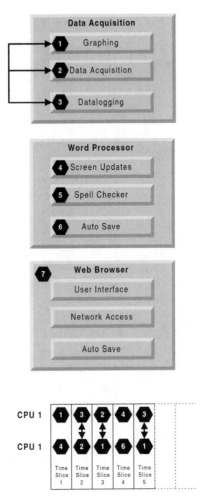

Figure 2-7
Parallelism with different execution priorities

2.5   Debugging

Powerful error detection functions are indispensable for a graphical programming language. LabVIEW offers a wealth of comfortable debugging functions (see Figure 2-8), such as `breakpoints`, `probes`, and `step-in/over/out` functions, for subprograms that greatly facilitate debugging of complex applications.

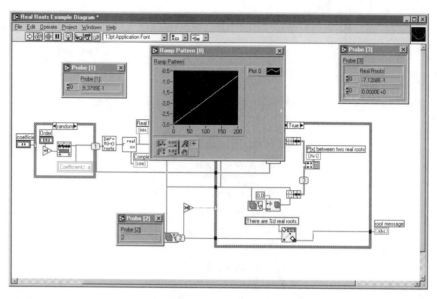

Figure 2-8
Debugging tools in LabVIEW

Breakpoints are used to interrupt the program execution at any point if you need to gain insight into the program state. Probes can be used to view actual data, generated during run time. Step-in/over/out functions can be used to skip or exit subprograms or to jump within subVIs.

An excellent way to follow the program flow visually is offered by the highlighting function. This is an excellent tool to analyze the data flow by means of wandering points (bubbles) and individual momentary values.

In addition, the profiler can be used to analyze the memory utilization, the frequency of subprogram calls, and the runtime behavior of subVIs. The profiler is particularly helpful to find *hot spots* (points in the program that handle an extremely high work load) by analyzing the tabular subVI statistics (mean, minimum, and maximum values of memory utilization, subVI visits, i.e., number of sequences). Figure 2-9 shows a profiler snapshot (a snapshot of a LabVIEW application, including subVIs).

As applications grow and involve more developers, the management of multiple versions of the same application can become very difficult. The graphical differencing in a LabVIEW tool can help to track changes in an application by comparing multiple versions of a VI. The differences can be found in both the front panel and block diagram, as well as other VI attributes, such as settings in VI Setup (see Figure 2-10).

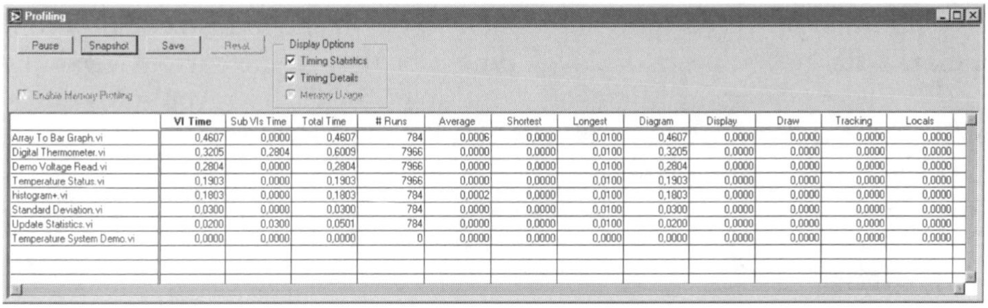

Figure 2-9
LabVIEW profiler

Figure 2-10
Differences dialog box

2.6 Platform Independence

Despite short innovation cycles, no single platform has yet established itself as the standard. One of the reasons can surely be attributed to various benefits and drawbacks of the relevant systems. Therefore, it becomes even more important for a software solution to support the most popular computer systems and to allow the use of their specific properties. This means that the user may expect from a powerful environment that a program developed on a given platform runs in another hardware environment without time-consuming modification. Of course, limitations are imposed by the hardware-specific properties, e.g., DLLs that can be supported in Microsoft Windows only, but not on Apple Macintoshes or HP workstations, while a TCP/IP protocol exists on several platforms so that it also has to be portable. LabVIEW meets most of these requirements. The development of this graphical compiler has shown that one may even expect a "native version" on new platforms, provided they prove themselves in the industry and become popular. With eleven years of development behind it, the current version of LabVIEW is available for Microsoft Windows 95/NT/3.X, Apple Macintosh 68k/PowerMac, PowerPC, Sun SPARC, and the HP-700 series.

2.7   Openness

Openness is a catchword frequently used in connection with software environments, and has become a kind of buzzword in the industry. Looking more closely, however, you will see that many systems praised as having this attribute do not live up to user expectations. Openness may have many meanings in the practical use of a software development environment, where each user has his or her individual preferences. Most expect the following characteristics, with the weighting being dependent on the specific application:

- It should support standardized system properties, e.g., OLE (ActiveX), DDE, DLL, TCP/IP, and shared libraries.

- It should allow users to embed their own special solutions as existing source code into the environment, e.g., by offering an open interface.

- It should support common I/O interfaces, such as RS-232, GPIB, PCI, AT bus, PCMCIA, and VXI.

- It should offer software drivers for such I/O interfaces for developers to allow shorter development times.

LabVIEW is open to a number of devices (serial, IEEE-488, VXI, etc.), bus systems (PCI, ISA/EISA, PCMCIA, MXI, etc.) and data acquisition cards (multi-I/O, A/D, D/A, CTR, Dig I/O, DSP, etc.). By supporting standardized system properties, it is possible to use any I/O interface, e.g., through DLLs (for instance, field buses normally used in process visualization and control, e.g., Interbus-S or CANBus).

2.8   Graphical Compiler

Of course, we must not forget the execution speed of the programs we create. While we had been limited to the rather slow interpreter mode only a few years ago, working with numerous BASIC dialects, a compiler is considered indispensable today. A compiler translates source code into more or less optimized machine code, thus offering quick program execution as an interpreter. As the power of hardware increased and faster processors were introduced, some interpreters had been sufficient for certain applications, but the typical industrial use of computers in fields that had been solely controlled by special equipment only a few years ago requires the power of an optimized compiler. The block diagrams created in LabVIEW are translated into optimized machine code by a "graphical" compiler so that the execution speed of LabVIEW programs is comparable to that of compiled C or Pascal programs.

2.9   Documentation

In times of total quality management, certification, and validation, appropriate documentation is essential. LabVIEW provides all the necessary documentation output facilities in a single integrated environment. In fact, users can print or export VI descriptions to formatted file formats such as Rich Text Format (RTF) and HyperText Markup Language (HTML). RTF can be imported into standard document-publishing software. Also, RTF files are the source for online help files. The HTML format represents the standard for online documents, especially those you intend to publish on the World Wide Web (see Figure 2-11).

2.10   Summary

In the search for a graphical programming language that would be usable by a large majority of application developers with various programming knowledge and experience, the LabVIEW system had been designed in the 1980s, based on a data flow model and enhanced by graphical control flow structures. LabVIEW avoids the limitations and problems inherent in the data flow approach because it enhances the data flow model by newer mechanisms like programming constructs or the options of a hierarchical, modular, and structured graphical syntax, without losing the natural readability of the data flow code. Modules (subVIs) similar to subprograms

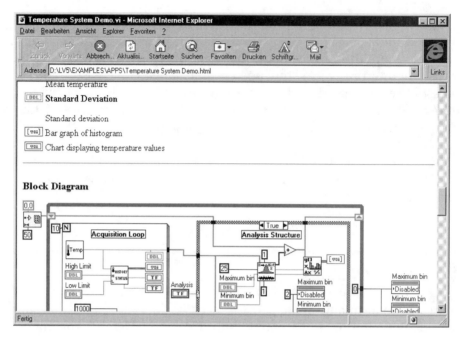

Figure 2-11
Documentation in HTML format

and represented in an icon-based notation hide the program complexity from developers, enabling them to master complex applications. The compiler integrated in the LabVIEW development system is probably unique in the world of graphical data flow systems. It allows extremely high execution speeds, comparable to textual programming environments based on standard compilers, e.g., C, C++, or Pascal.

3

Data Acquisition and Instrument Control in LabVIEW

... a great deal of our thinking is in terms of theories. The word "theory" derives from the Greek "Theoria," which has the same root as "theatre," in a word meaning "to view" or "to make a spectacle." Thus it might be said that a theory is primarily a form of insight, i.e., a way of looking at the world, and not a form of knowledge how the world is.

— David Bohm
Wholeness and the Implicate Order

Successful DAQ applications depend on the decisions you make when taking measurements. That is why any DAQ solution must deliver measurements, not estimates. This chapter describes the core purpose of LabVIEW: data acquisition and instrument control. You will become familiar with DAQ basics, DAQ drivers, and DAQ programming in LabVIEW. You will learn about the Virtual Instrumentation Software Architecture (VISA), the GPIB bus, VXI, PXI, USB, and several important standards.

3.1   LabVIEW and DAQ

The term *PC-based data acquisition* refers primarily to the acquisition of data both by means of PC-based plug-in cards, designed as classical multi-function cards, and external box systems (SCXI, VXI-DAQ, parallel port, etc.) and by means of PC-based instruments. A multifunction card is a PC plug-in card handling analog input and output capabilities, digital input and output ports, and one or several counters and timers. In turn, PC-based instruments combine the advantages of integration into industry-standard PC hosts and bus systems (PCI, VXI, PCMCIA, PC Card, etc.), offering the flexibility of stand-alone devices, for example, oscilloscopes, multimeters, functional generators, and spectrum and performance analyzers, based on plug-in cards.

    The possibilities and limitations of PC-assisted measurement value acquisition cards are still not fully understood, and a considerable amount of vagueness and misinterpretation is found in the user community. For this reason, it would appear sound to consider first the physical and technical measuring basics for the acquisition and processing of analog signals, including the underlying phenomena and sources of error. The following section is not limited to PC multifunction cards, but also includes external systems that may have various communication interfaces (e.g., PC card, RS-232/485/422, parallel port, and USB).

❑ 3.1.1   Common Sources of Error in A/D Cards for PCs

The market share of multifunction cards in the PC measurement technology increases every year. These cards can be used to acquire and output analog measuring values and to acquire or generate digital signals. This popularity can be mostly attributed to the flexibility these types of cards offer in connection with a computer and the corresponding software (e.g., LabVIEW and LabWindows). This is one reason why it is important for the potential user

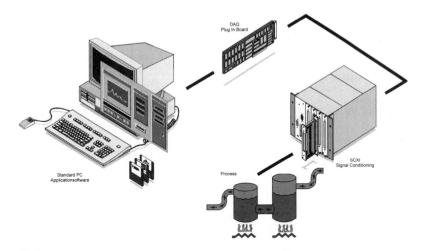

Figure 3-1

Main components of a data acquisition system, including signal acquisition, signal conditioning, A/D converter card, personal computer, and software

of such cards to analyze their specifications in greater detail. This section provides the newcomer to the field of computer-assisted measuring data acquisition with basic information regarding the selection of multifunction cards.

Before discussing the important characteristics of digitizing an analog signal, we will describe the main components (Figure 3-1) of a data acquisition system and illustrate the path a signal travels.

Figure 3-2 shows a block diagram depicting the structure of the analog input components of a typical data acquisition card. The input multiplexer (MUX) switches several input signals to one, single, programmable instrumentation amplifier (PGIA), which raises the input signal to the correct level. After amplification, the analog signal is held constant in a sample&hold step for a certain interval so that the A/D converter can digitize the analog value.

Most data acquisition cards store this digital value in a FIFO (First In First Out) buffer, which ensures that no data is lost in the event that it is not immediately transferred from the data acquisition card over the I/O channel of the PC into the main memory of the computer. Having a FIFO buffer is important, particularly when the card is to be used under an operating system subject to long interrupt delay times, for instance, Microsoft Windows.

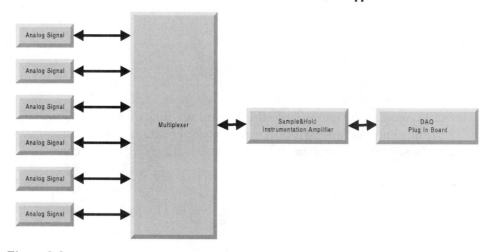

Figure 3-2

Analog input area of an A/D converter card, consisting of multiplexer, programmable instrumentation amplifier, A/D converter with sample&hold, and FIFO buffer

❏ 3.1.2   Analog Parameters

Almost all card specifications include information on number of channels, maximum sampling rate, resolution, amplification, and input range. These parameters are described below.

- **Number of channels** — In addition to multiplexers, amplifiers are switched in parallel to A/D converters. These amplifiers can be operated either in single-ended mode (depending on mass) or in differential mode (independent from mass). In the differential mode, only half as many input channels as in single-ended mode are available. On the other hand, interferences that could have an impact on the measuring lines are eliminated and earth circuits are avoided. In general, the single-ended mode is suitable when the input signals to be acquired are larger than 1 V, the connection lines from the signal source to the data acquisition card are shorter than 5 m, and if all input signals have the same mass reference point. If any one of these three conditions is not met, use the differential mode. The differential mode is also the preferred mode when signals have to be measured from a grounded signal source, when the ground potential lies on a potential other than that of the A/D converter card. In addition, there are also cards with pseudo-differential inputs. All inputs of this type of card have one common reference point, similar to the single-ended mode, but the reference point is not connected to ground. These data acquisition cards have the advantage that they possess many input channels and support noise compensation. Noise compensation is particularly useful when noise affects all input channels equally. Many data acquisition cards can be used in connection with external multiplexer cards (e.g., NI AMUX-64T) so that the number of available input channels can be expanded.

- **Sampling rate** — The sampling rate indicates how fast the input signal can be acquired and digitized. Most specifications state a sum sampling rate. This is the maximum sampling frequency that can be achieved when measuring one single channel. For example, when working with a 16-channel card equipped with only one single A/D converter, you get a maximum sampling rate of 1/16 of the maximum sum sampling rate per channel (when reading from all 16 channels). This is particularly important when you need to read several AC signals. It requires that the sampling rate be twice as high as the highest frequency occurring in this signal. But even if you want to acquire DC signals, you may want to work with a higher sampling rate to acquire and determine several values. This technique increases the acquisition accuracy, because the noise effect is reduced when the mean value is determined.

- **Resolution** — The resolution indicates the smallest voltage change in which the A/D converter can still digitize. It is indicated in bits and calculated by $\Delta U = 1/2^N$, where ΔU is the resolution and N is the number of bits. A reading range of 0–10 V and a resolution of 16 bits result in a smallest detectable voltage of 156.6 µV. The information stated in the specifications of some data acquisition cards is misleading about the actual resolution of these cards. With many cards, the specifications merely provide information on the resolution of the *Analog/Digital Converter* (*ADC*), without considering the linearity or the system noise. This means that there is a lack of information on the actual resolution of the data acquisition card. You need to know the value of the ADC's resolution in connection with the settling time and the integral/differential nonlinearity (INL/DNL) and the system noise to determine the real accuracy of a card.

- **Reading range and amplification** — The reading range and the amplification provide information on the available input ranges. Frequently, the two values are specified separately, which means that you have to calculate the actual input range from these two values with the following formula:

$$\text{input range} = \text{reading range}/\text{amplification}$$

Example: If a data acquisition card is equipped with one A/D converter with a reading range of ±10 V and an amplification of 100, you obtain an input range of ±100 mV. The more this input range can be adapted to the input signal range, the better the card's acquisition capability. Therefore, it is important to pay attention to the amplifications and reading ranges available with different cards.

☐ 3.1.3   Resolution, Accuracy, and Sources of Error

As mentioned earlier, information on the resolution of a specific ADC alone does not allow clear conclusions on the actual accuracy and quality of a data acquisition card. A number of factors reduce the specified resolution noticeably and thus falsify the measured values considerably. The most important sources of error are linearization errors INL and DNL (integral and differential nonlinearity) of the A/D converter, nonlinearity of the amplifier, amplification and offset errors, temperature drifts, parameter-drift phenomena due to age, and noise caused by external or immanent components and structures.

Figures 3-3 through 3-6 show an A/D-D/A transmission path with ideal transmission characteristics, potential error modes of the A/D-D/A transmission path (offset errors, amplification errors, linearization errors), INL (*integral nonlinearity*) or TNL (*total nonlinearity*) and missing codes at a quantization width of ≥ ±1 LSB (*least significant bit*) and DNL (*differential nonlinearity*).

Figure 3-3 shows an A/D-D/A transmission path with ideal transfer characteristic. A 3-bit (eight quantization steps) type of A/D and D/A converters were used for simplicity. The analog input signal is assumed to be ideal. The A/D and D/A converters are also ideal (no errors, no deviations). The center of the figure shows the input voltage of the A/D converter

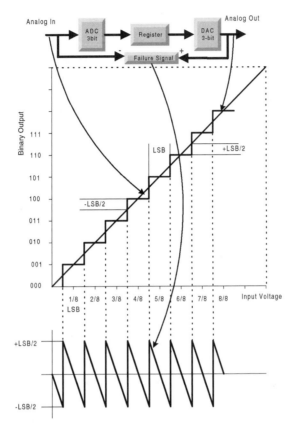

Figure 3-3

A/D-D/A transmission path with ideal transfer characteristic

and the output voltage of the D/A converter (as a typical ideal quantization step signal). The bottom part of the figure shows the difference between the input and output characteristics of the two signals in the form of a typical "sawtooth." We can see that the maximum deviation between the two signals can be ±0.5 LSB.

- **Amplifier specifications** — Amplifiers have a large impact on the accuracy of a data acquisition card and normally cause amplification and offset errors as well as nonlinearity errors. For high-quality A/D converter cards, capable of being calibrated, a higher amplification and offset accuracy are normally specified because both amplification and offset errors can be corrected by a simple

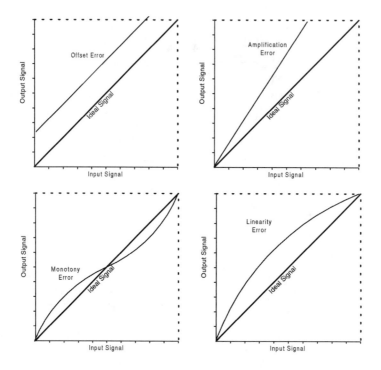

Figure 3-4

Potential errors on the A/D-D/A transmission path

two-point calibration. If the data acquisition system is to be used in an environment subject to temperature fluctuations, then the drift of the amplification and the offset must be studied before you can draw conclusions on the system stability. The amplifier drift is normally indicated in parts per million (ppm)/ºC, while offset errors are expressed in volts/ºC.

■ **INL errors** — Generally speaking, a linearity error describes the deviation of the output value from the ideal transmission function. If we were to subject the input of an A/D converter to a voltage extending over the entire input range, then the conversion result would ideally be represented as a stair (thin line in Figure 3-5). However, as the real conversion result deviates, the actual stair differs from the ideal characteristic. The maximum deviation between ideal and real characteristics is called INL. A good (low

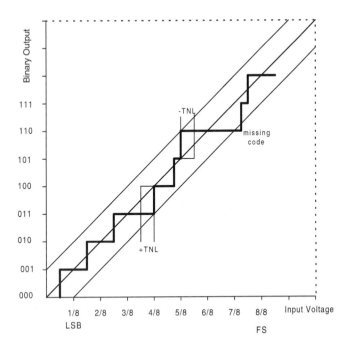

Figure 3-5

INL (TNL) and missing codes at a quantization width of $\geq \pm 1$ LSB

deviation) INL means that a data acquisition card can guarantee the exact conversion of analog input values into digitized output values over the entire input range.

- **DNL errors** — Ideally, one would want to have the digitized values in equal quantization steps. Because the quantization width can become unequally large in practice, we define DNL, which describes the deviation of the converter from the ideal quantization width. The DNL is stated in fractions of 1 LSB:

ideal quantization width = 1 LSB = input range / 2^N

where N represents the resolution in bits. DNL should ideally be smaller than ±1 LSB to be able to use all bit combinations. If DNL becomes larger than ±1 LSB, then certain bit combinations will simply

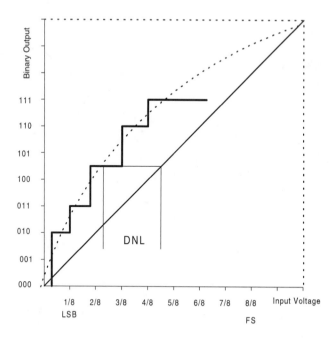

Figure 3-6

Representation of the DNL of a data acquisition system

not occur, i.e., the corresponding quantization step is skipped (see Figure 3-5). The result of this is that the resolution of the data acquisition card will be smaller than the specified resolution of the converter.

■ **Noise** — System noise is the parameter of a data acquisition system that is most difficult to predict because the amount of noise depends largely on the environment. Data acquisition card vendors specify noise in different ways. The most common method is that all inputs are connected to mass, then the amount of noise is measured. A data acquisition system should be checked for good design with regard to noise impact. It is not meaningful to generally assume that external, encapsulated data acquisition boxes are less sensitive to noise than the type of data acquisition card that is plugged into a computer. It is more important to ensure that the analog processing part is separated from the digital one on the board and that the analog part is equipped with additional screening. In many applications, the major amount of noise occurs directly at the signal sources or along the signal lines. This effect can be minimized by shortening the signal lines and/or by using a signal processing method in the immediate proximity of the signal source.

□ 3.1.4   Effective Sampling Rate

The sampling rate or conversion time of an A/D converter is normally specified in data sheets. However, the actual achievable throughput can be determined only if factors like the *settling time* of the converter, the availability of a FIFO buffer on the card, and the software used with the converter are taken into account.

- **Settling time** — The settling time indicates the time required by the output of the *programmable gain instrumentation amplifier (PGIA)* to obtain a transient state. We consider the following input signals as examples, assuming that one signal is at –5 V on channel 0 and the other one at +5 V on channel 1. In this worst case, the amplifier output must rise from –5 V to +5 V. Due to its limited bandwidth, the preamplifier requires a certain settling time to stabilize. If samples are taken at a sampling rate of 50 kHz, then the settling time of the measuring amplifier must be smaller than 20 μs (= 1/50 kHz). Unless this condition is maintained, we get wrong test results. In many data acquisition cards from various vendors, the required settling time extends as soon as higher amplification factors are used, leading to a lower sum sampling rate. Only a few vendors offer data acquisition cards with a PGIA that guarantees extremely short settling times, even with high amplification, regardless of the selected amplification to avoid sampling losses.

- **FIFO buffer** — Data acquisition cards that plug into a PC/XT or AT computer bus are capable of transmitting data acquired over the DMA and interrupt operation to the computer's main memory at relatively high speed. Nevertheless, it happens in practice that delay times arise during interrupt processing and DMA operation. To avoid data loss in the event of such delays, many data acquisition cards have a FIFO buffer to temporarily store data on its way out from the computer's main memory. While older data acquisition cards normally have a FIFO buffer with a capacity of 16 acquired and converted values, the FIFO buffer of modern cards can hold up to 1000 values. The popularity of Microsoft Windows, which causes major interrupt delays, requires expansion of this FIFO buffer.

❏ 3.1.5   Insulation

One major problem in measuring values acquired over a data acquisition card is the unacceptable grounding of this card when acquiring signals from grounded signal sources that lie on another earth potential. In this case, the voltage difference of the two earth potentials can cause a compensating current between them. This leads to two problems. First, an extremely high voltage difference can damage the data acquisition system. Second, the desired signal is measured, but the voltage resulting from the compensating current is added to the value. This voltage provoked by the compensating current normally contains noise, usually a 50-Hz noise from the voltage supply. If it is not possible to remove one of these earth sources, either on the data acquisition card by selecting the differential input operating mode or by removing the earth at the signal source, then this compensating current can be prevented by using a signal conditioning system (e.g., SCXI System from National Instruments). This type of signal connection produces a galvanic separation of the signal source from the measured acquisition system.

❏ 3.1.6   Measuring Thermocouples

We will use the measurement of thermocouples as an example to explain the above factors. We assume that thermocouples type T (Cu-constant), distributed on an extruder, are monitored, and they are to be operated in the 0–350ºC temperature range. We further assume the use of an A/D converter card with the specifications listed in Table 3-1.

Because this example uses thermocouples, the data acquisition system must be capable of detecting the temperature at the point where the other thermocouples are connected to the system. The data acquisition system can use this temperature directly through corresponding hardware or software for

Table 3-1   *Specifications of example data acquisition card*

Parameter	Specification
Resolution	12
Reading range	±10 V; 0–10 V
Programmable amplification	1, 10, 20, 50, 100, 500
A/D conversion time	10 µs
Relative ADC accuracy	0.8 LSB
Amplification error	0.05%
Offset	
Error before amplification	5.0 µV
Error after amplification	1.0 mV
Amplifier nonlinearity	0.01% FSR
System noise	1.0 LSB rms
Settling time	
Amplification = 1	10 µs
Amplification = 100	30 µs
Amplification = 500	50 µs

compensation. This method is also called *cold-junction compensation*. At 350ºC, a thermocouple type T generates a thermocouple voltage of approximately 18 mV, so we will operate our data acquisition system in the 0–10 V reading range, with an amplification of 500, to achieve an input range of 0–20 mV. At this amplification, the amplifier is specified with a maximum settling time of 50 µs. This means that we are limited to a maximum sampling rate of 20k samples/s.

To determine the accuracy of our measurement, we have to add all error sources that may potentially influence the measured value. For this reason, we will convert all specifications to error voltage and corresponding temper-

Table 3-2   *Calculation of measuring errors*

Parameter	Specification	Relative Input Error	Relative Input Error in C for Thermocouple Type T
Relative accuracy (incl. INL and DNL)	0.8 LSB	3.9 µV	0.10°C
Amplification error	0.05% of read value	0.05% of read value	0.05% of read value
Offset error			
Preamplification	5.0 µV	5.0 µV + 1.0 mV/ampl. = 7.0 µV	0.12°C
Postamplification	1.0 mV		
Amplifier nonlinearity	0.01% FSR	0.01% (20 mV) = 2 µV	0.05°C
System noise	1.0 LSB	4.88 µV rms -> 3 µV	0.07°C
Cold-junction error	0.2°C	0.2°C	0.2°C
Sensor temperature gradient	0.5°C	0.5°C	0.5°C
Sum			0.05% of read value + 1.04°C

ature. With an amplification of 500 and a reading range of 0–10 V, 1 LSB corresponds to a voltage of 20 mV/2^{12}, or 4.88 µV/°C. Table 3-2 shows how this is calculated. We also have to include the cold-junction error, which has a significant impact on the total error rate. This error is calculated by adding the error detected in the sensor used for cold-junction compensation during measured value acquisition to the temperature difference with regard to the contact points of the sensor versus the contact points of type T thermocouples we are using here.

Another important factor we should not forget is that the error calculation described above may not take the largest error source into account: the thermocouples used. Errors due to faulty connection lines of a thermocouple by using extension cords with different thermoelectric properties can cause an error of several °C. To obtain exact values, it is necessary to calibrate the entire system. This calibration removes many errors, including amplification and offset errors and the largest part of cold-junction compensation and thermocouple errors.

☐ 3.1.7   Summary and Current State of the Art

Both the resolution and the sampling rates of PC-based data acquisition systems have constantly increased during the past years. However, a detailed knowledge of the remaining parameters (e.g., long-term drift behavior, accuracy, differential nonlinearity) is important because resolution and sampling rate are not the only characteristics of a data acquisition system.

You can obtain this information by carefully selecting a suitable A/D converter board for the intended application based on its specifications. This means that you select a data acquisition system with a sufficient number of channels, a sufficient sampling rate to be able to acquire all signals in the resolution necessary to detect even minimal voltage changes that may potentially occur in the application, and an appropriate input range. Subsequently, you ensure that the INL and DNL specifications are smaller than 1 LSB and that the amplifier meets the requirements. Moreover, it is important to be able to set the PGIA to the required time and accuracy to make sure that the system can actually be operated in the necessary sampling rate. Care should be taken with poorly specified data acquisition cards. For instance, an A/D converter card may be specified with an ADC resolution of 12 bits, while it eventually works merely at an accuracy of 8 bits because of noise, high INL, missing codes, or insufficient settling time. Knowing which factors need careful attention when comparing card specifications, you should be able to select a suitable data acquisition card for your application.

Sufficient speed and sufficient buffer provide the only method to ensure proper oversampling at high accuracy and low proneness to interference. Conventional AT bus-based cards cannot handle high data rates due to the low bus bandwidth. Also, some PCI-based systems have performance problems due to the load sharing of the PCI bus (in general, there is only one PCI bridge per motherboard). A way out of this dilemma is demonstrated by high-speed data acquisition cards for modern buses. The following subsection provides a short insight into the power and limitation of the nonproprietary *Peripheral Component Interface Bus*.

Although the PCI bus offers superior performance compared with ISA, Micro Channel, EISA, and Nu-bus systems, a testing, measuring, data acquisition (DAQ) or control system based on PCI does not achieve the theoretically possible transmission rate of 132 Mbytes/s. The system performance is determined by two essential factors: the maximum data throughput rate of the components linked over the PCI bus (e.g., PCI plug-in cards) and the underlying computer system (processor type, memory components, and motherboard chip sets). The transmission rate of PCI bus components can be raised to that of the computer's main memory if DMA bus master capability is present, which may be implemented on PCI plug-in cards. The PCI's lack of DMA support has encouraged National Instruments to develop the *MITE* chip. This proprietary ASIC (Application-Specific Integrated Circuit) is capable of handling DMA block transfers. For instance, a capability called *scatter gather DMA* was implemented specially to handle measuring and control processes. This approach ensures that data transfers are executed between nonconsecutive memory blocks without the necessity to reprogram the DMA controller. Scatter gather DMA ensures that the DMA controller is reconfigured automatically by reading a series of concatenated data files located in buffers. These data files contain the source address of the data, the target address of the data, number of bytes to be transmitted, and the address of the next data file in the chain. The microprocessor needs to initialize the DMA controller to the first data file address only once at the beginning.

☐ 3.1.8   The DAQ Driver Concept

The use of DAQ-based systems is normally accompanied by a driver, which spares you from complicated hardware-related interface configurations and handling (i.e., special protocols and other software layers). The driver concept used in LabVIEW is based on three cornerstones (see Figure 3-7):

- **Hardware level** — This lowest level can consist of various component combinations. This level includes both the sensor-side connection and signal conditioning as well as acquisition plug-in cards (A/D multifunction card) or communication (RS-485, RS-232, parallel port) in the PC.

- **NI-DAQ driver software** — This medium level acts as connecting link between the hardware level and the application software (top level).

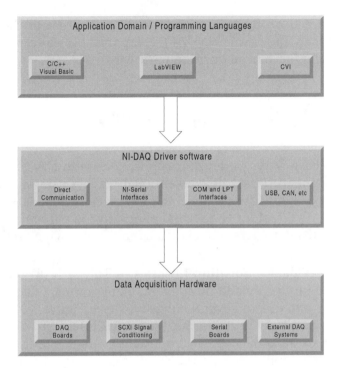

Figure 3-7

The NI-DAQ three-level model

- **Application software** — This language-independent top level allows development environments like LabVIEW, LabWindows/ VCI, Visual Basic (ComponentWorks), Excel VBA (Measure), and (Visual)C++ to access the hardware level effectively.

What does this concept mean for the development, maintenance, and portability of applications? The standardization of drivers for a large range of hardware components simplifies program development enormously. Assume we have written an application for a PC-based multifunction card. Now we want to run this application on a notebook with a parallel port device or a PC card hardware. Since the application was initially developed for the multifunction card in LabVIEW, we merely have to install (store) it when porting it to the notebook; we don't have to rewrite it. Note that although the communication path has changed from the original PC-based multifunction card to the parallel port or PCMCIA of the notebook, this has no impact on the LabVIEW program because it is already considered on the driver level (not transparent to the end user). The driver functionality of NI-DAQ compares basically to that of an operating system for data acquisition hardware.

❑ 3.1.9   DAQ Programming in LabVIEW

There are two basic ways to develop applications with DAQ hardware and software components: First, the conventional way — using the data acquisition libraries of LabVIEW; second, the new way, available since Version 4.1 — using DAQ Wizards. The second alternative is particularly useful for newcomers. Both alternatives are explained in the following sections.

- **LabVIEW data acquisition libraries** — The LabVIEW data acquisition library is divided into several main groups, which are again divided into one or several subgroups (see Figure 3-8). The main groups and their purposes are explained briefly in the following section because they are an important prerequisite to understanding the application examples. Basically, we distinguish the libraries by categories (*Basic* and *Advanced*) for users with basic knowledge and for advanced users.

- **Analog input** — These libraries (see Figure 3-9) contain functions concerning the A/D conversion. The choice ranges from single-point measurement to continuous multiple-point acquisition with sampling rates ranging from the megahertz range to the PCI bus.

- **Analog output** — These libraries (see Figure 3-10) are similar to the previous group, except that the functions they contain are used by the D/A converter card to output arbitrary or continuous curve forms.

- **Digital I/O** — The Digital I/O libraries (see Figure 3-11) allow you to set, reset, and read input and output ports in any port width. You can freely define the inputs and outputs. Moreover, handshake protocols are available for communication.

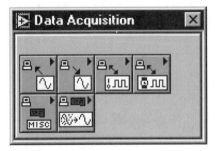

Figure 3-8

Main palette in the data acquisition library

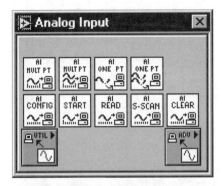

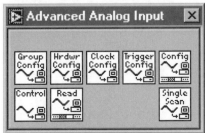

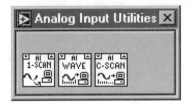

Figure 3-9

Analog input libraries

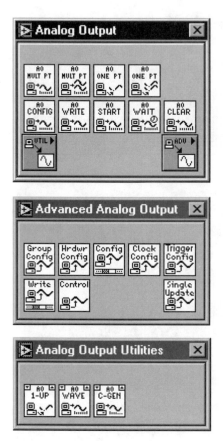

Figure 3-10

Analog output libraries

- **Other functions** — Additional functions (see Figure 3-12) mainly concern signal conditioning, counter/timer, calibration, and configuration. The signal conditioning routines allow you to adapt the input signals (e.g., by amplifying) to the voltage levels taken by the plug-in cards (e.g., ±5 V or ±10 V) by using the required signal conditioning hardware (e.g., SCXI). *Timers* and *Counters* are important components relating to the functionality of measurement value acquisition cards. The system uses these components to define the sampling rate and the time window for sampling and for evaluation of counter pulses. Additional fields of application include event, time, and frequency acquisition, where triggering can be realized in different ways.

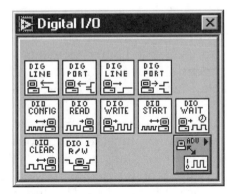

Figure 3-11

Digital I/O libraries

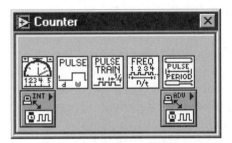

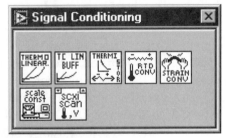

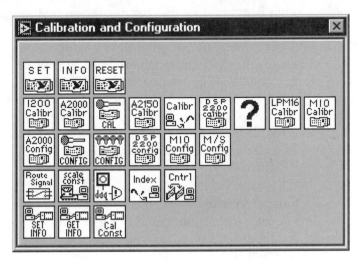

Figure 3-12

Other libraries

A series of functions are available for calibration of data acquisition systems (e.g., offset, amplification, and correction of characteristics). Professional test engineers can use these library elements very efficiently. They allow rational development of powerful, reliable, and stable applications. The number of available functions allows different solution options. This choice of ways to the goal — the final application — is not necessarily simple for the less experienced developer or for the newcomer. For this reason, National Instruments has introduced powerful tools to generate complete virtual instruments for data acquisition. Working with easily usable point-and-click configurations, users with little or no experience can quickly generate virtual measuring devices. This technology is explained in the following sections.

☐ 3.1.10   DAQ Wizards

Since the introduction of LabVIEW Version 4.0 in January 1996, National Instruments has worked extensively to improve the data acquisition system development. It became evident from user feedback that the development of DAQ configuration tools required some improvement. This need was satisfied with the new DAQ Wizard technology introduced in Version 4.1. Experts consider this technology advanced that a patent has been applied for and is pending. What benefits does this configuration methodology offer for users?

The *Custom DAQ Application Wizard* helps you specify systems for data acquisition in a series of question-and-answer dialog boxes. You select a number of operating elements, e.g., analog inputs, digital I/Os, and file inputs and outputs, and use them to form your application-specific data acquisition system. Once you have defined your system, LabVIEW generates a fully executable program with the relevant LabVIEW components, i.e., with panels, diagrams, and icon/connectors. This program can be edited to your needs (Figure 3-13).

- **The DAQ Channel Wizard** — Formerly, developers of data acquisition systems spent a lot of time defining signal types, connections, transmission equations, and unit conversions before they could start working on the actual system development. For instance, when using thermocouples, it was necessary to calculate a cold-junction compensation (CJC) to calculate the actual temperature values from

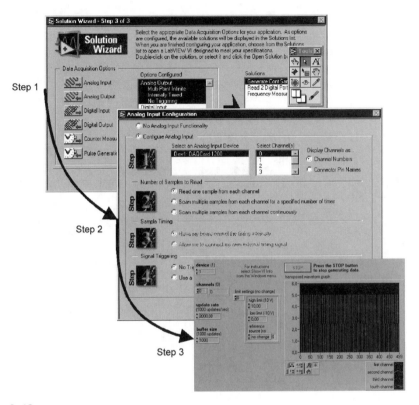

Figure 3-13

Solution Wizard in three steps

a lower-level input voltage (µV). Moreover, it was necessary to state interpolation developments, which depend on the type of the temperature sensor used. Conversion into the correct units (°C, °F, K) required additional program codes.

With the new DAQ Channel Wizard, all these complicated steps are reduced to defining input signals by entering a name and a description, specifying the measuring instrument, state conversion, and scaling factors, and defining CJC values and unit conversion, as shown in Figures 3-14 and 3-15. The name of the input channel is used later as reference across the entire application. This makes the exchange of data much more transparent for the user.

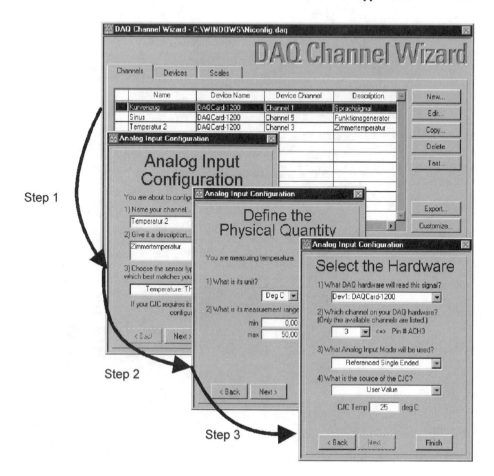

Figure 3-14

DAQ Channel Wizard in three steps

- **The DAQ Solutions Gallery** — The DAQ Solutions Gallery (see Figure 3-16) is a library containing solution suggestions for the most commonly used applications for data acquisition. You search this library for an application that comes closest to your application. Typical examples are oscilloscopes, multimeters, function generators, data loggers, temperature acquisition systems, spectrum analyzers, arbitrary generators, etc. The LabVIEW block diagrams can be thought of as templates that you can adapt easily to different problems.

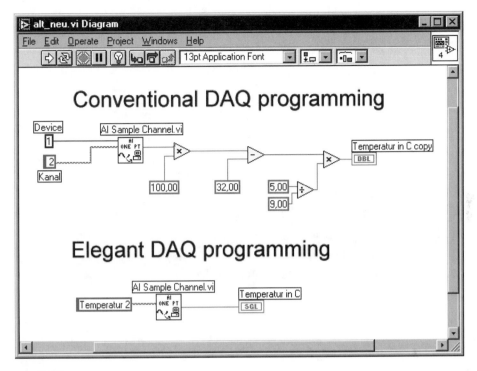

Figure 3-15

Conventional and DAQ channel-based LabVIEW programming

☐ 3.1.11 PC-Based Instruments

Stand-alone measuring devices have been available for several centuries and are now facing fierce competition by PC-based instruments based on PCI plug-in cards. In contrast to conventional stand-alone measuring devices, PC-based instruments introduce almost no overhead because important hardware components are already available in the host PC. The PCI bus offers the required bandwidth to fully utilize the motherboard memory. The production of PC-based instruments is extremely cost efficient because casing, power supply, memory, operating elements, displays, and a large part of the firmware are not required. Thanks to low storage cost (small dimensions) and the flexibility in developing new MMIs and new customer-specific features, a PC-based instrument can be adapted to changing market situations much more easily. Important development steps

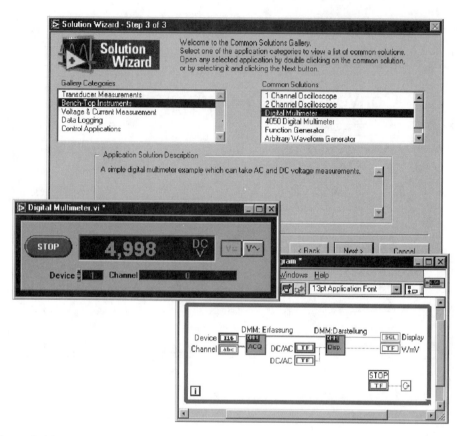

Figure 3-16

DAQ Solutions Gallery

can be ignored (e.g., almost the entire mechanical constructions) so that the
time to market of such products is unbeatable.

National Instruments is one of the pioneers in the field of measuring
devices based on plug-in cards. National Instruments is the only manufacturer capable of offering all important instruments based on DAQ cards.

By using the new technologies (DAQ Wizards, DAQ Channel, and DAQ
Solution Wizards) in connection with new form-factor-reducing ASICs, it is
now possible to implement data acquisition devices (PC-based instruments)
in fields of use that have been traditionally controlled by stand-alone devices.
The new series of PC-based instruments includes the DAQScope™ oscilloscope, the DAQArb™ arbitrary function generator, and the DAQMeter™
digital multimeter (DMM). Together with the related software, PC-based

instruments are particularly suitable for stand-alone use and for integration into existing systems for data acquisition purposes. The DAQScope family is available in PCI-5102, AT-5102, and DAQCard-5102. The DAQCard-5102 is the first PCMCIA card of its kind; it can be used as a two-channel oscilloscope. The DAQArb product line includes the PCI-5411 and the AT-5411 cards, and the PCMCIA DAQCard-4050 is used for DAQMeter.

DAQMeter DAQCard-4050

The DAQCard-4050 is a full-feature digital multimeter (DMM) for hand-held and notebook computers with a Type II PC Card (PCMCIA) slot. The card features accurate 5½-digit DC voltage, true root mean square (*true-rms*) AC voltage, and resistance (ohms) measurements. This multimeter handles AC/DC voltage, resistance, and (with the optional current shunt accessory) current measurements. The card can also be used as a continuity tester or for diode testing. The signal connection is based on 4-wire technique over safety banana jacks and complies with international safety standards, including UL 3111 and IEC 1010. The input voltage protection of the multimeter is ± 250 VDC or 250 true-rms in all input ranges. The DAQCard-4050 is fully configurable through software and runs under MS Windows 3.1x/95/NT and other operating systems (Figure 3-17).

- **Reading rates** — Three different reading rates (10, 50, or 60 readings per second) can be selected in all measuring areas and measuring modes of the DAQCard-4050. Highest accuracy is achieved by selecting a rate of 10 readings per second.

- **AC/DC voltage range** — Five different input voltage ranges (±20 mV, ±200 mV, ±2.0 V, ± 25 V, ± 250 V) are available for measuring DC voltage. This allows you to measure an input voltage to the specified accuracy, as long as the input value is within the range from 10 to 100% of the selected input voltage range. In AC voltage ranges, the DAQCard-4050 measures the AC-coupled true-rms (or t-rms) value of a signal. With the t-rms method, a small AC signal is measured in the presence of a large DC offset. A high DC presence of up to 250 V can be faded out in every input voltage range. The DAQCard-4050 is capable of processing signals with a crest factor of up to 10, where an additional inaccuracy of 1% FSR has to be expected with a crest factor of 2–5 and 2% FSR with a crest factor of 5–10. Tables 3-3 through 3-5 list the accuracy of input ranges.

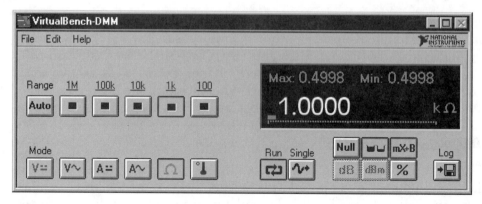

Figure 3-17

Panel of a virtual multimeter (VMM), based on a DAQ meter card

Table 3-3   *Accuracy of DC input voltage ranges*

Input Voltage Range ±	Read Error in % ±	Range Error in % ±	Temperature Coefficient ± (% of measured value + % of range)/C ±	Overvoltage Range ±	Input Resistance
20.0000 mV	0.01%	0.01%	0.0004% + 0.05%	22.0000 mV	>1 GΩ
200.000 mV	0.01%	0.01%	0.0004% + 0.005%	220.000 mV	>1 GΩ
2.00000 V	0.01%	0.005%	0.0004% + 0.0005%	2.20000 V	>1 GΩ
25.0000 V	0.05%	0.002%	0.0006% + 0.0007%	27.0000 V	1 MΩ
250.000 V	0.05%	0.001%	0.0006% + 0.0007%	250.000 V	1 MΩ

Table 3-4   *Accuracy of AC input voltage ranges referred to a 1-kHz sine signal*

Input Voltage Range ±	Read Error in % ±	Range Error in % ±	Temperature Coefficient ± (% of measured value + % of range)/C	Overvoltage Range ±	Input Resistance
20.0000 mV	0.3%	0.25%	0.01% + 0.05%	22.0000 m true-rms	> 1 MΩ
200.000 mV	0.3%	0.25%	0.01% + 0.05%	220.000 m true-rms	> 1 MΩ
2.00000 V	0.4%	0.4%	0.01% + 0.08%	2.20000 true-rms	> 1 MΩ
25.0000 V	0.4%	0.4%	0.01% + 0.08%	27.0000 true-rms	1 MΩ
250.000 V	0.3%	0.04%	0.01% + 0.008%	250.000 true-rms	1 MΩ

- **Resistance** — The DAQCard-4050 has five basic input ranges (200 Ω, 2.0 kΩ, 20 kΩ, 200 kΩ, 2 MΩ) and an extended range for two-wire resistance measurement. With the extended range, measurements of up to 20 MΩ are possible. In the extended ohms range, a 1 MΩ resistor is added in parallel with the test resistor.

- **Continuity testing** — The DAQCard-4050 multimeter makes continuity measurements by comparing the measured resistance between the two probes to a set value. To perform continuity measurements on a circuit, you set the DAQCard-4050 to the 200-Ω range and compare the measured value to a low resistance value (10 Ω is typical).

- **Diode testing** — The DAQCard-4050 multimeter measures the forward drop across a diode, up to 2 V. The multimeter biases the diode with 100 µA and measures the resulting voltage drop. Diode measurements are made with a fixed range of 2.0 V to an accuracy of 0.01%.

Table 3-5   *Accuracy of resistance input ranges*

Input Voltage Range ±	Read Error in % ±	Range Error in % ±	Temperature Coefficient ± (% of measured value + % of range)/C	Test Current
200.000 Ω	0.05%	0.1%	0.0015% + 0.045%	100 µA
2.00000 kΩ	0.01%	0.01%	0.0015% + 0.0045%	100 µA
20.0000 kΩ	0.01%	0.01%	0.0015% + 0.00045%	100 µA
200.000 kΩ	0.01%	0.02%	0.007% + 0.0045%	1 µA
2.00000 MΩ	0.01%	0.001%	0.007% + 0.00045%	1 µA
Extended range	see note		0.05% + 0.00045%	1 µA

Note: The accuracy of measurement in the extended range varies in line with the value of the resistance to be measured. The accuracy is 0.05% for a resistance of up to 1 MΩ, 0.1% for a resistance between 1 MΩ and 20 MΩ, and 0.2% for a resistance >20 MΩ.

- **Measuring current** — The DAQCard-4050 with the optional CSM Series current shunt accessories measures both AC and DC current. The CSM accessories include a precision resistor that converts the current through the shunt into a voltage measurable by the DAQCard-4050. To determine the amperage, calculation with the $I = U/R_{shunt}$ formula is done subsequently.

Its size, weight, and low power consumption make the DAQCard-4050 ideal for portable measurements and data logging with hand-held and notebook computers.

DAQScope 5102

The DAQScope 5102 digital oscilloscopes are a family of dual-channel 20 MS/s digital oscilloscopes for use with PCI, PXI, CompactPCI, PCMCIA, and ISA bus computers. These plug-in cards have two analog inputs and one analog trigger — CH0, CH1, and TRIG, respectively. They allow simultaneous sampling of two channels with signal sampling rates from 1 kS/s to 20 MS/s in real time and 1 GS/s in random-interleaved sampling method. The maximum vertical resolution is 8 bits at a bandwidth of 15 MHz. The on-board memory can accommodate up to 663,552 measured values. This corresponds to an acquisition time of 33 ms at a maximum real-time sampling of 20 MHz. Four different input voltage ranges (±5 V, ±1 V, ±0.25 V, ±50 mV) can be selected, which can be extended by the use of a 1:10 (±50 V, ±10 V, ±2.5 V, ±0.5 V) or 1:100 (±500 V, ±100 V, ±25 V, ±5 V) scan head. The AC/DC link can be selected in the software so that it is easy to acquire small AC voltages with a high DC share by fading out the DC share. The input resistance is 1 MΩ ± 1% in parallel to 30 pF ± 15 pF. A programmable frequency output is integrated to align scan heads or for other applications. This frequency output is TTL-compatible and can be operated with two different time bases (1.25 MHz and 7.16 MHz). These allow output of a rectangular signal in the range from 19 Hz to 2.38 MHz through a corresponding splitting ratio (3...65535).

In addition, DAQScope 5102 offers pre-trigger and post-trigger acquisition modes. In *post-trigger* acquisition mode, the device acquires a requested number of samples after the start trigger occurs. The desired number of samples are stored in on-board memory or (in the case of PCI-5102 and PXI-5102) are transferred to PC memory in real time. The PCI and PXI-5102 can acquire a maximum of 16 million samples if the computer has at least 16 Mbytes free. These two cards are capable of PCI bus mastering. They can sustain data transfer from board to system memory at the maximum sample rate, using the PCI MITE bus-master interface ASIC. The MITE is specifically designed for PCI-based DMA transfers. On the AT-5102 and DAQCard 5102, data transfer takes place after the acquisition ends, for up to 663,000 samples.

In *pre-trigger* acquisition mode, the card acquires a predetermined number of samples, the pre-trigger scan count, before the trigger event can be recognized. A start trigger initiates the acquisition. After satisfying the pre-trigger scan count requirement, the hardware keeps acquiring data and stores it in a circular buffer implemented in on-board memory. The size of this buffer is equal to the pre-trigger scan count. As new data is acquired, the hardware overwrites the oldest points in the circular buffer until the stop trigger occurs. When the stop trigger occurs, the desired number of post-trigger scans is acquired and the acquisition terminates.

Two digital trigger inputs are available at the programmable function inputs/outputs (PFI1/PFI2). These inputs can be configured through the software to the acquisition of rising/dropping flank of the digital trigger signal. Channel 1, channel 2, and an independent analog trigger input can be used for analog triggering. Five different trigger modes can be selected for analog triggering, where an upper and lower threshold value is defined via software:

- **Below-low-level analog triggering** — Triggering occurs when the value at the specified trigger input is less than the lower threshold value.

- **Above-high-level analog triggering** — Triggering occurs as soon as the value at the specified trigger input is greater than the upper threshold value.

- **Inside-region analog triggering** — Triggering occurs when the value at the specified trigger input is within the range between the lower and the upper threshold values.

- **High-hysteresis analog triggering** — Triggering occurs as soon as the value at the specified trigger input is greater than the upper threshold value. The hysteresis is specified by entering the lower threshold value.

- **Low-hysteresis analog triggering** — Triggering occurs when the value measured at the specified trigger input is less than the lower threshold value.

Tables 3-6 and 3-7 list the maximum number of values in pre- and post-trigger acquisition modes.

The *RTSI* bus (*Real-Time System Integration* bus, a PCI/ISA bus version) sends timing and triggering signals to other boards to synchronize multiple devices including other computer-based instruments or DAQ boards. The RTSI bus has seven bidirectional trigger lines and one bidirectional clock signal. The PXI-5102 uses the PXI trigger bus (six triggers and one clock) to route timing and triggering signals. Any of the seven RTSI or six PXI trigger lines can be programmed as inputs to provide start trigger, stop trigger, and scan clock signals from a master board. Similarly, a master board can output its internal start trigger, stop trigger, scan clock, and analog trigger circuit output (ATC_OUT) signals on any of the trigger lines. The RTSI bus clock

Table 3-6   *Maximum number of measured values in post-trigger acquisition mode*

Number of Channels	PCI-5102		AT-5102 and DAQCard 5102	
	Min	**Max**	**Min**	**Max**
One	360	16,777,088*	360	663,000
Two	180	16,777,088*	180	331,500

* Depending on available RAM

Table 3-7   *Maximum number of measured values in pre-trigger acquisition mode*

Number of Channels	PCI-5102		AT-5102 and DAQCard 5102	
	Min	**Max**	**Min**	**Max**
One	360 10	663,000 16,777,088*	360 10	663,000 – post-trigger measuring values 663,000 – pre-trigger measuring values
Two	180 5	331,500 16,777,088*	180 5	441,500 – post-trigger measuring values 331,500 – pre-trigger measuring values

* Depending on available RAM

line is a special clock line on the RTSI bus that can carry only the timebase of the master board to the slave board.

VirtualBench-Scope software is shipped with all 5102 oscilloscopes. VirtualBench-Scope is a "soft" front panel that controls the 5102 with no programming required. All hardware features of the 5102 are accessible by the software. You use VirtualBench-Scope just as you use stand-alone instruments, but you benefit from the processing, display, and storage capabilities of computers. VirtualBench-Scope can save waveform data to disk, generate reports, and perform statistical measurements. Figure 3-18 shows an example of a virtual scope panel.

If you want to build an automated test application or to integrate the 5102 card in your test software, use the 5102 instrument driver. The instrument driver works with LabVIEW and BridgeVIEW, LabWindows/CVI, Microsoft Visual C++, Borland Delphi, and Visual Basic. The 5102 cards use the NI-DAQ driver software as the low-level hardware and operating system interface. The instruments work with Windows NT/95/3.x. Using the 5102 digital oscilloscopes, you combine the power and capability of a stand-alone oscilloscope with the flexibility and benefits of a computer to create highly capable virtual instrument solutions that can leverage off the Internet, Pentium processing power, and greater high-speed data storage. The DAQCard-5102 is a PCMCIA (PC Card) plug-and-play instrument, which means that you can easily "hot insert" cards, and you can build portable test systems because of the card's small size and low power consumption.

DAQArb 5411

The National Instruments PCI-5411 and AT-5411 are full-featured 40 MS/s arbitrary waveform generators for use with PCI or ISA computers. In Arb mode, you first define the waveform and then store the digital representation of the waveform in the 5411 on-board memory. Two memory options are available — one for storing up to 2 million digital waveform samples and the other for storing up to 8 million samples. Different methods exist for defining the waveform, including using an oscilloscope to digitize and store a physical signal or using VirtualBench-ARB to define the waveform mathematically, with standard functions, or to draw the waveform on the screen. When Arb mode is used to generate signals, waveform sequencing, looping, and linking techniques contribute to the arrangement of the digital waveform data. Waveforms are produced with 12-bit vertical resolution and at a maximum update rate of 40 MS/s.

The 5411Arb permits up to 8000 different waveform segments to be defined in on-board memory. Each segment can be looped up to 65,535 times to form a waveform stage. Waveform linking creates a complete waveform by connecting various stages together. Up to 292 stages are permitted in Arb mode. Waveform linking and looping is an essential feature for generating very long arbitrary waveforms using minimal memory.

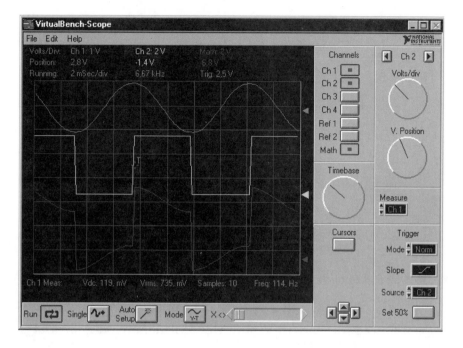

Figure 3-18

Virtual scope panel, based on a DAQMeter card

Using the card in DDS mode is more suitable for generating standard waveforms that are repetitive in nature, such as sine, TTL, square, and triangular waveforms. DDS mode limits the number of waveform samples stored to 16,384 and the number of waveform stages to 340. By using DDS, the 5411Arb generates signals with high-frequency accuracy, temperature stability, wideband tuning, and fast, phase-continuous frequency switching.

Analog voltages generated by the 12-bit DAC pass through a lowpass filter and are preamplified before a 10-dB attenuator. Through software, you select an output impedance of either 50 or 75 Ω. Most applications require a load impedance of 50 Ω, but 75 Ω is necessary for video testing. The SYNC output is a TTL version of the sine wave generated by the DAC and has a maximum frequency of 16 MHz. SYNC is useful as a high-frequency resolution, software-programmable clock source.

DAQArb 5411 cards feature *phase-lock loops* (PLLs) to synchronize with other National Instruments DAQ plug-in cards or with external hardware that supports TTL-compatible SYNC input. Moreover, the card's Dig Out connector is a 50-pin very high density (VHD) SCSI connector for access to

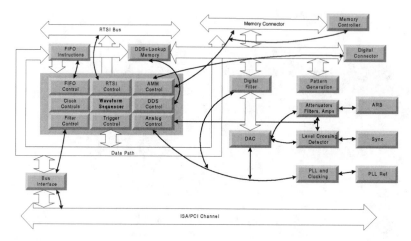

Figure 3-19

DAQArb 5411 block diagram

the external trigger input, digital pattern outputs, digital pattern clock output, marker output, and ±5-V power output. Figure 3-19 illustrates the elements of the DAQArb 5411.

❑ 3.1.12　LabVIEW and DAQ – Summary

The above sections provide an overview of the ways DAQ applications can be programmed in LabVIEW. There is a large number of application notes, application solutions, DAQ course and training material, and other information available from National Instruments dealing with a wide range of DAQ aspects. There is almost no limit to using DAQ-based hardware and software. DAQ-based products have established themselves well next to conventional measuring devices, now being more than a mere alternative to these systems.

3.2　LabVIEW and VISA

VISA (Virtual Instrument Standard Architecture) is a uniform, vendor-independent basis for the development, integration, and interoperability of software components like device drivers, soft front panels, and application software. Currently, work is in progress for formal standardization according to IEEE 1226.5. Though the VXI*plug&play* Systems Alliance specified VISA, it is not limited to the VXI bus; it is rather a uniform approach, containing guidelines for future bus systems and communication technologies. This section describes the motivation and the possibilities based on a LabVIEW example.

❏ 3.2.1 History

In September 1993, a group of VXI product vendors announced the formation of the VXI*plug&play* Systems Alliance. The founding members of this organization, all of whom are leaders in VXI technology and products, are GenRad, National Instruments, Racal Instruments, Tektronix, and Wavetek. Since its formation, many additional VXI vendors have joined the alliance. The VXI*plug&play* Systems Alliance is an open organization: its membership is open to both vendors and users of VXI technology.

The objective of the VXI*plug&play* Systems Alliance is to increase ease of use for end users of VXI technology. The Alliance members share a common commitment to end-user success with open, multivendor VXI systems, and a common vision for multivendor system architecture, including both hardware and system-level software.

The intention of the member companies is to accomplish major improvements in ease of use by endorsing and implementing common standards and practices in both hardware and software beyond the scope of the VXIbus specifications. Both formal and de facto standards are used to define complete system frameworks. These standard frameworks give end users true "plug and play" interoperability at both the hardware and system software level.

To understand a standard architecture and methodology for the development of instrument drivers, it is useful to understand some of the history of modern instrument driver software. A first step in this process is to answer the question, "What is an instrument driver?" It is not a new concept. Instrumentation users have been writing their own instrument drivers for years. An instrument driver, in the simplest definition, is a piece of software that handles the details of controlling and communicating with a specific instrument.

The task of programming instruments in a test system has always been a major concern for end users and a major cost for the overall system development. Many users know that programming can often be the most time-consuming part of developing a system. The developer spends much valuable time learning the specific programming requirements of each instrument in the system, including the undocumented or unexpected features.

"Rack and stack" IEEE-488 (or GPIB) instruments (see Section 3.3) have long been the most widely used type of instruments for multivendor systems. Almost all GPIB instruments are designed for interactive use through a physical front panel and also offer remote control capability via a GPIB port on the back of the instrument. Documenting an instrument command set in the user manual, along with some example program listings, has traditionally been the standard method for an instrument vendor to assist the end user in programming the instrument. These documentation methods have served the industry well for many years, but this approach still places the responsibility for writing the program code on users, many of whom may end up writing very similar application programs.

In the early days of computer-assisted instrumentation systems, the interpreted BASIC language was a widely used programming environment. To control the instrument from the computer, users would typically use I/O statements throughout their application program to send and receive the appropriate command and data strings to and from the various instruments in the system. The BASIC language, with its built-in formatted I/O capabilities, matched well with the formatted ASCII command and data strings used to control message-based GPIB instruments. Over the years, special-purpose versions of BASIC or similar languages with ever more powerful tools oriented specifically toward message-based instrument programming appeared in the market and enjoyed various levels of success. Throughout the 1980s, computer-assisted instrumentation became more widespread. Several key factors helped this spread to occur and influenced the evolution of instrument drivers.

The personal computer revolution enabled more users to take advantage of automated computer-assisted systems. Not only did many more scientists and engineers become comfortable using computers, but also the lower cost of PC-based systems made such technology applicable to many more applications. Also during this time, computer programming technology made tremendous strides (see Chapter 2). Compiled languages such as Pascal and C offered many benefits over interpreted BASIC, including faster execution speed and more powerful program and data structures. In addition, these new languages offered the ability to make programs more modular by organizing them into separate pieces that could be developed, maintained, and managed as independent, but related, software objects.

As computer-assisted instrumentation became more widespread and successful, many users began to realize that they often used the same instruments in a variety of systems. In addition, maintaining or enhancing existing systems often entailed replacing older instruments with newer or less expensive models. These realizations led to two identifiable trends in instrumentation applications: standard instrument command sets and modular instrument driver software.

If the same commands work for multiple instruments, regardless of the manufacturer, users can interchange or upgrade instruments and reduce the amount of changes to their application programs. In particular, many of the installed base of users who had substantial investments in BASIC or other software environments that did not easily lend themselves to software modularity lobbied for this approach. Through the mid- to late 1980s, many standards organizations, including the IEEE, worked on this objective with little progress. The IEEE 488.2 specification (see Section 3.3.3) more carefully defined the operation of instruments but did not address the issue of standard command sets.

Finally, in 1990, the SCPI Consortium (see Section 3.3.4) was formed. This organization, many of whose members were also involved in the VXIbus Consortium and are now involved in the VXI*plug&play* Systems Alliance, approved a specification for standardized commands for message-based programmable instruments.

In the early 1980s, as users gained experience with computer-assisted instrumentation and programming tools improved, they began to make their code more modular and to reuse the portions that control specific instruments. Over the years, many users developed their own guidelines for writing, documenting, and managing the test software their companies developed. In some cases, these internally developed software standards evolved over a number of years and represent tens or even hundreds of man-years of development.

At the same time, instrumentation suppliers began to deliver powerful, PC-based software tools that combined the latest advances in general-purpose PC software with instrumentation-specific tools. Libraries of pre-written software routines for particular instruments, known as *instrument driver libraries*, appeared in the market and began to be used for the first time. These initial instrument drivers were often written by the software supplier, rather than the instrument vendor, and the list of available drivers was small but growing. Numerous software suppliers entered the market and promoted their particular application software tools along with the associated libraries of instrument drivers.

Through the mid-1980s until today, more and more users have taken advantage of instrument driver concepts, whether internally designed or through a commercial software product. While the benefits were real, one drawback was that each software package had a unique approach to instrument drivers and unique instrument drivers. As years passed and the marketplace evaluated this variety of approaches, the natural process of competition helped the market refine the good ideas and discard the others. It also allowed the developers of successful products to continue the invest-ment needed to continuously improve and refine the product itself and its associated instrument drivers, take advantage of the latest technology, and maintain compatibility with the rapidly evolving PC market.

By the early 1990s, both personal computers and instrument driver technology had become a mainstream technology for instrumentation users. Though software products from different vendors still had unique instru-ment drivers, the more mature packages generally all featured a very long list of available instrument drivers. Some instrument vendors had recognized the value an instrument driver adds to their instrument and were actively involved in writing and promoting instrument drivers. Other instrument vendors, on the other hand, continued to simply document their command sets in the instrument manual and leave the instrument driver up to someone else, often the software supplier.

The most successful instrument driver concepts have always distributed instrument drivers in source code and provided end users with access to the same tools developers use to write drivers. With this philosophy, new instrument drivers were often easily developed by end users by modifying an existing driver for another instrument. End users, in general, had come to view the availability of an instrument driver as an important factor in the choice of a particular instrument. However, users still had access to standard instrument driver development tools and source code for other instruments.

With this situation in mind, the VXI*plug&play* Systems Alliance identified an initial set of guiding principles to drive the activities of the alliance and come up with solutions:

- Maximize ease of use and performance

- Maintain long-term compatibility with the installed base

- Maintain multivendor open architectures

- Maximize multiplatform capability

- Maximize expandability and modularity in frameworks

- Maximize software reuse

- Standardize the use of system software elements

- Treat instrument drivers as part of the instrument

- Accommodate established standards, both de facto and formal

- Maximize cooperative support of end users

☐ 3.2.2   VISA from the Users' Perspective

An important consideration for instrument drivers is how they perform I/O to and from instruments. In the VXI*plug&play* architecture, the I/O interface is provided by a separate layer of software that is standard and available on numerous platforms.

This principle means that VISA is similar to most existing I/O libraries: The standard provides a set of core functions that you can use to control arbitrary device types, regardless of the underlying physical interface. In addition, there is a small collection of interface-specific functions offering you all benefits of a specific interface. One may regard VISA as a type of functional superset of existing I/O software products offered by leading measurement technology manufacturers. The GPIB library NI-488.2 of National Instruments contains approximately 60 functions, and NI's NI-VXI library offers about 130 functions, while Hewlett-Packard's SICL library has more than 100 functions. Compared to this, VISA requires only about 90 operations to provide the functionality of these products and more (see Figure 3-20).

For use on MS Windows PCs, VISA ships in the form of a Dynamic Link Library (DLL). For UNIX systems, VISA is available as Shared Library (sl). VISA uses a language-independent syntax, which can be addressed from the popular text and graphic programming environments. These are text-based languages or development environments like MS C, C++, Visual Basic, LabWindows, and graphical languages like LabVIEW and HP VEE. All device drivers based on the VISA standard are available in source code for LabVIEW and LabWindows.

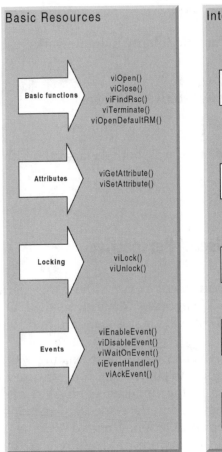
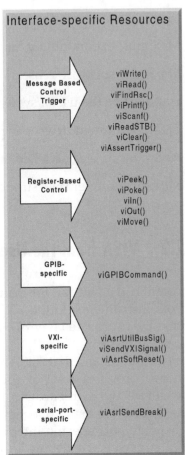

Figure 3-20

A subset of the VISA command set

To fully comply with the VXI*plug&play* standard, a device has to be shipped with the following components:

- Interactive user interface (soft front panel) as executable program

- C library, containing an ANSI C device driver (.c, .h), a function panel file (.fp), and a dynamic link library (.dll)

- Knowledge base, containing the specification of a measuring device

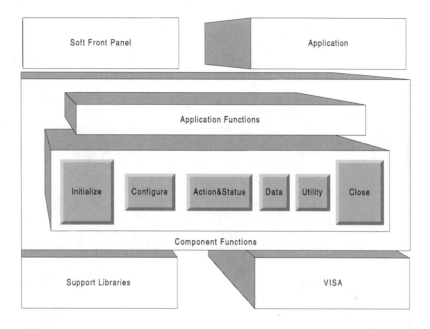

Figure 3-21

Device driver model based on VISA

Figure 3-21 illustrates a device driver based on VISA.

Basically, there are two approaches for a VISA model: first, an external approach to show how the driver adapts to the software environment (as closed structure) and second, an internal approach to show the functionality of a driver.

The external approach considers the driver as the functional core, and an application program builds on this core. To support the software developer, there is an interactive front panel for direct testing and understanding the individual driver and device functions. The driver itself relies on I/O routines, which handle the actual conversion for the connected interface bus. In addition, there is a feature to integrate other software components (e.g., additional libraries).

The internal organization of a measuring device driver consists of six different functional groups, accommodating the functions on the lowest level: initialization, termination, configuration, action sequence/status, data processing, and additional functions. On this level, but still within the core, are application functions, containing meaningful measuring sequences (horizontal function chains) or functional groups (vertical function chains).

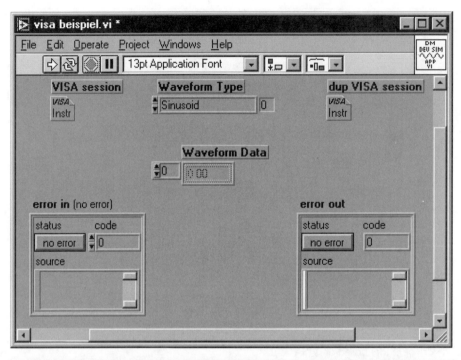

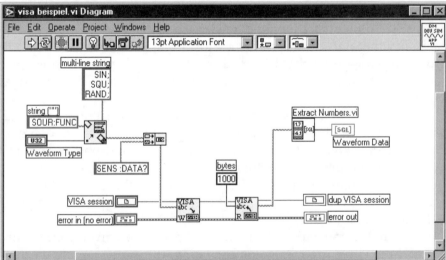

Figure 3-22

Graphical implementation of a sample program in LabVIEW (panel and diagram)

This structure is maintained for all device drivers so that once you understand the principle of one driver, you can integrate any other driver intuitively, which saves you a lot of time.

❏ 3.2.3   VISA in Real-World Applications

By now you may probably ask the question: "How does a VISA program look in practice?" Figure 3-22 shows the VISA syntax to control a GPIB instrument in LabVIEW. This example includes the block diagram and the front panel. The block diagram shows two important functions: read and write.

In this practical example, the Pick Line and Append function produces a control string, composed of two substrings. Using the generated string and an additional string, a Concatenate function produces a Write string for the VISA Write VI to address a connected multifunction device (in this case, a curve generator). The read signal with a length of 1000 measured values is read through the VISA Read function and subsequently visualized.

❏ 3.2.4   VISA in Distributed Systems

As one of the important VISA standardization goals is transparent communication in distributed measuring systems, we consider the measuring arrangement shown in Figure 3-23 to illustrate this capability. This example emphasizes the following points:

- The system is considered as one single unit, although the measuring hardware is composed of three different types.

- The entire communication between the two applications and the devices is handled by the VISA Resources Manager, which is connected to the VISA Device Resources.

- Applications do not have to know the physical location of the device to be controlled.

Next-generation VISA will respond increasingly to the requirements in distributed measurement architectures. Plans also include additional object-oriented enhancement and OLE support.

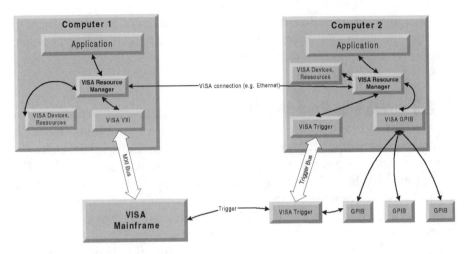

Figure 3-23

Using VISA in a distributed measuring architecture

☐ 3.2.5   LabVIEW and VISA – Summary

The tremendous success of the VXI*plug&play* Systems Alliance shows how important it is today to encourage manufacturers to adopt (de facto) standards. VISA represents a milestone in the history of I/O driver software standardization. It is a testament to the value of cooperation among product vendors who wish to establish more specific definitions regarding the interoperability of their products and services.

3.3   LabVIEW and the GPIB Bus

Bus systems are of increasing importance in modern computer-assisted measurement and test technologies: They offer standardized interfaces to interconnect components with different functions in almost any arbitrary constellation. Today, this is the only way to achieve the high degree of flexibility and modularity required by modern measurement and automation systems, while harnessing the benefits of modern PC technology at the same time.

GPIB, without a doubt, is the most widely used bus standard for interfacing test and measurement instruments with computers. Many years ago, instrument vendors and users recognized the innovation in interfacing technology that GPIB offered for instrumentation. This section discusses the basics of this important interface and its capabilities. Understanding this concept is necessary for the efficient use of the GPIB libraries built in LabVIEW and for the real-world applications described in later sections.

❏ 3.3.1 History

Hewlett-Packard developed the *General Purpose Interface Bus* (*GPIB*) in the late 1960s to facilitate communication between computers and instruments. The Institute of Electrical and Electronic Engineers (*IEEE*) standardized GPIB in 1975, and it became known as the IEEE-488 standard. In 1977, the International Electrotechnical Commission (*IEC*) adopted it as an international standard and renamed it *IEC 625-1*. Today, this bus system is known under the following names:

- General-Purpose Interface Bus (GPIB)

- Hewlett-Packard Interface Bus (HPIB)

- IEEE-488 Bus

- IEC Bus

GPIB's original purpose was to provide computer control of test and measurement instruments. However, its use has expanded beyond these applications into other areas, such as computer-to-computer communication and control of multimeters, scanners, and oscilloscopes.

GPIB is a digital, 24-conductor parallel bus. It consists of eight data lines, five bus management lines (ATN, EOI, IFC, REN, and SRQ), three handshake lines, and eight ground lines. GPIB uses an 8-bit parallel, byte-serial, asynchronous data transfer scheme. In other words, whole bytes are sequentially moved across the bus at a speed determined by the slowest participant in the transfer. Because GPIB sends data in bytes, the messages transferred are frequently encoded as ASCII character strings. Your computer can only perform GPIB communication if it has a GPIB board (or external GPIB box) and the proper drivers installed.

You can have many instruments and computers connected to the same GPIB bus. Every device, including the computer interface board, must have a unique GPIB address between 0 and 30 so that the data source and destinations can be identified by this number. Address 0 is normally assigned to the GPIB interface board. Instruments connected to the bus can use addresses 1 through 30. The GPIB has one controller, usually your computer, that controls the bus management functions. To transfer instrument commands and data on the bus, the controller addresses one *talker* and one or more *listeners*. The data strings are then sent across the bus from the talker to the listener(s). The LabVIEW GPIB VIs automatically handle the addressing and most other bus management functions, saving you the hassle of low-level programming.

Although using GPIB is one way to bring data into a computer, it is fundamentally different from performing data acquisition, even though both use boards that plug into the computer. Using a special protocol, GPIB talks to another computer or instrument to bring in data acquired by that device, while data acquisition involves connecting a signal directly to a DAQ board in the computer.

To use GPIB as part of your virtual instrumentation system, you need a GPIB board or external box, a GPIB cable, LabVIEW and a computer, and an IEEE-488 compatible instrument with which to communicate (or another computer containing a GPIB board). You also need to install the GPIB driver software on your computer, according to the directions that accompany LabVIEW or the board.

A GPIB board is used to control and communicate with one or more external instruments that have a GPIB interface. All HP instruments support it as well as thousands of others. GPIB was updated and standardized by the IEEE and was duly named IEEE 488.2 (see Section 3.3.3). Some nice features about GPIB are:

- It transfers data in parallel, one byte at a time.

- The hardware takes care of handshaking, timing, etc.

- Several instruments (up to 15) can be strung together on one bus.

- Data transfer is fast: 800 Kbytes/s or more.

You can obtain a plug-in GPIB board for almost every platform and bus as well as external interfaces that will convert your serial port, parallel port, Ethernet connection, or SCSI port to GPIB.

Another feature that makes GPIB popular is that the computer and instrument talk to each other using simple, intuitive ASCII commands.

A multitude of GPIB instrument drivers, available free of charge, makes communication with an external instrument as simple as using a few subVIs that do all the ASCII commands for you.

There are two palettes for GPIB communication: one is called *GPIB* and the other, chock-full of more functions, *GPIB 488.2*. GPIB 488.2, also called *HS488* or *IEEE 488.2*, is the latest "version" of the standard. 488.2 establishes much tighter rules about the communications and command sets of instruments in an attempt to make programming easier and more uniform for different GPIB devices. Most GPIB devices will still work if you use LabVIEW's GPIB palette. If you need to adhere to the IEEE 488.2 standard, though, you should use the GPIB 488.2 palette.

Note that GPIB VIs will be replaced by the functions in the VISA palette. VISA VIs are more general-purpose functions that can communicate with other instrument types, such as VXI.

All GPIB instruments have an address, which is a number between 0 and 30, uniquely identifying the instrument. Sometimes instruments will also have a secondary address. In all the GPIB VIs, you must provide the address of the instrument (in string format) in order to talk to it. This makes it easy to use the same VIs to talk to different devices on the same GPIB bus. You can enclose the addresses in a `case` structure, for example.

The majority of GPIB communication involves initializing, sending data commands, perhaps reading back a response, triggering the instrument, and closing the communication.

❏ 3.3.2   IEEE 488.1 – The Traditional GPIB Standard

The IEEE-488 bus was developed to connect and control programmable instruments and to provide a standard interface for communication between instruments from different sources. Almost any instrument can be used with the IEEE-488 specification, because it says nothing about the function of the instrument itself or about the form of the instrument's data. Instead, the specification defines a separate component, the interface, that can be added to the instrument. The signals passing into the interface from the IEEE-488 bus and from the instrument are defined in the standard. The instrument does not have complete control over the interface. Often the bus controller tells the interface what to do. The active controller performs the bus control functions for all the bus instruments.

At power-up time, the IEEE-488 interface that is programmed to be the system controller becomes the active controller in charge. The system controller has several unique capabilities, including the ability to send Interface Clear (IFC) and Remote Enable (REN) commands. IFC clears all device interfaces and returns control to the system controller. REN allows devices to respond to bus data once they are addressed to listen. The system controller may optionally pass control to another controller, which then becomes the active controller.

Three types of devices can be connected to the IEEE-488 bus:

- Talkers

- Listeners

- Controllers

Some devices include more than one of these functions. The standard allows a maximum of 15 devices to be connected on the same bus. A minimum system consists of one controller and one talker or listener device (i.e., an HP 700 with an IEEE-488 interface and a voltmeter). Figure 3-24 shows a typical GPIB system.

It is possible to have several controllers on the bus, but only one may be active at any given time. The active controller may pass control to another controller, which in turn can pass it back or on to another controller. A listener is a device that can receive data from the bus when instructed by the controller, and a talker transmits data on to the bus when instructed. The controller can set up a talker and a group of listeners so that it is possible to send data between groups of devices as well. The GPIB bus management lines handle the following tasks:

- **IFC** (Interface Clear) — The IFC signal is asserted only by the system controller in order to initialize all device interfaces to a known state. After releasing IFC, the system controller is the active controller.

- **ATN** (Attention) — The ATN signal is asserted by the controller to indicate that it is placing an address or control byte on the data bus. ATN is released to allow the assigned talker to place status or data on the data bus. The controller regains control by reasserting ATN; this is normally done synchronously with the handshake to avoid confusion between control and data bytes.

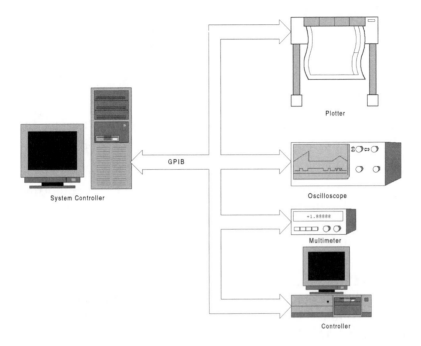

Figure 3-24

Typical GPIB system

- **REN** (Remote Enable) — The REN signal is asserted only by the system controller. Its assertion does not place devices into remote control mode; REN enables a device to go into remote mode only when addressed to listen. When in remote mode, a device should ignore its local front panel controls.

- **EOI** (End Or Identify) — The EOI signal has two uses. A talker may assert EOI simultaneously with the last byte of data to indicate end-of-data. The controller may assert EOI along with ATN to initiate a parallel poll. Although many devices do not use parallel poll, all devices should use EOI to end transfers.

- **SRQ** (Service Request) — The SRQ line is like an interrupt: it may be asserted by any device to request the controller to take some action. The controller must determine which device is asserting SRQ by conducting a serial poll. The requesting device releases SRQ when it is polled.

- **DAV** (Data Valid) — The DAV handshake line is asserted by the talker to indicate that a data or control byte has been placed on the data lines and has had the minimum specified stabilizing time. The byte can now be safely accepted by the devices.

- **NRFD** (Not Ready For Data) — The NRFD handshake line is asserted by a listener to indicate it is not yet ready for the next data or control byte. The controller will not see NRFD released (i.e., ready for data) until all devices have released it.

- **NDAC** (Not Data Accepted) — The NDAC handshake line is asserted by a listener to indicate it has not yet accepted the data or control byte on the data lines. The controller will not see NDAC released (i.e., data accepted) until all devices have released it.

Within a GPIB interface, the ATN line is used to distinguish between device data and dedicated interface messages. Device data (like instrument programming commands and returned measurement results) are always sent with ATN = 0. The interface messages are sent with ATN = 1 and serve a particular purpose, which is clearly defined by the GPIB standard. For example, a controller uses interface messages to address a talker or one or more listeners, to execute a serial poll when a service request (SRQ) occurs, or to execute a "device clear" action.

Note that the GPIB bus uses what is called "negative logic": A message being sent over a GPIB line is assumed to be true (1) when the electrical voltage level is low. Similarly, the message is false (0) when the electrical level is high. For example, when the data valid (DAV) line is low, the corresponding DAV message is true (DAV = 1).

Additionally, the GPIB handshake uses a "wired OR" concept for the NRFD and NDAC lines. This means that an NRFD line will only go high when none of the devices pulls the line low. In other words, the NRFD line will go high only when all devices send NRFD high (none pulls low).

NRFD (Not Ready For Data) is the negation of RFD (Ready For Data). As a result, a source will not receive the RFD = true message until all acceptors send the RFD message true. Therefore, it is said that an acceptor sends the RFD message "passive true." A similar concept applies to the Not Data Accept (NDAC) line, which carries the Data Accept (DAC) message. Figure 3-25 describes this concept.

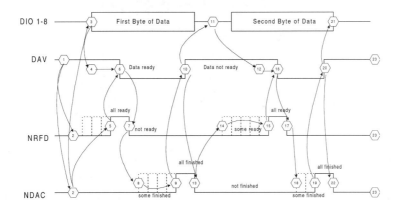

Figure 3-25

Three-wire standard GPIB handshake

- The data sender (controller, talker) checks the NRFD line to ensure that all devices are ready to receive. This is true when the NRFD line is not in active state (i.e., when the line is low).

- The data sender places the first data byte to the data bus over lines DIO1–DIO8 and sends DAV. Data present on the bus is now valid and can be accepted by the receiver.

- The data receiver (listeners) activates NRFD, setting it to low, to indicate that it is currently busy and not available for additional data.

- The data sender will wait until all devices have accepted data. Subsequently the NDAC line is set to high. This means that the slowest device on the bus dictates how fast data is transmitted.

- The data sender sets DAV to high (i.e., the data becomes invalid).

- The data sender places the second data byte to the data bus and waits until the NRFD line is high. Once all devices are ready to accept a new data byte, the sender can declare the second data byte to be valid and set DAV to low.

This means that data transfer is asynchronous at the speed of the slowest device. We also see that this method allows addressing of several acceptors at various transmission speeds in one single transfer. This allows fast devices to communicate with both fast and slow ones.

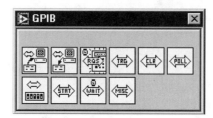

Figure 3-26

Conventional IEEE 488.1 LabVIEW library elements

   In summary, the IEEE-488.1 standard greatly simplifies the interconnection of programmable instruments by clearly describing mechanical,
hardware, and electrical protocol specifications. This was the first standard
to allow instruments from different vendors to connect over a standard
cable. Notice that this standard does not address data formats, status
reporting, message exchange protocol, common configuration commands,
or device specific commands. LabVIEW features ten basic VIs for communication as specified by the IEEE 488.1 standard (Figure 3-26).

❑ 3.3.3   IEEE 488.2 – The New GPIB Standard

The IEEE-488.2 standard enhances and strengthens the IEEE-488.1 standard
by specifying data formats, status reporting, error handling, controller
functionality, and common instrument commands. Another main
advantage is that the IEEE 488.2 standard remains compatible with the
original IEEE 488.1 standard. Although the IEEE 488.2 standard has a
greater impact on instruments that on Controllers, several new requirements and optional improvements for Controllers make them a necessary
component of new test systems. The following sections discuss the IEEE
488.2 requirements and recommendations.

Minimum Requirements of IEEE 488.2

IEEE 488.2 defines a set of minimum requirements for the traditional IEEE
488.1 interface. Essentially, all devices must be able to send and receive data,
support SRQ, and respond to a `Device Clear` function.

Message Transfer Protocol

IEEE 488.2 defines the formats of commands that are sent to devices and the
formats and encoding of replies received from devices. It also specifies error

handling and procedures for exceptional cases. In addition, it introduces the concept of *Forgiving Listening and Precise Talking*. Consider the following example to understand this concept: A function generator can output frequencies of 1, 5, 10, 50, and 100 Hz. Forgiving Listening requires the function generator to accept commands in data formats of 54.576, 5.0E+1, +50, 50, or +0.523E+2. However, to act as a precise talker, the device has to exactly maintain a specific reply protocol.

Data Coding and Data Formats

Basically, the 7-bit ASCII code is specified for all device messages. An 8-bit binary code is valid for integer values, and an IEEE754 format is valid for binary floating-point numbers.

Uniform Commands

All devices have to run some uniform operations on the bus (e.g., status queries, device identification, or synchronization). For instance, IEEE 488.2 specifies a set of common commands, which all devices have to understand, regardless of manufacturer and functionality. Table 3-8 lists the mandatory commands and queries as specified in IEEE 488.2.

Status Messages

IEEE 488.2 builds on the 1-byte state byte register defined in IEEE 488.1 and expands it. Formerly, only bit 6 (RQS bit) of the state byte register had been defined in the IEEE 488.1 specification, while all other bits could be optionally chosen by the manufacturer. The RSQ bit remains as specified in IEEE 488.1, but a new Event Status Bit (ESB) and a Message Available Bit (MAB) were added. Bits 0 through 3 and 7 are optional. The EBS bit indicates whether one of the standard events defined in the standard event status register has occurred. The MAB bit indicates whether a message is available in the output queue. IEEE 488.2 also defines a series of requests to the controller, including a set of interface options according to IEEE 488.1, bus control sequences, and bus protocols. In addition, an IEEE 488.2 controller has to provide functions to allow an application program to do the following:

- hold the IFC line >100 μs on true

- detect false/true change in the SRQ line

- monitor the SRQ line status

- set the REN line

- monitor and set the EOI line

- send interface commands specified in IEEE 488.1

- set a time-out for I/O transactions

- check individual bits of the status register

- receive and send common commands as specified in IEEE 488.2

IEEE 488.2 Control Sequences

The IEEE 488.2 standard defines common bus states and how devices should respond to standard messages. If an IEEE 488 instrument was designed for certain command sequences and/or bus conditions used by an IEEE 488 Controller from one vendor, the instrument might not work properly with a Controller from another vendor. With IEEE 488.2 control sequences, both Controller and Talker/Listener devices have a set of rules to follow to ensure complete compatibility among vendors. Table 3-9 shows the 15 required control sequences and four optional control sequences defined in the IEEE 488.2 standard Controllers. All of these control sequences are part of the LabVIEW GPIB 488.2 library.

IEEE 488.2 Protocols

IEEE 488.2 protocols are high-level routines that reduce development time because they use a proven algorithm, combining a number of control sequences to execute the most common test system operations. IEEE 488.2 defines two required protocols and six optional protocols as shown in Table 3-10. One of the most important protocols, FINDLSTN, uses the IEEE 488.2 Controller capability of monitoring bus lines to identify which devices are present on the bus. An application program might use FINDLSTN at the beginning of the test sequence to ensure proper system configuration and to create a valid list of active GPIB devices. The Controller implements the FINDLSTN protocol by issuing particular listen addresses, one at a time, and then monitoring the NDAC handshake line to determine if devices exist at those addresses. The result of the FINDLSTN protocol is a list of addresses for all the located devices.

Another important protocol, FINDRQS, provides an efficient mechanism for locating and polling devices that request service. It uses the IEEE 488.2 Controller capability of sensing the FALSE to TRUE transition of the SRQ line.

Table 3-8   Mandatory IEEE 488.2 commands

Mnemonics	Group	Description
*IDN?	System Data	Identification request
*RST	Internal Operations	Reset
*TST?	Internal Operations	Self-test request
*OPC	Synchronization	Operation complete
*OPC?	Synchronization	Operation complete request
*WAI	Synchronization	Wait for completion
*CLS	Synchronization	Clear status
*ESE	Status and Event	Event status enable
*ESE?	Status and Event	Event status enable request
*ESR?	Status and Event	Event status register request
*SRE	Status and Event	Service request enable
*SRE?	Status and Event	Service request enable request
*STB?	Status and Event	Read status byte request

An application program may prioritize the device list so that the more critical devices receive service first. If the SRQ line is asserted, calling FINDRQS immediately results in increased program efficiency and throughput.

For compatibility, LabVIEW contains two independent GPIB libraries. The GPIB library was briefly described in Section 3.3.2. Figure 3-27 shows the GPIB IEEE 488.2 library, including both mandatory and optional commands (see Table 3-9). IEEE 488.2 requires compatible hardware. The LabVIEW GPIB IEEE 488.2 library contains all control sequences specified in IEEE 488.2, i.e., 15 mandatory and four optional control sequences.

Goals of HS488 Standardization

HS488 is a high-speed version of the traditional GPIB handshake protocol. The goals of this development effort were to increase overall I/O throughput, maintain compatibility with existing instruments, and preserve the advantages of multivendor interoperability.

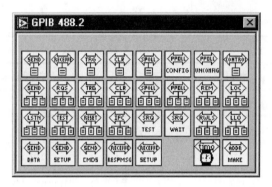

Figure 3-27

LabVIEW IEEE 488.2 library elements

Table 3-9 *IEEE 488.2 mandatory and optional control sequences*

Control Sequence	Description	Type
SEND COMMAND	Send command (ATN true)	mandatory
SEND SETUP	Set address to send data	mandatory
SEND DATA BYTES	Send data (ATN false)	mandatory
SEND	Send program messages	mandatory
RECEIVE SETUP	Set address to receive data	mandatory
RECEIVE RESPONSE MESSAGE	Receive data (ATN false)	mandatory
RECEIVE	Receive reply messages	mandatory
SEND IFC	Activate IFC line	mandatory
DEVICE CLEAR	Set device to DCAS status	mandatory
ENABLE LOCAL CONTROLS	Set device to local status	mandatory
ENABLE REMOTE	Set device to remote status	mandatory

Table 3-9 (Continued)

Control Sequence	Description	Type
SET RWLS	Set device to remote status with local lockout status activated	mandatory
SEND LLO	Set device to local lockout status	mandatory
READ STATUS BYTE	Read IEEE 488.1 status byte	mandatory
TRIGGER	Send group Trigger (GET) command	mandatory
PASS CONTROL	Pass controller function to other device	optional
PERFORM PARALLEL POLL	Perform parallel poll	optional
PARALLEL POLL CONFIGURE	Enable parallel polling capability of device	optional
PARALLEL POLL UNCONFIGURE	Disable parallel polling capability of device	optional

Compatibility is critical because of the sheer volume of computer-controlled GPIB instruments being used in the industry today. Devices implementing the new high-speed protocol must be able to coexist and communicate with the huge installed base of existing instruments. This new protocol must be able to work with standard GPIB cables and not impose any further restrictions on cabling distances. Ideally, an improved protocol should be transparent to the application program to protect the industry's investment in existing application code.

Addressing the overall I/O performance, two issues are paramount. First, the raw transfer rate must be increased significantly — preferably an order of magnitude — but at least to the theoretical limit that the cable medium will support. Second, this increase in throughput cannot be offset by additional overhead in setting up the transfer.

Table 3-10   *IEEE 488.2 controller protocols*

Protocol	Name	Type
RESET	Reset System	mandatory
FINDRQS	Find Device Requesting Service	optional
ALLSPOLL	Serial Poll All Device	mandatory
PASSCTL	Pass Control	optional
REQUESTCTL	Request Control	optional
FINDLSTN	Find Listeners	optional
SETADD	Set Address	optional
TESTSYS	Self-Test System	optional

The HS488 Protocol

HS488 meets the requirements for successfully deploying a high-speed extension to the GPIB standard — speed, compatibility, and transparency in existing systems. By using the "wink" signal to initialize data transfers, HS488 maintains compatibility with existing instruments by reverting to the traditional three-wire handshake if any instrument involved in the transfer is not HS488-compliant. Because this protocol is implemented in hardware, HS488 works with existing GPIB applications and does not impose additional software overhead for configuring HS488 transfers. HS488 listeners also use an input buffer to ensure that no data is lost if they assert either the NDAC or NRFD signals to force the talker to pause data transfer or revert back to the IEEE 488.1 handshake. HS488 increases the maximum data throughput of a GPIB system to 8 Mbytes/s and works with existing cables, the same number of devices, cable lengths, and cabling configurations as the original specification. Table 3-10 lists the controller protocols used in HS488.

HS488 Data Transfer Flow Control

The listener may assert NDAC to temporarily prevent more bytes from being transmitted, or assert NRFD to force the talker to use the three-wire handshake. Through these methods, the listener can limit the average transfer rate. However, the listener must have an input buffer that can accept short bursts of data at the maximum rate because by the time NDAC or NRFD propagates back to the talker, the talker may have already sent another byte.

You can configure the required settling and hold times, depending on the total length of cable and number of devices in your system. Between two devices and 2 m of cable, HS488 can transfer data up to 8 Mbytes/s. For a fully equipped system with 15 devices and 15 m of cable, HS488 transfer rates can reach 1.5 Mbytes/s.

HS488 controllers always use the standard IEEE 488.1 three-wire handshake to transfer GPIB commands.

Standard IEEE 488.1 Handshake

The standard IEEE 488.1 three-wire handshake requires the talker to assert the Data Valid (DAV) signal to indicate to the listener that a data byte is available and for the listener to unassert the NDAC signal when it has accepted that byte. This means that a byte cannot transfer in less than the time it takes for the following events to occur:

- The DAV signal propagates to all listeners;

- The listeners accept the bytes and assert NDAC;

- The NDAC signal propagates back to the talker; and

- The talker gives a settling time before reasserting DAV.

HS488 Handshake

HS488 increases system throughput by removing propagation delays associated with the three-wire handshake. To enable the HS488 handshake, the talker pulses the NRFD signal line after the controller has addressed all listeners. If the listener is HS488-compliant, then the transfer occurs using the HS488 handshake. Once HS488 is enabled, the talker places a byte on the GPIB DIO lines, waits for a preprogrammed settling time, asserts DAV, waits for a preprogrammed hold time, unasserts DAV, and drives the next data byte on the DIO lines. The listener keeps NDAC unasserted and must accept the byte within the specified hold time. A byte must transfer in the time set by the settling time and hold time, without waiting for any signals to propagate along the GPIB cable. Figure 3-28 is a schematic view of the HS488 handshake.

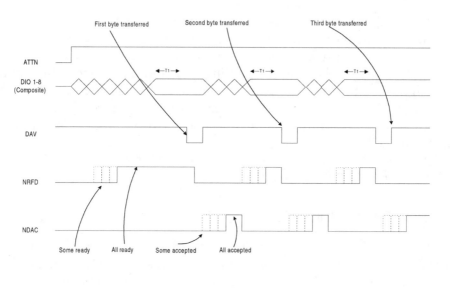

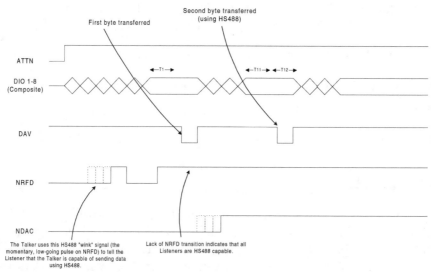

Figure 3-28

HS488 handshake

HS488 Protocol Initialization

The premise behind HS488 is to eliminate much of the propagation delays associated with the IEEE 488.1 handshake. Instead of waiting for listeners to assert NDAC or NRFD, the talker assumes that HS488 listeners are always ready for data and always accepting data. Once the listeners are addressed and ready for data, they unassert the NRFD signal. After the HS488 active talker detects that all devices are ready for data, it asserts NRFD, then unasserts it after a predetermined time. This process is known as the *HS488 wink*. Addressed HS488 listeners interpret the NRFD wink to signify that the active talker is HS488-compliant. If all listeners can use HS488, the data transfer will use the HS488 protocol until one of the listeners exits HS488 to use normal IEEE 488.1 handshaking. If one or more listeners cannot use HS488, they ignore the wink, and the entire transfer takes place using the IEEE 488.1 handshake.

HS488 Data Transfer

HS488 employs the same proven, high-speed, data streaming techniques used with VME, PCI, and Fast SCSI. Once HS488 is enabled for the transfer, the talker places the first data byte on the GPIB DIO lines and asserts DAV. If the talker sees the NRFD signal unasserted and the NDAC signal asserted, there is at least one active listener on the GPIB and all active listeners are ready to receive data. At this point in the handshake, HS488 listeners know that the talker is HS488-compliant. Next, the HS488 listeners unassert NDAC to signal to the talker that they are HS488-compliant. The talker then begins transferring bytes using the HS488 noninterlocked handshake protocol. Once HS488 listeners are enabled, they always accept data and are always ready for data.

The talker places a second data byte on the DIO lines, asserts and then unasserts DAV. The listener keeps NDAC unasserted and accepts the byte. The talker continues transmitting bytes using the noninterlocked handshake. As long as the talker detects NDAC and NRFD both high, it continually sends data bytes, asserting DAV for each byte.

☐ 3.3.4   Standard Commands for Programmable Instrumentation

The *Standard Commands for Programmable Instrumentation (SCPI)* defines a standard set of commands to control programmable test and measurement devices in instrumentation systems. SCPI was initially developed by Hewlett-Packard. Starting with families of languages for oscilloscopes, meters, RF sources, etc., HP began to develop a common command set based on signal parameters and a universal instrument block diagram. The internal standard was first known as *HP-SL*, then as *TMSL*, and finally as *SCPI*. Simultaneously, Tektronix developed *ADIF*, a data interchange format combining raw waveform data with contextual information about how a measurement was made. These developments laid the foundation for the SCPI Consortium. The Consortium's founding members include Brüel & Kjaer, Fluke, Hewlett-Packard, Keithley Instruments, National Instruments, Tektronix, Philips, Racal-Dana, and Wavetek. Figure 3-29 shows the GPIB standard structure, with SCPI on top of the hierarchy.

SCPI is a major advance in providing a standard instrument vocabulary and provides the following benefits to users of programmable instrumentation:

- Shorter programming time (no learning of unique instrument vocabularies)

- More understandable and maintainable test programs

- Greatly increased likelihood of instrument interchangeability

The system controller sends commands to one or more instruments over the bus. These commands are called *program messages*. Instruments may send reply messages back to the controller. The reply message may be a measurement result, instrument setting, error message, etc. When a program message directly generates a reply, it is called a *query*. In the past, the commands that controlled a particular device function varied among instruments with similar capabilities. SCPI provides a uniform and consistent language to control test and measurement instruments. The same commands and replies control corresponding instrument functions in SCPI equipment, regardless of the vendor or the type of instrument.

SCPI provides a high degree of consistency among instruments. The command to measure a frequency is the same whether the measurement is made by an oscilloscope or a counter. The set of commands to control multimeters from two vendors differs only in places where the underlying hardware is different. Thus, instruments from different vendors can be expected to be interchangeable in many applications.

One of the fundamental objectives of SCPI is to provide an easy way to perform simple operations. The MEASure command is the easiest way to configure and read data from an instrument. For example, when the program message ":MEAS:VOLT:AC?" is received by a voltmeter, the meter will select settings and configure its circuitry for an AC voltage measurement, initiate the measurement, and return the results to the system controller.

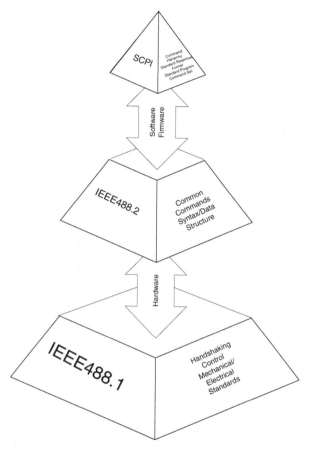

Figure 3-29

GPIB standard structure

The commands in the measure family are "signal oriented" in the sense that they specify a measurement in terms of the desired result and in terms of the signal to be measured. You can specify characteristics of the signal measurement, such as expected signal value or the resolution of the measurement, by adding parameters to the command. Based on these parameters, the instrument selects suitable settings. For example, you might send the following command to a voltmeter:

:MEASure:VOLTage:AC? 20, 0.001

This command instructs the meter to configure itself to make an AC voltage measurement on a signal of around 20 V with a 0.001-V resolution. Portions of the command in lowercase type are optional. The question mark at the end of the command instructs the voltmeter to return the measured value to the controller. The heart of the SCPI standard is the *Command Reference* — a grouped list of definitions for all the program messages. These definitions precisely specify the syntax and semantics for every SCPI message. Instrument functions covered by the standard may only be controlled through SCPI commands. This restriction does not mean that SCPI equipment manufacturers are limited to those functions already defined in SCPI. SCPI was designed with a modular structure that permits commands controlling new functions to be added at any time.

Measure commands provide the highest level of compatibility between instruments because the measurement is specified entirely in terms of signal parameters. No knowledge of the hardware components of the instrument is required to perform the measurement. The `:MEAS:VOLT:AC` command applies equally well to any instrument that can make an AC voltage reading: a digital multimeter, digitizing oscilloscope, or phase-angle voltmeter.

Some measurements require direct control over an instrument's hardware. To provide this control, SCPI contains command subsystems that control particular instrument functions and settings. These commands trade interchangeability for fine control. The ability to configure instruments and make measurements with different degrees of control is a major benefit of SCPI. To define the commands for this control, SCPI uses a generalized model of a programmable instrument. Commands to control a particular instrument's hardware make reference to this generalized block diagram, rather than the specific instrument's hardware diagram.

This model controls the way instrument functionality is divided among the SCPI command subtrees. This block diagram permits commands to be placed within SCPI in a consistent manner for all types of instruments. Each block in the model is a major area of signal processing functionality and generally corresponds to major SCPI command subtrees.

Another advantage of SCPI is the possibility to reuse instrument firmware. This applies both to reuse of existing code and to development of common code components for new functions.

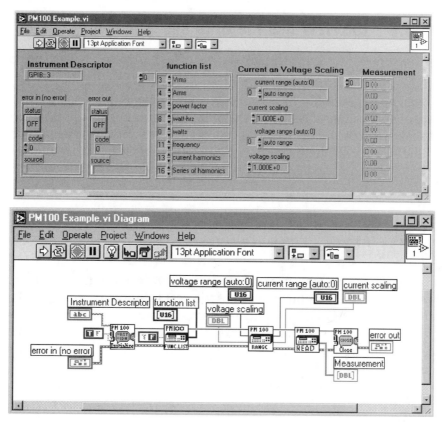

Figure 3-30
VISA driver for Voltech PM100 power analyzer

❑ 3.3.5   Sample Application – LabVIEW VISA Driver for an IEC Bus Device

This section briefly describes how you can use a LabVIEW VISA driver for an IEC bus device.

Figure 3-30 shows how a VISA device driver is set up. All supported functions of the measuring device are summarized by topic in virtual instruments. By combining device functions contained in various VIs, you quickly obtain a device driver that features the required performance characteristics. This example initializes the performance analyzer, defines the list of functions, scales the power and voltage ranges, and finally reads the desired measuring values.

❑ 3.3.6   LabVIEW and the GPIB Bus – Summary

GPIB, without doubt, is the most widely used bus standard for interfacing test and measurement instruments with computers. GPIB plays a unique role. GPIB was designed specifically for simplicity of implementation and use in a wide variety of applications in stable or changing configurations. It uses a three-wire handshake protocol that is well understood and has proven to be very reliable. With the new HS488 high-speed version, GPIB offers several advantages that have made it the dominant bus in test and measurement applications.

3.4   LabVIEW and VXI

This section introduces an exciting, fast-growing platform for instrumentation systems. VXI combines some of the best technology from GPIB instruments, plug-in DAQ boards, and modern computers. To better understand this technology, the standard behind it, and the reason why it has been so successful during the past years, we describe some historical information and a small real-world sample application.

❑ 3.4.1   History

VXI stands for *VME eXtensions for Instrumentation*. The VME standard was developed by Motorola and a number of other companies in the late 1970s. It has been widely accepted as a backplane standard for many electronic platforms. It defines the electrical and mechanical backplane characteristics that allow a wide variety of companies to develop products to work in a mix-and-match fashion to develop electronics systems.

The *VXIbus*, an acronym for *VMEbus eXtensions for Instrumentation*, is an instrumentation standard for instrument-on-a-card systems. First introduced by the VXIbus Consortium in 1987 and based on the VME-bus (IEEE 1014) standard, the VXIbus is a fast-growing platform for instrumentation systems. VXI consists of a mainframe chassis with slots that hold modular instruments on plug-in boards. A variety of instrument and mainframe sizes is available from numerous vendors, and you can also use VME modules in VXI systems. VXI has a wide array of uses in traditional test and measurement and *ATE* (*automated test equipment*) applications. VXI's popularity is also growing as a platform for data acquisition and analysis in research and industrial control applications. Many users are migrating to VXI by integrating it into existing systems along with traditional GPIB instruments and/or plug-in DAQ boards.

VXI*plug&play* is a name used in conjunction with VXI products that have additional standardized features beyond the scope of the baseline specifications. VXI*plug&play*-compatible instruments include standardized software, which provides soft front panels, instrument drivers, and installation routines to take full advantage of instrument capabilities and make your programming task as easy as possible. LabVIEW software for VXI is fully compatible with the VXI*plug&play* specifications.

❑ 3.4.2   Motivation

The five test companies that formed the VXI Consortium felt there was a need to develop a standard with which they (and anyone else) could develop instruments that would:

- Be capable of handling demanding electronic test problems

- Be open to all

- Contain interpretable modules

- Reduce the size of current instrumentation systems

- Increase the speed of ATE systems.

To accomplish these goals, the Consortium chose the VME standard because of its many desirable characteristics. It was a well-recognized standard, providing a platform that accomplished the downsizing and speed but was designed primarily for high-speed digital communications. While the problems of developing sophisticated test equipment required these characteristics, developers also wanted to be able to handle low-level analog signals as well as very high frequency signals in the radio frequency (RF) and microwave frequency range.

To allow low-level signals to coexist on a backplane with high-speed digital and RF or microwave signals, the VME standard had to be enhanced to allow for shielding, both on the backplane and between the individual modules themselves. To provide for this shielding, the VXI specification increased the spacing between adjacent boards from 0.8 inch (the VME standard) to 1.2 inches.

In addition, instrumentation modules required more power and a greater variety of available power.

Another very important aspect is that VXI actually has two paradigms for instrument control. The first paradigm is *message-based* control, which is very similar to the message-based control methodology used for years with GPIB instruments. With this paradigm, the instrument is controlled by fairly high level ASCII command and data strings, which are communicated to and from all instruments in a standard way and interpreted by each instrument in a device-specific way.

The other paradigm available in VXI is *register-based* control, which is a new paradigm for instrument control. With register-based control, the instrument is controlled through very low level peeks and pokes of individual binary registers. Because there is no overhead for the instrument to interpret a message string and because the software can directly access binary registers and data on the instrument without any formatting or interpretation, register-based control can, in some cases, offer higher performance than a message-based approach. However, because each register-based device has its own unique set of registers that operate in a unique way for each instrument, the software to control a register-based module is device specific. Register-based control software is similar in concept to the "device driver" software modules required for individual computer plug-in boards such as network interfaces, display adapters, and disk controllers.

Because VXI has two paradigms available for instrument control, it promises the benefit of higher performance than GPIB technology. A particular instrument can use one or the other of these paradigms or a mixture of both on the same module. Some of the VXI performance potential lies in register-based programming, but higher performance is also available through the use of other unique new VXI concepts, such as fast data channels, shared memory, and distributed, multiprocessor architectures that work with both message-based and register-based devices.

❏ 3.4.3 VXI System Variants

Figure 3-31 illustrates how you can set up your VXI system in a wide range of hardware topologies; the most important are:

- **GPIB-VXI** — Allows you to integrate a VXI system into an existing GPIB system.

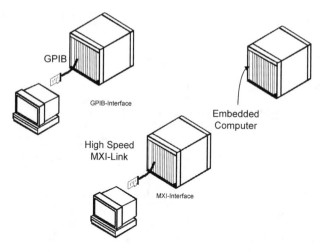

Figure 3-31
VXI system variants

- **VXI-MXI** — Allows you to connect the VXI bus to an external computer.

- **VXIpc** — Allows you to integrate a computer to function as local microprocessor within a VXI mainframe.

- **VXI-DAQ** — These plug-in modules use the VXI trigger bus to synchronize several instrument modules in a VXI system.

☐ 3.4.4 Accessing VXI Modules

You can access VXI modules in two ways. As described above, a VXI system can contain both message-based and register-based modules. Register-based modules are programmed at a low level by exchanging binary data. The code is written to or read directly from the module registers. The advantage of this communication type is speed. The drawback is that there are many types of modules, and writing to and reading from their registers is different for each module (see Figure 3-32).

In contrast to this, message-based modules communicate on a higher level, using ASCII characters. The ASCII character strings vary slightly from one module to another, but you do not have to access registers to pass them on. In addition, many message-based modules use the SCPI format so that the commands are identical for all devices.

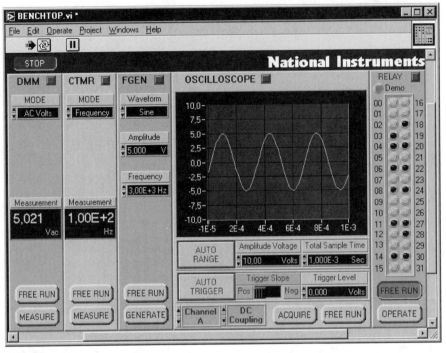

Figure 3-32

Front panel of a VXI system, consisting of five VXI devices

The following example illustrates the difference. In either case, a message should be executed with a DMM and the value should be returned:

- **Message-based** — A controller sends the MEAS:VOLT? ASCII string to DMM. The DMM interprets this string and triggers a measurement. Then it reads the binary information from ADC, converts it into ASCII characters, and returns it to the controller.

- **Register-based** — The module can map its internal register to the VXI bus. This means that the controller can easily manipulate these registers to trigger a measurement and to read the binary measured value from the same registers. This variant saves two process steps: The ASCII character string does not have to be parsed, and data does not have to be converted from binary to ASCII.

Figure 3-33 illustrates the two types.

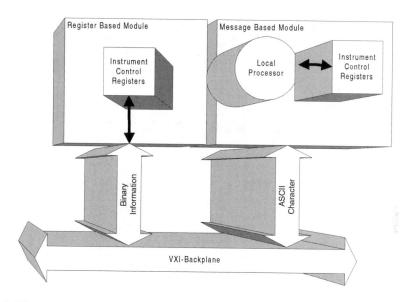

Figure 3-33

Message-based and register-based VXI communication types

The system configuration is divided into three categories: VXI bus control over GPIB, MXI, or VXIpc. A VXI system may, for example, consist of a VXI mainframe, which is connected to an external controller over the GPIB bus. The GPIB controller in the PC uses GPIB to talk to a GPIB interface module, installed in the VXI mainframe, which handles conversion between the GPIB protocol and a VXI word serial protocol.

The configuration tools include VXIInit to initialize the system, a Resource Manager for automatic configuration, VXIEdit to change the controller's start parameters, and the VXI Resource Editor to integrate VME modules.

The interactive programs can be used to learn and experiment with communication paths and functions. They are available in graphical and text-based forms, providing the traditional VXI functions and VISA library functions.

The NI-VXI library contains more than 120 functions for all VXI bus communication and configuration needs (see Figure 3-34). It is identical for all computer platforms, which means that it is source-code compatible.

NI-VXI contains libraries to communicate with message-based and register-based modules. In addition, the library contains functions for signals, interrupts and triggers. National Instruments' hardware and software supports each trigger protocol defined for VXI, including ASYNC, SYNC, START/STOP, and SEMI-SYNC. There are several ways to respond

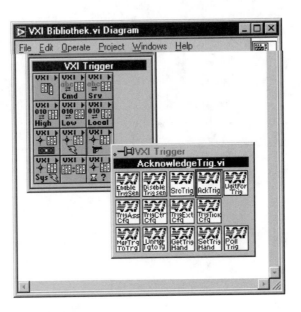

Figure 3-34

VXI libraries

to signals, interrupts, and triggers. You can use NI-VXI functions either to wait for an event to occur or to install an interrupt handler to have an interrupt routine automatically called for you when an event occurs.

NI-VXI function calls are identical on all computer platforms supporting this standard. LabVIEW accesses the NI-VXI API (Application Programming Interface) and offers the VXI library based on this API. The VISA library represents an API interface as defined by the VXI*plug&play* Alliance, which is independent of system and vendor. When you use VISA functions to write your programs, you don't have to worry about the underlying hardware (e.g., GPIB, MXI, Embedded Controller).

Figure 3-35 shows two examples. In the top example, a :MEAS:VOLT? command in SCPI format is sent to a message-based module with the logical address 1 to run a message and output a measuring value. Subsequently, the value is read, stored in the read buffer, and displayed on the front panel.

In the second example, value 29 is written to the register (A24 address space with 0x0 offset) of a register-based module with logical address 2. Subsequently, a word is read from the same address space with offset 0x2 and displayed on the front panel. Note that the names of VISA functions are identical on different platforms to facilitate programming in VISA.

Message based programming with VISA

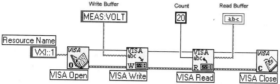

Register based programming with VISA

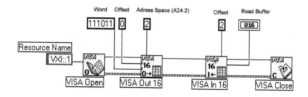

Figure 3-35

Message-based and register-based data flow diagrams using VISA functions

☐ 3.4.5　LabVIEW and VXI – Summary

This section provided a brief description of the VXI standard and VXI products. It showed that VXI combines the best technology from GPIB instruments, modular plug-in DAQ boards, and modern computers. VXI instruments have the ability to communicate at very high speeds, using the best technology from both GPIB instruments and plug-in DAQ boards.

3.5　LabVIEW and PXI

Today's test challenges demand instrumentation at PC prices. From a test engineer's perspective, test complexity is increasing, customers are expecting higher quality, profit margins are decreasing, and time to market is more crucial than ever before. In a nutshell, engineers today must test more for less money in less time. The new PXI architecture leverages the mainstream technologies to increase performance and flexibility, decrease costs, and improve ease of use for instrumentation applications. In this section, we introduce the PXI philosophy and its benefits for LabVIEW users.

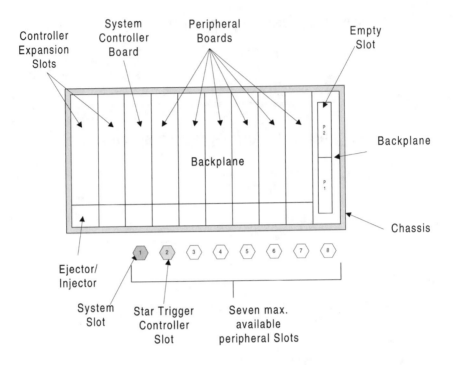

Figure 3-36

Example of a 3U PXI system

☐ 3.5.1   What Is PXI?

The *PXI* (*PCI eXtensions for Instrumentation*) specification defines a modular instrumentation platform. The only difference between PXI and mainstream desktop PCs is that the computer and plug-in instrument modules are repackaged in an industrial form factor with numerous expansion slots. PXI takes advantage of the *PCI* (*Peripheral Components Interconnect*) bus and extends the CompactPCI (a fast-growing standard for industrial computers based on PCI in a rugged industrial form factor) specification with instrument-specific features to create a platform for modular instrumentation at PC prices (see Figure 3-36).

   In contrast to desktop PCs the PXI specification designates system-level guidelines for clean, reliable power; forced air cooling; electromagnetic and environmental compatibility; and multimodule timing and synchronization.

❑ 3.5.2 Mechanical Features

PXI provides mechanical features that make PXI systems ideal for industrial environments and easy to integrate. The rugged Eurocard packaging system and high-performance IEC connectors defined in the CompactPCI specification are also used in PXI. PXI adds specific cooling and environmental requirements. Finally, two-way interoperability with standard CompactPCI systems is offered through the PXI specification.

The following PXI mechanical features are shared with CompactPCI:

- **High-performance connector system** — PXI employs the same advanced pin-in-socket connector system called out by CompactPCI. These highly dense (2-mm pitch) impedance-matched connectors are defined by the International Electrotechnical Commission (IEC-1076) and offer the best possible electrical performance under all conditions. These connectors have seen widespread use in high-performance applications, particularly in the telecommunications field.

- **Eurocard mechanical packaging and form factors** — The mechanical aspects of PXI and CompactPCI are governed by Eurocard specifications (ANSI 310-C, IEC 297, and IEEE 1101.1), all of which have a long history of application in industrial environments. A small (3U = 100 mm × 160 mm) and a large (6U = 233.35 mm × 160 mm) form factor are supported. Figure 3-37 shows the two primary form factors and the associated interface connectors for PXI peripheral boards. The most recent additions to the Eurocard specifications (IEEE 1101.10 and 1101.11) address electromagnetic compatibility, user-defined mechanical keying, and other packaging issues that apply to PXI systems. These electronics packaging standards define compact, rugged systems that can withstand harsh industrial environments in rack-mount installations.

All PXI features are implemented on the J2 connector of a 3U board and may selectively be used by peripheral boards. PXI-compatible backplanes must implement the complete PXI feature set. 6U PXI boards and PXI chassis only need to implement connectors J1 and J2. Future additions to the PXI specification may define the pinouts for connectors J3, J4, and J5 for

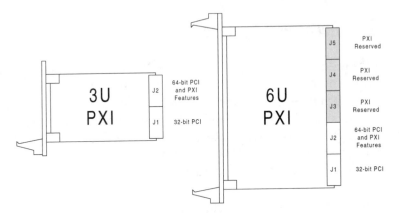

Figure 3-37
PXI 3U and 6U form factors

additional functionality in 6U. Note that any 3U peripheral board can work in a 6U chassis by means of a simple adapter panel.

PXI defines the system slot location to be on the far left end of the bus segment, as shown in the system diagram of Figure 3-36. This defined arrangement is a subset of the numerous possible configurations allowed by CompactPCI (a CompactPCI system slot may be located in any single position on a backplane). Defining a single location for the system slot simplifies integration and increases the degree of compatibility between controllers and chassis from multiple vendors. Furthermore, the PXI specification stipulates that the System Controller board should expand to the left into what are defined as controller expansion slots. These expansion slots do not have any CompactPCI connectors associated with them on the backplane and are basically expansion space. Expanding to the left prevents system controllers from using up valuable peripheral slots.

- **Additional electronic packaging specifications** — All mechanical specifications defined in the CompactPCI specification apply directly to PXI systems; however, PXI does include additional requirements that simplify system integration. As discussed above, the system slot in a PXI chassis must be located in the leftmost slot, and controllers should be designed to expand to the left to avoid using up peripheral slots. The airflow direction for required forced-cooling of PXI boards is defined to flow from the bottom to the top of a board. The PXI specification recommends complete environmental testing, including temperature, humidity, vibration, and shock for all PXI products, and requires documentation of test results. Operating and storage temperature ratings are required for all PXI products.

- **Interoperability with CompactPCI** — An important feature offered by PXI is that it maintains interoperability with standard CompactPCI products. Many PXI-compatible systems may require components that do not implement PXI-specific features. For example, a user may want to use a standard CompactPCI network interface card in a PXI chassis. Likewise, some users may choose to use a PXI-compatible plug-in card in a standard CompactPCI chassis. In this case, the user will not be able to implement PXI-specific functions but will still be able to use the plug-in card's basic functions. Note that interoperability between PXI-compatible products and certain application-specific implementations of CompactPCI (other sub-buses) is not guaranteed. Of course, both CompactPCI and PXI leverage the PCI Local Bus, thus ensuring software and electrical compatibility, as depicted in Figure 3-38.

❑ 3.5.3 Electrical Features

Many instrumentation applications require system timing capabilities that cannot be implemented directly across standard ISA, PCI, or CompactPCI backplanes. PXI adds a dedicated system reference clock, bused trigger lines, star triggers, and slot-to-slot local buses to address the need for advanced timing, synchronization, and sideband communication. PXI adds these instrumentation features while maintaining all of the advantages of the PCI bus. Finally, PXI offers three more peripheral slots per bus segment than desktop PCI for a total of seven.

- **System reference clock** — PXI defines the means to distribute a 10-MHz system reference clock to all peripheral devices in a system. This reference clock can be used for synchronization of multiple cards in a measurement or control system. The implementation of the reference clock on the backplane is strictly defined. As a result, the low-skew qualities afforded by this reference clock make it ideal for qualifying individual clock edges of trigger bus signals for sophisticated trigger protocols.

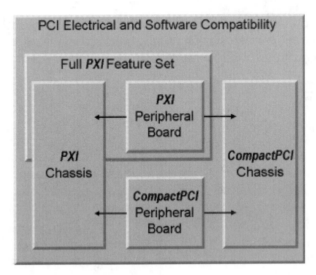

Figure 3-38
Interoperability between PXI and CompactPCI

- **Trigger bus** — PXI defines eight highly flexible bused trigger lines that may be used in a variety of ways. For example, triggers can be used to synchronize the operation of several different PXI peripheral boards. In other applications, one board can control carefully timed sequences of operations performed on other boards in the system. Triggers may also be passed from one board to another, allowing deterministic responses to asynchronous external events that are being monitored or controlled. The number of triggers that a particular application requires varies with the complexity and number of events involved.

- **Star trigger** — The PXI star trigger bus offers ultra-high performance synchronization features to users of PXI systems. The star trigger bus implements a dedicated trigger line between the first peripheral slot (adjacent to the system slot) and the other peripheral slots. An optional star trigger controller can be installed in this slot to provide very precise trigger signals to other peripheral boards. Systems that don't require this advanced trigger can install any standard peripheral board in this slot. Note that the star trigger can be used to communicate information back to the star trigger controller, as in the case of reporting a slot's status, as well as responding to information provided by the controlling slot.

PXI's star trigger architecture gives two unique advantages in augmenting the bused trigger lines. The first advantage is a guarantee of a unique trigger line for each card in the system. For large systems, this approach eliminates the need to combine multiple card functions on a single trigger line or to artificially limit the number of trigger times available. The second advantage is the low-skew connection from a single trigger point. The PXI backplane defines specific layout requirements such that the star trigger lines provide matched propagation time from the star trigger slot to each card for very precise trigger relationships between each card.

■ **Local bus** — The PXI local bus is a daisy-chained bus that connects each peripheral slot with its adjacent peripheral slots to the left and right. Thus, a given peripheral slot's right local bus connects to the adjacent slot's left local bus and so on. Each local bus is 13 lines wide and can be used to pass analog signals between cards or to provide a high-speed sideband communication path that does not affect the PCI bandwidth.

Local bus signals may range from high-speed TTL signals to analog signals as high as 42 V. Keying of adjacent boards is implemented by initialization software that prohibits the use of incompatible boards. Boards are required to initialize their local bus pins in a high-impedance state and can only activate local bus functionality after configuration software has determined that adjacent boards are compatible. This method provides a flexible means for defining local bus functionality that is not limited by hardware keying. The local bus lines for the leftmost peripheral slot on a PXI backplane are used for the star trigger, as represented in the local bus schematic in Figure 3-39.

■ **Peripheral component interconnect (PCI) features** — PXI offers the same performance features defined by the desktop PCI specification with one notable exception. A PXI system can have up to eight slots per segment (one system slot plus seven peripheral slots), whereas most desktop PCI systems only offer three or four available peripheral slots. Multiple-segment PXI systems built using PCI-PCI bridges offer this increased number of slots per segment, making very high slot count systems possible (256-slot theoretical maximum). The capability to have additional peripheral slots is defined in the CompactPCI specification on which PXI draws. Otherwise, all of PCI's features transfer to PXI:

- 33-MHz performance

- 32- and 64-bit data transfers

- 132 Mbytes/s (32-bit) and 264 Mbytes/s (64-bit) peak data rates

- System expansion via PCI-PCI bridges

- 3.3-V migration

- Plug-and-play capability

☐ 3.5.4   Software Considerations

Like other bus architectures, PXI defines standards that allow products from multiple vendors to work together at the hardware-interface level. Unlike many other specifications, however, PXI defines software requirements in addition to electrical requirements to further ease integration. These requirements include the support of standard operating system frameworks such as Windows NT and 95 (WIN32). Appropriate configuration information and software drivers for all peripheral devices are also required. Clearly, the PXI software specification is motivated by the benefits achieved through leveraging existing desktop software technology.

The PXI specification presents software frameworks for PXI systems, including Microsoft Windows NT and 95. A PXI controller operating in either framework must support the currently available operating system and must support future upgrades. The controller must also support a defined set of industry-standard application-programming interfaces, including LabVIEW, LabWindows/CVI, Visual Basic, Visual C/C++, and Borland Turbo C++.

PXI requires that all peripheral cards have device driver software that runs in the appropriate framework. Hardware vendors for other industrial buses that do not have software standards often do not provide any software drivers for their devices. The customer is often only given a manual, which describes how to write software to control the device. The cost to the customer, in terms of engineering effort, to support these devices can be enormous. PXI removes this burden by requiring that manufacturers, rather than customers, develop this software.

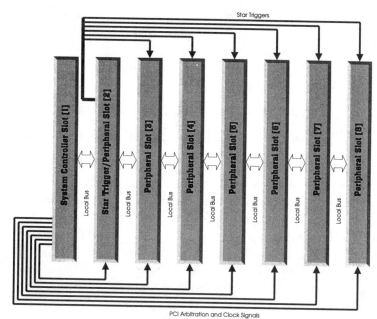

Figure 3-39

PXI local bus routing

❏ 3.5.5 LabVIEW — PXI, VXI, and GPIB

PXI fills the void that exists between low-cost desktop PCs and relatively more expensive GPIB- and VXI-based test systems. Just over 20 years ago, GPIB came into existence as a multivendor instrumentation standard. Ten years ago, VXI emerged as a multivendor standard for high-end modular instrumentation. Today, the main direct application of the PC as the core technology for instrument and text applications has been formalized by the PXI specification. This new, PC-based instrumentation platform can create opportunities for both vendors and users to take advantage of modular instrumentation at PC prices. The software applications and operating systems that run on PXI systems are already familiar to end users, as they are already in use on common desktop PCI computers.

As mentioned earlier, the PXI specification leverages mainstream off-the-shelf software. All PXI components are supported by LabVIEW, similarly to plug-in boards, external GPIB instruments, or other I/O systems. PXI systems that control external GPIB, VXI, and serial instruments should use the VISA software standard. The VISA standard facilitates using PXI, GPIB, and VXI in hybrid systems and paves the way for GPIB and VXI vendors to offer PXI versions of their products with minimal investment in software.

3.6   LabVIEW and the Universal Serial Bus

The obvious lack of flexibility in reconfiguring PC-based (test and measurement) systems has been acknowledged as an obstacle to their further deployment. The combination of user-friendly graphical interfaces and the hardware and software mechanisms associated with new-generation bus architectures, such as PCI, plug-and-play ISA, and PC-Cards, has made computers less confrontational and easier to reconfigure.

The concept of plug-and-play was designed to address many of the problems associated with installing new peripherals (I/O and DMA address as well as IRQ conflicts, which can create serious problems that can cause hardware and software to fail or even crash) on PC systems, but, in essence, plug-and-play is really only applicable for components installed inside the PC itself. As a result, external peripherals such as modems, scanners, printers, and test and measurement devices are very poorly addressed by current technology. In the last four years, the information technology industry developed two new PnP-capable mainstream serial buses, the *universal serial bus* (*USB*) and the *IEEE1394 bus* (called *Firewire*).

This section introduces a cost-effective new way to build easy-to-use instrumentation systems based on the USB.

❑ 3.6.1   History

The addition of external peripherals continues to be constrained by port availability. The lack of a bidirectional, low-cost, low-to-mid-speed peripheral bus has held back the creative proliferation of peripherals such as test and measurement devices, telephone/fax/modem adapters, answering machines, scanners, PDAs, keyboards, mice, etc. Existing interconnects are optimized for one or two point products. As each new function or capability is added to the PC, a new interface has been defined to address this need.

In 1995, a consortium of seven companies (Compaq, Digital Equipment Corp., IBM, Intel, Microsoft, NEC, and Northern Telecom) recognized the need for a new, simple, and efficient method of connecting peripherals to desktop PCs. In the following years they developed the USB.

With the advent of Windows 98 and Windows NT 5.0, there will be a boost for sophisticated serial technologies like USB and Firewire.

❑ 3.6.2   Specifications and Features

USB is one of the newest technologies for peripheral interfaces for PCs. Offering significant enhancements in ease of use and performance, USB is positioned to replace standard parallel and serial ports. USB data acquisition and GPIB devices are plug-and-play capable and are both installed and configured by the operating system without users having to change I/O resources or restart their computer.

The basic design of the USB features the following specifications:

- PC serial expansion bus

- Built-in power distribution for low-power devices

- Low speed, bus-powered devices

- Single master, host-driven bus

- Support for daisy chaining through a tiered star multi-drop topology

- Plug-and-play operation (hot plug)

- Sophisticated device enumeration

- USB transactions detailed by devices

- Sophisticated, three-layer protocol

- CRC packet error checks

- TDM scheduling of bus

- Dual speed: 1.5 and 12 Mbits/s

- Up to 5 meters per cable segment

- Four-wire serial bus

- Up to 127 devices

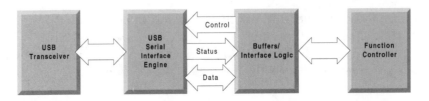

Figure 3-40

USB interface components

- Two-wire differential signaling

- Both isochronous and asynchronous data transfers

- All transactions originate from host

- 8-, 16-, 32-, 64-byte maximum data packet sizes

- Four packets: Token, Data, Handshake, Special

- Transaction setup by token packet

The Universal Serial Bus connects USB devices with the USB host. The USB physical interconnect is a tiered star topology. A hub is at the center of each star. Each wire segment is a point-to-point connection between the host and a hub or function — or a hub connected to another hub or function. There is only one host on any USB system. The USB interface to the host computer system is referred to as the host controller (see Figure 3-40). The host controller may be implemented in a combination of hardware, firmware, or software. A root hub is integrated within the host system to provide one or more attachment points. USB devices are hubs that provide additional attachment points to the USB, functions that provide capabilities to the system; for example, an ISDN connection, a digital joystick, or speakers.

Several attributes of the USB contribute to its robustness:

- Signal integrity using differential drivers, receivers, and shielding

- CRC protection over control and data fields

- Detection of attach and detach and system-level configuration of resources

- Self-recovery in protocol, using time-outs for lost or broken packets

■ Flow control for streaming data to ensure isochrony and hardware buffer management

■ Data and control pipe constructs for ensuring independence from adverse interactions between functions

☐ 3.6.3   LabVIEW and USB – Summary

The test and measurement industry is leveraging the newest mainstream technologies of a growing speed. National Instruments is one of the first companies to introduce data acquisition devices based on USB.

The new USB data acquisition devices include the NI-DAQ driver software; the GPIB interface includes NI-488.2M driver software. Users can easily run LabVIEW applications developed on other National Instruments DAQ or GPIB hardware with similar functionality on the new USB products without rewriting their applications.

NI-DAQ and NI-488.2M driver software will work with USB computers equipped with both OEM Service Release (OSR) Version 2.1, also known as Windows 95 Version 4.00.1111, and manufacturer-installed USB operating system device driver, UHCD.SYS. Windows 98 and Windows NT 5.0 will support USB out of the box.

The Universal Serial Bus has many advantages over existing technology, and one of the prime reasons for its development is the constant demand for cheaper PC and test and measurement solutions. From a mechanical point of view, the design is very simple. This should mean that implementing USB in the PC architecture is relatively easy at a moderate cost overhead. As a result, it would not be unreasonable to expect that, before long, USB will be a standard feature on practically all new PCs.

4

Commercial Communication Applications in LabVIEW

All things come out of the one, and the one out of all things. Change, that is the only thing in the world which is unchanging.

— Heraclitus
The Art and Thought of Heraclitus

*T*here are a number of ways to exchange data between programs or processes. Protocol-based communication and accessing common files (file sharing) are only two examples of interapplication communication. Both have advantages and drawbacks, which we describe in this chapter.

4.1   File Sharing

One of the simplest methods for using a common "data pool" is file sharing. File sharing is the only communication option when other communication methods are not available (i.e., when a process does not offer any other communication mechanism).

Common access to network files is straightforward because the drivers required for this access are provided by the network operating systems. It is sufficient to set the desired file name and the full path in the application to allow file access.

File sharing is frequently the simplest method when the hardware on the destination computer does not provide sufficient prerequisites to transfer processes that demand high resources, for example, very large LabVIEW VIs or large Excel spreadsheet tables, at an acceptable speed.

It is often desirable to design work flows sequentially. Particularly when managing measuring data, a clear separation is generally made between test management and measuring data acquisition and evaluation (e.g., in the form of logs and graphics).

One example of this job sharing is the data logger function of a universal test bench at Quelle.

An Access97-based database is responsible for test preparation (master data management, test management, LabVIEW application management, access to Oracle and SQL server databases, etc.).

In an initial step, the master data and the data relating to the test have to be input in Access. Subsequently, the LabVIEW test application is started. Access supplies LabVIEW with an ASCII file that contains all relevant test data. Once the LabVIEW application has been started, the Access database application closes automatically and LabVIEW takes over, loading the master data and control data and running the virtual instruments according to the control record.

In addition to collecting test data on a number of parameters, such as pressure, temperature, resistance, power, voltage, wattage, etc., LabVIEW assumes the complete control of the test bench.

In industrial practice, the Microsoft Excel spreadsheet software is frequently used to create logs. The main reason for this use is Excel's popularity, which makes it a widely used Office component offering many possibilities.

In the case of the universal test bench at Quelle, Microsoft Excel is used to handle the entire evaluation of complex test sequences (see Figure 4-1).

The Excel evaluation and creation of logs is based on a LabVIEW file, composed of Access master data and relevant test data as well as process data and instrument data. The tester can choose to evaluate the test results immediately or at a later stage. In the latter case, the LabVIEW VI closes automatically, without starting Excel, while in the first case, Excel starts and LabVIEW exits.

Both applications are active concurrently only for a few seconds. This means that each application can use the full computer power. This is important,

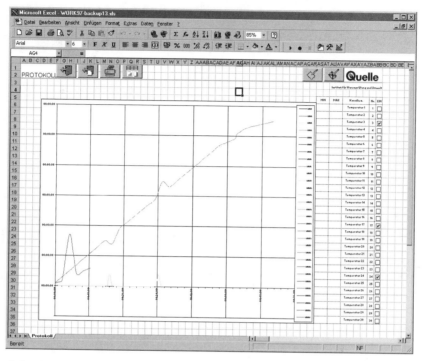

Figure 4-1

Evaluation and visualization template of the universal test bench at Quelle

particularly when a large number of channels are scanned at a moderate sampling rate, perhaps over several hundred hours.

The Excel evaluation software consists of a set of *VBA (Visual Basic for Applications)* modules handling the management of data rows, graphics, logging, and printing (see Figure 4-2).

Although VBA is a popular front-end and back-end application development language, programming in VBA produces very cryptic codes compared with LabVIEW. To be able to exchange data and programs with partner applications, it is necessary to find a common basis. Due to the dominating position of the Microsoft Office package in the market, Visual Basic (for Applications) has become a de facto standard for commercial applications during the past years. Even application developers pursuing a puristic approach (e.g., LabVIEW, C++, Prolog, Smalltalk) are well advised not to neglect the capabilities Visual Basic can offer and to consider this programming language as an alternative for "part-based" (best of all worlds) development. Although LabVIEW offers basically all

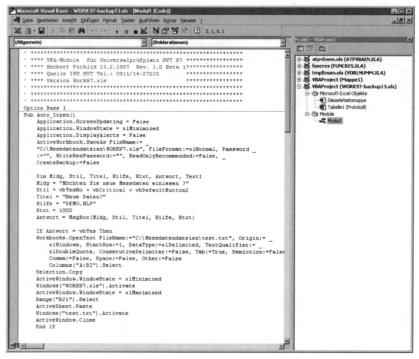

Figure 4-2

The VBA development interface

prerequisites to penetrate all domains of the application development, its coexistence with Visual Basic is meaningful to allow a smooth integration into Microsoft Office/Backoffice applications, covering the commercial aspects of data processing.

4.2   Communication Protocols

For communication between programs or between processes, the programs or processes must use a common communications language, referred to as a *protocol*. A communication protocol lets you specify the data that you want to send or receive and the location of the destination or source, without having to worry about what the underlying operating system has to do to make sure data gets there. The protocol translates your commands into data that network drivers can accept, and you don't have to worry about the deeper layers of that network communication. The network drivers then take care of transferring data between the sender and the receiver across the network as appropriate.

Several networking protocols have emerged as accepted standards for communication because the development of hardware and software has seen many different variants during the past decades (mainframes, workstations, PCs, Macs, MVS, VMS, CP-M, MS-DOS, MacOS, UNIX, Windows (NT), Assembler, Basic, C(++), Pascal, Prolog, Smalltalk, LabVIEW, etc.). In general, one protocol is not compatible with a different protocol. Thus, in communication applications, one of the first things you must do is decide which protocol to use both on stand-alone computers and in computer networks.

If you want to communicate with an existing, off-the-shelf application (e.g., MS Office, desktop publishing, graphics software, browser), then you have to work within the protocols supported by that application.

When you are actually writing the application, you have more flexibility in choosing a protocol. Factors that affect your protocol choice include the type of machines the processes will run on, the kind of hardware network you have available, and the complexity of the communication that your application will need. You may also want to include strategic aspects in your consideration; for instance, the following issues: Will the protocol be used in the future? Does the protocol have routing capabilities? Is the protocol powerful, reliable, stable, and so forth?

Several protocols are built into LabVIEW, some of which are specific to a certain type of computer. LabVIEW uses various protocols to communicate between computers. Some of the proprietary protocols explained below are mentioned only for reasons of completeness.

PPC and AppleEvents are two protocols for the Apple Macintosh, in addition to the usual standard protocols, to let processes communicate. *Program-to-Program Communication* (PPC) is a Macintosh intrinsic protocol allowing the transfer of data blocks between different processes, where the generated virtual instruments can act both as clients and as servers. Compared with *AppleEvents*, PPC-based low-level VIs are a simpler and faster communication option.

AppleEvents have a much more complex structure and are similar to *DDE* (*Dynamic Data Exchange*) known in the Windows world. AppleEvents is another protocol based on different topics offered by server applications and requests made by the client to the server. AppleEvents function on stand-alone computers and in networked environments. Apple offers a very large library with commands and messages. Although they allow complex and powerful communication applications, they burden the developer with a lengthy training curve. Given that Apple Macintosh computers are used mainly in graphic arts, desktop publishing, musical, and similar applications, while scraping a bare living in industrial and other commercial areas, we will not describe Apple-specific communication forms in detail.

In addition to more exotic communication forms, such as Named Pipes (known in the UNIX world), AppleEvents (Mac world), and System Exec VIs (virtual instruments capable of handling system calls, which are available under UNIX and Windows), the proprietary Windows-based communication forms, Dynamic Data Exchange (DDE, Net-DDE) and *OLE* (*Object Linking and Embedding*), clearly dominate in all fields of use.

To ensure platform-independent communication in the future, you should use standard protocols that support routing, such as the *TCP* (*Transmission Control Protocol*) standard Internet protocol, and *UDP* (*User Datagram Protocol*), another protocol built into the LabVIEW TCP/IP suite (unless you need to use other protocols for specific reasons).

The following section describes the client/server model in brief to explain the world of network protocols.

❑ 4.2.1 The Client/Server Model

The client/server model is a common model for networked applications. In the client/server model, one set of processes (clients) requests services from another set of processes (servers). A server can be a LabVIEW test value acquisition application, which serves a requesting client (e.g., LabVIEW VI, MS Office application, Internet browser).

For example, in your application you could set up a dedicated computer for acquiring measurements from the real world. The computer acts as a server when it provides data to other computers on request. It acts as a client when it requests another application, such as a database program (e.g., Oracle, SQL Server, Informix), to record the data that it acquires. In LabVIEW, you can use client and server applications with all protocols except Macintosh AppleEvents. If you need server capabilities on the Macintosh, use either TCP, UDP, or PPC.

Figure 4-3 shows a panel of a server VI that generates various curve forms, optionally imposes them with noise, and then provides them globally in a TCP/IP-based network.

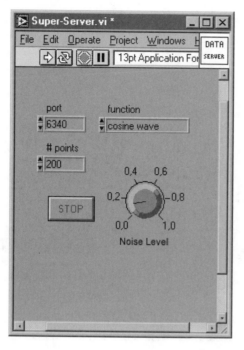

Figure 4-3

Panel of a server application providing curves overlaid with white noise in the network

The panel shown in Figure 4-4 displays the user interface of a client application, which visualizes the curve generated by the server application. This communication requires common port addresses and the server's IP address. The next section describes the TCP/IP communication in more detail.

To better understand how client/server communication works, it is useful to study Figure 4-5 closely. The TCP Open Connection is an IP function that takes two communication parameters, port address and IP address, to establish a connection to the server application. The TCP Read function is used to read the data stream available in the network, which is subsequently visualized by Type Cast and displayed in a chart.

If a break condition is met, for example, a time-out or when the user requested the program to exit, then the client program sends an acknowledge message to the server application informing it with a Write command that all data has arrived, then it sends a TCP Close command to close the communication. A simple error handler handles the error management in this virtual instrument.

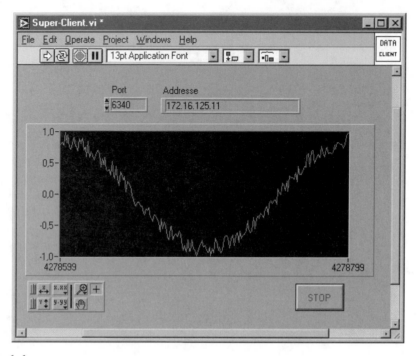

Figure 4-4

Panel of a client application displaying the noise-infested curve generated by the server application

☐ 4.2.2   TCP/IP and UDP in LabVIEW

TCP/IP is a suite of communication protocols, originating from the ARPANET, developed for the Defense Advanced Research Projects Agency (DARPA) at the end of the 1960s. Subsequent experiments demonstrated that the ARPANET protocols were not suitable for running over multiple networks. This observation led to more research on protocols, culminating with the design of the TCP/IP model and protocols. By 1990, the ARPANET had been overtaken by newer networks that it itself had spawned, so it was shut down and dismantled.

In 1984, the National Science Foundation (NSF) began designing a high-speed successor to the ARPANET that would be open to all university research groups. To have something concrete to start with, NSF decided to build a backbone network to connect its supercomputer centers. This *NSFNET*

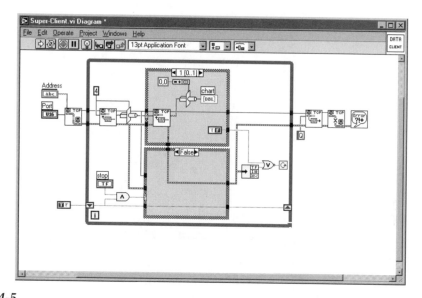

Figure 4-5

Diagram of a client application that visualizes the noise-infested curve generated by the server application

was based on a totally different software technology: It spoke TCP/IP right from the start, making it the first TCP/IP WAN. NSFNET was also the first internet to use top-level domains (the domains used in the Internet today): the generic domains *gov, mil, edu, com, org, net,* and geographic domains, for example, *uk, au, at, ch, de*. Since its development, TCP/IP has become widely accepted.

The name "TCP/IP" comes from two of the best-known protocols of the suite, the *Transmission Control Protocol (TCP)* and the *Internet Protocol (IP)*. TCP, IP, and the *User Datagram Protocol (UDP)* are the basic tools for network communication today.

TCP/IP enables communication over single networks or multiple interconnected networks, which are known as an *internetwork* or *internet*. The individual networks can be separated by great geographical distances. TCP/IP routes data from one network or internet computer to another. Because TCP/IP is available on most computers, it can transfer information between diverse systems.

To let applications within an internet communicate smoothly, a set of protocol layers, known as the *TCP/IP Reference Model*, is required. At the bottom of the stack is the *physical layer*; it is concerned with the hardware, for

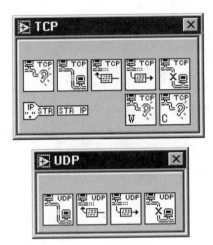

Figure 4-6

LabVIEW TCP and UDP libraries

example, Ethernet, Fast Ethernet, Gigabit Ethernet, Token Ring, Arcnet, FDDI, Sonet, ISDN, RS-232, ATM. Next comes the *internet layer*, the linchpin that holds the whole architecture together. Its job is to permit hosts to send packets to any network and have them travel independently to the destination (potentially on a different network). The layer above the internet layer is the *transport layer* (where TCP acts). It is designed to allow peer entities on the source and destination hosts to carry on a conversation. On top of the stack is the *application layer*. It contains all the higher-level protocols (e.g., FTP, SMTP, HTTP). Figure 4-6 illustrates how LabVIEW presents TCP and UDP libraries.

The IP transmits data across the network. This low-level protocol takes data of a limited size and sends it as a *datagram* across the network. IP is rarely used directly by applications because it does not guarantee that the data will arrive at the other end. Also, when you send several datagrams, they sometimes arrive out of order or are delivered multiple times, depending on how the network transfer occurs. UDP, which is built on top of IP, has similar problems.

TCP is a higher-level protocol that uses IP to transfer data. TCP breaks data into components that IP can manage. It also provides error detection and ensures that data arrives in order without duplication. For these reasons, TCP is usually the best choice for network applications.

Each host or node on an IP network has a unique 32-bit internet address. This address identifies the network on the internet to which the host is attached and the specific computer on that network. You use this address to identify the sender or receiver of data. IP places the address in the datagram headers so that each datagram is routed correctly.

One way of describing this 32-bit address is the *IP dotted decimal notation*, which divides the 32-bit address into four 8-bit numbers. The address is written as the four integers, separated by decimal points; for instance, 130.127.78.240.

Another — much more elegant — way of using the 32-bit address is by names that are mapped to the IP address. Network drivers usually perform this mapping by consulting a local hosts file that contains name-to-address mappings or by consulting a larger database using the *Domain Name System* (*DNS*) to query other computer systems for the address of a given name. Your network configuration dictates the exact mechanism for this process, which is known as *hostname resolution*. For instance, WINS is the most popular mapping server in Windows domains.

Considering the current explosion of network nodes, directories, and files, the use of domain models is no longer suitable because it gets increasingly difficult to obtain an overview on network resources. Large systems can no longer be managed economically. During the coming years, there is no doubt that the current domain models will be replaced successively by intelligent distributed file, directory, and resource systems. This will offer sophisticated dynamic address and resource management on many platforms.

☐ 4.2.3   DDE and OLE or ActiveX

Dynamic Data Exchange (DDE) is a protocol for exchanging data between Windows applications. DDE is most appropriate for communication with standard off-the-shelf applications (e.g., Microsoft Excel).

Since Version 4.0, LabVIEW for Windows 95/NT offers OLE (Object Linking and Embedding) automation mechanisms. This interapplication communication technology is part of Microsoft's OLE 2.0 specification (see Figure 4-7).

The *NT OLE* automation protocol was designed as a successor of the popular DDE protocol, which has strongly influenced communication between Windows applications. While DDE allows only asynchronous communication between Windows applications, OLE offers synchronous communication that allows development of more reliable applications.

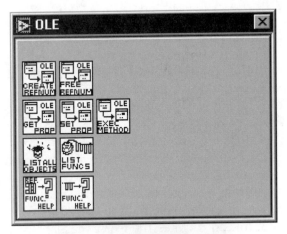

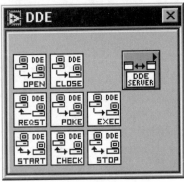

Figure 4-7

LabVIEW DDE and OLE libraries

Today, DDE is less popular among software developers, so this aged communication standard will not be further discussed here.

OLE automation mechanisms allow LabVIEW to access objects residing on automation servers. LabVIEW itself is not an OLE (automation) server or container application.

In OLE context, OLE objects are characterized by data abstractions exported by applications. You can manipulate these objects by using OLE-capable Windows applications. *Linking* and *embedding* are the two methods available to access OLE objects. You can use OLE automation to make functions and methods of one application available to another application. You can then access these functions and methods, which are normally grouped in objects.

In March 1996, Microsoft renamed its OLE standard to *ActiveX*. Basically, ActiveX can be divided into three different components:

- ActiveX documents

- ActiveX automation

- ActiveX controls/containers

Generally, these three methods are based on an object-oriented software architecture, defined by the *COM (Common Object Model)* standard. Let's look at these three ActiveX components more closely.

ActiveX documents are the logical continuation of the initial OLE standard, allowing integration of various documents within the same environment.

ActiveX automation as the next step in the evolution of dynamic data exchange represents new options for flexible data and command inter-operability between various applications on different systems in local, networked, and WAN environments. For instance, it allows you to call functions, methods, and objects from remote applications and to manipulate them locally. ActiveX communication always functions according to the client/server principle. The server application supplies its properties on the operating system level for use by other applications. Today, a number of applications support ActiveX automation and optimum communication with LabVIEW front-ends.

ActiveX controls or *containers* represent a further step in the development of Visual Basic VBX tools based on the COM standard. ActiveX controls build on the Windows 32-bit technology, in contrast to the 16-bit VBX controls. ActiveX controls are control elements that can be embedded in ActiveX containers. One of the new ActiveX features, *visual editing*, allows direct activation of objects within a document without having to exit the calling application and including processes like in-place editing and display. Visual editing facilitates the integration of applications from various vendors. For instance, *HiQ* (a mathematics and analysis and visualization software from National Instruments) allows you to visually edit a Microsoft Excel spreadsheet embedded in a HiQ ActiveX container. The most commonly used methods for data transfer between applications are cut/copy and paste. On the other hand, drag and drop is the most natural way to let selected objects communicate between applications. ActiveX supports this drag-and-drop functionality. For instance, you can transfer data from Excel to HiQ by dragging and dropping.

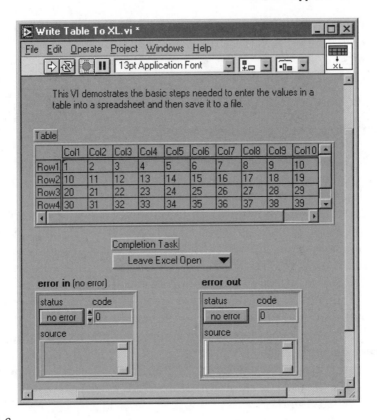

Figure 4-8

Panel of an OLE application that transfers a table generated in LabVIEW to an Excel worksheet and then saves it

To understand how OLE communication works, let's look at the following example of a LabVIEW/Excel communication.

Figure 4-8 shows the panel of an OLE application that transfers a table generated in LabVIEW into an Excel worksheet and then saves it. To better understand this application, we will look at the underlying diagram more closely.

In Figure 4-9, we can see three main blocks: The left block concerns preparatory steps for communication (i.e., opening the worksheet and table). The block in the center concerns the assignment of row and column labels and inserting the table into the specified area. The block on the right closes the Excel documents and exits Excel. One of the most important subVIs in this application is Set Cell Value (see Figure 4-10).

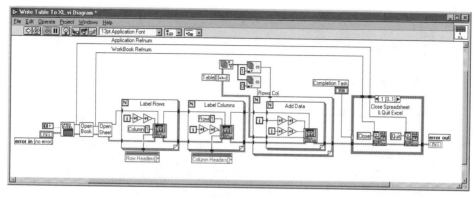

Figure 4-9

LabVIEW diagram of an OLE application that transfers a table generated in LabVIEW into an Excel worksheet and then saves it

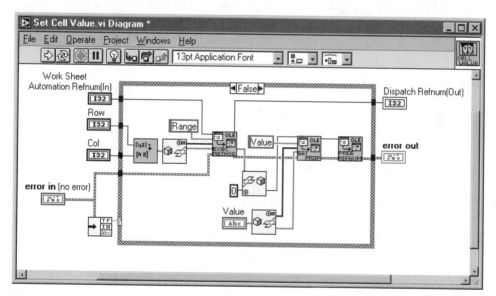

Figure 4-10

Sample VI (front panel, block diagram connector)

The Set Cell Value subVI makes sure LabVIEW data is converted into cell values Excel can understand. To do this, it uses the flatten function, and data so converted is known as *flattened data*. Two functions, OLE EXEC Method and OLE Set Prop, force the subVI to first set the range and then to transmit the value generated by the LabVIEW application

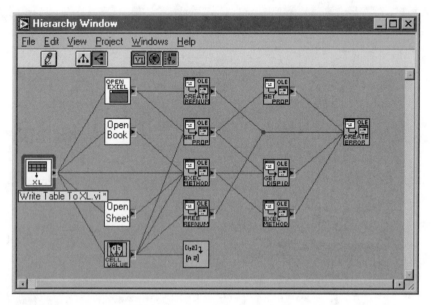

Figure 4-11

LabVIEW VI hierarchy of an OLE application

into this area. An additional function, OLE Free Refnum, releases the reference number previously defined by the system.

The hierarchical overview shown in Figure 4-11 helps in understanding the structure of this OLE example.

4.3   OPC – OLE for Process Control

While there is no lack of matching software drivers for industry standard hardware in the field of office automation, the situation is totally different in automation technology. Considerable time and cost have to be deployed to build an automation system to ensure the exchange of information in systems or devices from multiple vendors. On this basis, the OPC Foundation specified an open interface to solve this problem. *OPC (OLE for Process Control)* is based on the Microsoft OLE/COM and DCOM technology, which allows control hardware and software from different vendors to interoperate so that users are relieved from tedious driver programming and arduous connection to a selected bus system.

❑ 4.3.1 History

A number of automation system manufacturers and system integrators have developed special interfaces for their devices and control systems to allow data access within a predefined time or real time. Microsoft DDE (Dynamic Data Exchange) had become the de facto standard interface several years ago. This standard had been used by many devices in the automation industry. To exchange data between unequal systems, the DDE protocol and additions like NetDDE had been widely used, for instance, to transfer a spreadsheet table into a word processor. However, users soon realized the limits of DDE with regard to capability and reliability of data transfer. A few years ago, Microsoft replaced DDE by OLE (Object Linking and Embedding) to provide a more powerful technology, ensuring a more stable and reliable data exchange.

OLE is based on COM (Component Object Model). COM is a standard that defines how to build software components that interoperate in a reliable way. COM provides functions that enable you to build components that are distributed, secure, scalable, and reusable. COM also enables cross-platform support, robust versioning, programming language independence, and transparent remote operations. *DCOM (Distributed Component Object Model)* supports communication between clients and components that are located on different computers. This communication is identical to that between clients and components residing on the same computer. However, to use this client/server structure effectively, it is necessary that all vendors use the OLE standard.

Today, a number of major automation hardware and software vendors cooperate with Microsoft to agree on specifications for the new OLE for Process Control (OPC) standard. OPC is a standard for transmission protocols based on the OLE technology to allow interoperability between field devices, automation/control systems, and commercial applications from different manufacturers.

Version 1.0 of the OPC Specification defines standards for interfaces like objects, methods, and properties. It also specifies standards to access data residing on application servers. Developers use the OPC Specification to design application servers or clients capable of exchanging information in real time between different systems (DCS [Distributed Control Systems], SCADA [Supervisory Control and Data Acquisition], PLC [Programmable Logic Controllers], popular I/O systems, smart field devices, and networks).

❑ 4.3.2   Benefits of OPC

With wide industry acceptance, OPC will provide an unprecedented way to integrate different vendor hardware components in industrial automation across an entire organization in a plug-and-play software environment. Moreover, there is a high potential to reduce system integration costs because all software and hardware components use one single standardized transmission protocol.

Initially, automation system vendors will probably use the OPC technology to develop interface servers for hardware I/O. Instead of the original I/O driver software, an OPC server will be shipped with the new hardware. OPC client applications, for instance MMI (Man Machine Interface) software, will be required to exchange data with OPC servers. The single, compact OPC interface allows software vendors to expand the functionality of their programs instead of having to develop a number of different drivers for their hardware devices.

The OPC technology will extend beyond hardware I/O to more complex control and production systems. Considering that DCS, SCADA, MMI, planning and maintenance tools, and a series of other applications can be used as OPC clients and servers for different applications to exchange data, users can concentrate on their actual applications without having to worry about system integration problems.

❑ 4.3.3   The OPC Foundation

Leading manufacturers in the field of industrial automation have recently formed the *OPC Foundation*. This independent nonprofit organization promotes the development and improvement of the OPC Specification to ensure successful proliferation of the OPC technology on a global level.

The charter of the OPC Foundation is to develop an open and simple interface standard based on the properties of OLE/COM and DCOM technologies to improve the interplay between automation/control applications, field systems/devices, and commercial applications. To ensure the success of the OPC technology, a steering committee and various work groups (consumer advice, technical and marketing committees) have been established.

Some industrial automation vendors have already communicated to the OPC Foundation that they intend to manufacture and distribute OPC-compliant products. Anyone interested in improving and promoting the

OPC Specification can become a member of the Foundation. The Foundation offers memberships with and without voting rights.

☐ 4.3.4   LabVIEW/BridgeVIEW and OPC

National Instruments is a founding member of the OPC Foundation and participated in developing the specification for an open and simple data transfer system standard based on Microsoft's ActiveX technology.

LabVIEW and BridgeVIEW (this package is described in the next chapter) are truly open products regarding their communication properties. They support all types of I/O signals and communication protocols, and they ensure that users working in these environments can access all bus systems. To achieve this, the two development systems support not only mechanisms like OLE, DDE, TCP/IP and DLL, but they also implement future technologies as described in the OPC Specification.

Both LabVIEW and BridgeVIEW use OPC server toolkits to create servers for a wide range of automation components (SPC, field buses, RS-232/ RS-422/RS-485-based devices, etc.). Today, LabVIEW and BridgeVIEW run in many installations as OPC clients to access and manipulate objects, methods, and functions residing on OPC servers. The first practical implementation of this type was demonstrated at the CeBIT 1997 trade fair. It demonstrated an OPC-based connection of a BridgeVIEW OPC client to a Siemens S7 OPC server.

Figure 4-12 shows possible locations for OPC interface software in industrial automation environments.

4.4   Measuring and Process Control and the Internet

New communicative media like the Internet and the World Wide Web (WWW) are gaining ground and have become an integral part of the engineering workplace. The following section describes the LabVIEW development software and LabWindows/CVI to provide an overview of the possibilities and uses of the Internet and the WWW in measuring and process control applications.

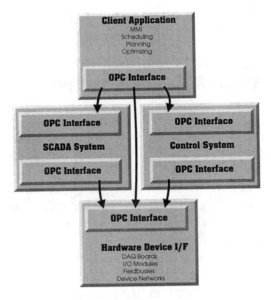

Figure 4-12

OPC interface software layers in an industrial automation environment

☐ 4.4.1   Terminology

Worldwide networking over the Internet, corporate data exchange over LANs or WANs, and their linkage over the Internet are the beginning of a new communicative information age. Catchwords like *connectivity*, *data highway*, *intranet*, *global village*, just to name a few, characterize this trend. On the other hand, users are confused about the meaning of all these terms. For this reason, it appears worthwhile to define a few basic terms of Internet communication and both the Internet concept and its potential for the measuring and process control industry.

The following list defines the most important terms in the context of this chapter. The Glossary at the end of this book contains a comprehensive list of terms and definitions.

- **Client** — In the client/server paradigm, a program running on a computer system that requests services from another program over a network connection with the WWW. The client computer uses a browser to simplify the access on URLs and to represent HTML pages for the user.

- **Common Gateway Interface (CGI)** — A technology used to create dynamic WWW documents. CGI programs run only on server computers. They are used, for instance, to send more recent HTML pages to the client (browser) or to create HTML pages dynamically.

- **FTP (File Transfer Protocol)** — A protocol and Internet service to transfer complete files from one computer to another. FTP includes special commands for file management.

- **HTML (Hypertext Markup Language)** — A document standard specified for the WWW. It defines a simple logical structure including titles, headings, paragraphs, lists, forms, tables, and mathematical equations, as well as a language to specify hypertext links.

- **HTTP (Hypertext Transfer Protocol)** — The standard mechanism used in the WWW to transfer documents between server and client systems.

- **Internet** — Name given to the collection of interconnected networks spanning the globe and running the Internet Protocol (IP). The Internet is formed of commercial and public networks, private or corporate networks, and governmental or publicly funded networks. In the Internet, the IP runs over a variety of underlying technologies. The development of the Internet, its promotion, and technical support are organized under the auspices of the Internet Society, a nonprofit, professional membership organization. Also (when not capitalized in this book), a set of physical networks interconnected by additional hardware systems (e.g., routers).

- **Intranet** — A private or corporate network used only by authorized members of that organization. Most corporate intranets use the protocols and software tools applied in the Internet to provide a universal service within their networks.

- **JPEG/GIF** — Common image file formats to represent images in the WWW.

- **URL (Uniform Resource Locator)** — A syntactic form used to identify a page of information in the WWW. A URL contains several logical parts, including the specification of the protocol to be used for the transfer or for the access and, in the case of a remote document, the actual Internet location of the document, that is, the Internet name of the host where the document is stored and its complete file name. URLs can also include complex search commands (e.g., to search databases).

- **Web server** — In the client/server paradigm, a program that provides services upon request from another program, called the client, over a network connection within the WWW. A server serves client requests by sending data in the form of HTML pages, images (JPEG, GIF), sound, or video. Web servers can respond dynamically to user inputs, for instance, to display new pages or perform certain computations.

- **World Wide Web (WWW)** — Name of a project developed at CERN, Geneva, to design and develop a series of concepts, communications protocols, and software to support the interlinking of various types of information according to the hypertext and hypermedia paradigm. The result is a series of protocols and reference implementations that have been endorsed and enriched by the Internet community and widely adopted outside CERN. The interconnection of thousands of WWW servers over the Internet creates what is called the WWW information space.

❏ 4.4.2 Internet Applications

The Internet is primarily used as a huge repository of contents, a kind of world-spanning multimedia information source. The rapid growth can surely be ascribed to the development of the WWW and the intuitive graphical browsers like NCSA Mosaic, Netscape Navigator or Microsoft Internet Explorer. The use is very easy because all information is available interactively through HTML. For this reason, the number of companies using the same protocols as in the Internet (TCP/IP, HTTP, FTP) to allow their staff easy access to internal corporate information is growing continually. A corporate network is normally called an *intranet*. In addition to these classical fields of use, the Internet becomes increasingly interesting for the transmission and evaluation of acquired data.

Particularly in the age of the Internet, the significance of virtual instruments increases with regard to distributed modular applications, opening entirely new horizons in the sense of a communicative measuring and process control technology. Which new possibilities can the Internet and the WWW offer for virtual instruments today?

Users working in the LabVIEW development environment can use the new Internet toolkit and the Internet to create measurement and process control applications. A classical field of use for test engineers is, for example, remote data monitoring of a test running in a laboratory. Also, long-term experiments in field studies where various data sets are measured can be monitored conveniently at a local computer. The acquired data can then be evaluated automatically on another computer over the Internet. Using the intranet structure, test engineers can, for instance, automatically control the results of production tests on their computers in the development department and add changes as needed during the test run.

The benefits of this approach are obvious: Developers can use existing network resources, such as standard TCP/IP and Ethernet adapters, without the need to build special network topologies. In addition, because most users are familiar with the Internet, access to remote virtual instruments is straightforward. All users have to do is to start their preferred web browser and enter a URL.

❑ 4.4.3   The Internet Toolkit for LabVIEW

The *Internet Developer Toolkit* can be used to provide user interfaces of virtual instruments, developed with LabVIEW, as HTML pages in the Internet. Virtual instruments can send e-mail messages automatically or receive and send FTP files. Although the toolkits for LabVIEW and LabWindows/VCI are implemented differently, they work in a similar way. Both provide tools to convert the front panels of active virtual instruments into JPEG image files. These image files are then transmitted to the computer that sent a corresponding request. In addition, these toolkits offer mechanisms allowing users to interact with the virtual instrument. Based on the graphical G programming language of LabVIEW, user-specific CGIs (Common Gateway Interfaces) can be created to evaluate user inputs. The LabVIEW web server receives a user's request, passes it on to a CGI-VI for processing, and returns a reply to the requesting user. To process common requests, the LabVIEW toolkit contains a collection of various CGI-VIs and a sample VI for users to implement their own CGIs in LabVIEW.

☐ 4.4.4   HTTP Server on LabVIEW Basis

Because the graphical data flow philosophy of LabVIEW and its structure provide an ideal integrative environment for various processes or VIs, the LabVIEW HTTP server was implemented entirely as a self-contained VI. This is particularly useful because image files of existing LabVIEW VIs can be sent in their source code format; they do not have to be modified. In addition to HTTP server VIs, the LabVIEW developer obtains a collection of VIs to develop e-mail and FTP clients. During installation of the Internet Toolkit, a home directory is set up to accommodate all HTML documents, images, icons, and CGI programs of the server.

Probably the most common use of the HTTP server is to create image files of active front panels and to send them to all clients connected to the server. As this alone may not provide sufficient information, you can use a standard HTML editor to create attractive HTML pages with additional information, images, or links to other web sites. A front panel is connected to the active VI through a URL (e.g., to the computer of that VI). Figure 4-13 shows an example of using the Netscape Navigator browser.

Listing 4-1 contains the HTML source code for the page shown in Figure 4-13.

Listing 4-1

HTML source code

```
<html>
<head>
   <title></title>
   <meta name="Author" content="">
   <meta name="GENERATOR" content="Mozilla/2.01Gold (Win32)">
</head>
<body>
<img src="http://130.164.158.174/.monitor?SCXI  XE "SCXI"  +Client
XE "Client"  .VI&re-fresh=1.0lifespan=60.0" border=0 height=405
width=625>
</body>
</html>
```

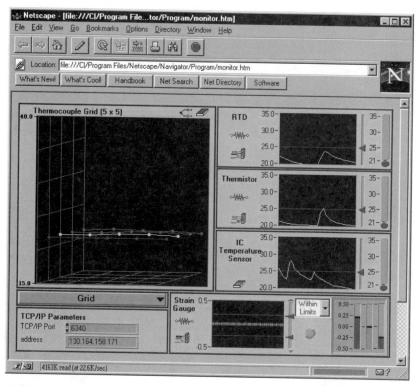

Figure 4-13

Netscape browser interface

The front panel of a specific virtual instrument is referenced with the img src tag. It makes sure the user interface is converted into an image file. The refresh parameter is used to define how often the image is to be updated (every second in this example). The lifespan parameter specifies the animation time (60 seconds in this example) before the browser terminates the connection to the server.

The Internet Toolkit for LabVIEW contains several CGI-VIs to handle various user requests. Most CGI programs existing in the Internet were written in C or Perl, but any modern programming language is basically suitable. When programming CGIs based on LabVIEW, the developer has the additional benefit that the programs are portable because, in addition to the Windows NT/95/3.1 platforms, LabVIEW is also available for the Macintosh, Sun Solaris and HP-UX. One of the CGI sample programs included in the LabVIEW package is described step by step in the following

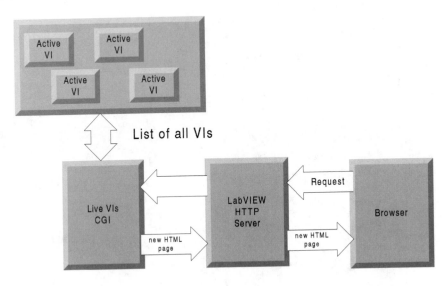

Figure 4-14

LabVIEW HTTP server and CGI-VI in an Internet/intranet environment

section. In this example, we use a CGI called livevis, which is used to build an HTML page that dynamically shows a list of all currently active CGIs. The user can then select and view a VI.

- **Client** — Sends a request to the server to obtain a page that lists all active VIs. The page can be requested either by entering the URL of that page in the browser's address field by integrating the URL in the HTML document as a hyperlink. In our example, the request could look like this:

  ```
  <http://IP address/cgi-bin/livevis.vi>
  ```

  where the cgi-bin directory is in the home directory of the server computer identified by IP address.

- **Server** — Receives the request and executes the LiveVIs CGI program.

- **LiveVIs CGI** — Finds all running LabVIEW VIs in the memory and dynamically generates an HTML page listing the names of these VIs.

- **Server** — Receives the HTML page created by the CGI program and returns it to the requesting client.

- **Client** — When a VI is selected from the list, sends a new request to the server to display the VIs selected by the user.

In addition to the HTTP server applications described in this example, the Internet Toolkit allows you to send e-mail messages and to transfer files via FTP automatically. The e-mail functions included in the toolkit allow you to send messages with attached files, similar to a classical mail system. The FTP libraries include high-level functions, which connect to the FTP server, enter login and password identification, transfer the desired file, and terminate the connection in automatic steps. Moreover, the FTP libraries include functions that allow you to execute these operations manually, and functions for usual FTP operations, such as change directory, create/delete directory, list file structures, and rename files.

❑ 4.4.5 Outlook

The ongoing commercialization of the Internet and its broad acceptance in the manufacturing industry give rise to new ways for measurement and process control applications. The LabVIEW and LabWindows/CVI development environments have proven themselves for many years and have a large installed base. They now offer new toolkits to create virtual instruments that fully integrate the Internet as an important communication medium. This paves the way for a totally new generation of virtual devices, particularly with regard to distributed modular applications for the measurement and process control industry.

4.5 LabVIEW Port I/O under Windows NT 4.0

LabVIEW Port I/O VIs make a trivial matter the access to the I/O bus of PCs under the Windows 3.x and Windows 95 operating systems. In contrast, however, the Windows NT architecture renders direct access to hardware more difficult. This is due mainly to the concept and positioning of the microkernel operating system as a reliable, sturdy, and stable SMP (Symmetric Multiprocessing) multitasking operating system, particularly for mission-critical tasks. It prevents programmers from using hardware-specific codes that violate protected memory locations. While this practice ensures much more reliable and stable applications, considerable effort is necessary to access specific hardware when there is actual need to do so, for instance, to address simple, low-level plug-in cards. To better understand this problem, we first discuss some important aspects of the NT architecture.

❏ 4.5.1   NT Architecture

The Windows NT operating system is divided into the *user mode* and the *kernel mode.* The user mode is characterized by applications, such as logon process and various clients, and by its protected subsystems (e.g., Security, Win32, Posix, and other subsystems).

To prevent direct access to hardware, the Microsoft NT team headed by Dave Cutler developed an intermediate layer called *NT Executive.* This is the portion of Windows NT that runs in kernel mode. It provides process structure, interprocess communication, memory management, object management, thread scheduling, interrupt processing, I/O capabilities, networking, and object security. Application programming interfaces (APIs) and other features are provided in user-mode protected subsystems.

❏ 4.5.2   NT Executive Components

Each component of the NT Executive is modular in its own way. It can be taken out and replaced with something else, as long as all the interfaces are handled properly. Now, let's go over each component of the Executive:

- **Object Manager** — Creates, deletes and manipulates objects that are basically system resources.

- **Security Reference Monitor** — Enforces the security of different objects. For instance, it monitors operating system resources and protects objects during run time. It also has some sophisticated mechanisms to audit services waiting for execution and available resources.

- **Process Manager** — Creates and terminates processes and threads. It can stop and restart processes and threads during run time. It also manages information on all NT processes.

- **Local Procedure Call** (LPC) — Is the message-passing facility for the client and the server within a computer node (consisting of one or several processors). LPC represents a localized, optimized version of the Remote Procedure Call (RPC), an industry-standard communication mechanism.

- **Virtual Memory Manager** — Implements memory management and paging. It makes sure that each process obtains a large, protected, and private address space (which cannot be accessed by other processes). Once that available main memory is used up, the Virtual Memory Manager provides mechanisms to free space by swapping data on additional media (e.g., hard disk or mass storage).

- **Kernel** — Responds to interrupts and exceptions. It schedules threads and synchronizes activities. It creates basic objects that can be used by Executive components above it to create higher objects.

- **I/O System** — Is made up of many subcomponents responsible for processing input and output for various devices. The subcomponents are as follows: The *I/O Manager* provides input/output facilities and maintains a model for NT Executive I/O. *File System* is an NT driver that accepts file-oriented I/O requests for various devices. The *Network Redirector* and the *Network Server* are network drivers responsible for sending and receiving remote I/O requests from the network. The *Cache Manager* speeds up I/O performance by storing the most recently read information into the memory. It also uses paging to write in the background to "write without wait." The *NT Device Drivers* are low-level drivers that directly manipulate hardware to write output to or retrieve input from a physical device or a network.

- **Hardware Abstraction Layer** (HAL) — not to be confused with HAL, the computer character in Stanley Kubrick's famous *2001: A Space Odyssey*. Exists between the hardware and the rest of the NT Executive components to provide an interface between the hardware and the operating system. This way, the NT Executive can maintain portability by calling HAL, rather than deal with any kind of hardware changes by itself. If you want to port Windows NT to another type of computer, all you have to do is rewrite HAL without having to worry about the rest of the operating system.

The services of these individual compartments of the NT Executive components are called *native services*.

☐ 4.5.3   Xon-Software Device Driver

Windows NT is a portable operating system, optimized to minimize the code overhead to support all kinds of processor and I/O hardware. Still, a minimum amount of processor-specific code (e.g., Intel Pentium, DEC Alpha, Power PC) is required and is normally accommodated deep down at the NT OS kernel level, while a small part of the code is held in the Virtual Memory Manager. The components listed above, particularly Kernel, hide processor architectures and memory management differences from the remaining operating system parts. The cross-platform part of the NT code is totally hidden in HAL and is provided by the relevant processor supplier. Although device drivers contain device-specific code, they avoid processor and platform dependencies by directly accessing the I/O hardware that was specifically designed and developed for each one (e.g., PCI hardware).

This situation means that you need a device driver to be able to manipulate low-level I/Os directly. The Munich, Germany, based Xon-Software developed a driver to solve the problem described above and implied in Figure 4-15.

When installing the Xon-Software package, copy the `Xon_PortIO.sys` driver to the following directory:

```
C:\WINNT\SYSTEM32\DRIVERS\XOn_PortIO.sys
```

All system drivers are located in this path. To activate the driver, you need to add an entry in the NT Registry. Select **Programs** and **Administrative Tools** from the **Start** bar to open the Registry Editor (Figure 4-16) and edit the database entries as in Listing 4-2.

Listing 4-2

Adding a driver entry in the Registry (sample, 32 bytes in the range from 0x320 to 0x33f)

```
HKEY_LOCAL_MACHINE
   SYSTEM
   CurrentControlSet
   Services
   XOn_PortIO                             [new Key]
   ErrorControl    0x00000001 (1)         [DWORD]
   Group           "Extended Base"        [STRING]
   Start           0x00000002 (2)         [DWORD]
   Type            0x00000001 (1)         [DWORD]
   Parameters                             [SubKey]
   IOPortAddress   0x00000320             [DWORD]
   IOPortCount     0x00000020             [DWORD]
```

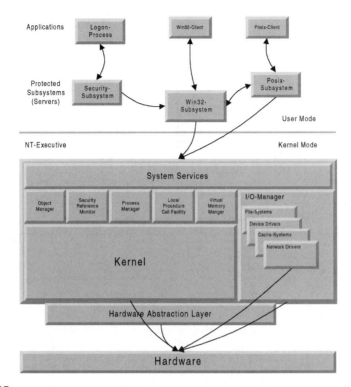

Figure 4-15

Architecture of the Microsoft Windows NT micro-kernel (incomplete)

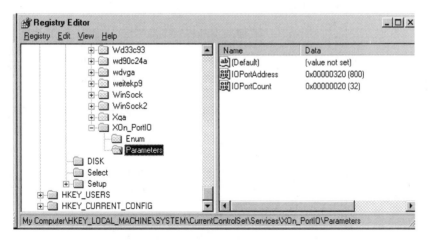

Figure 4-16

The Windows NT Registry Editor

Table 4-1   *NT startup options*

Start	Designation	Meaning
0	Boot	When booting the computer, e.g., hard disk driver
1	System	During NT start
2	Automatic	After NT start
3	Manual	Manual start
4	Disabled	Not active

If your Registry contains an older driver, you can stop and deactivate this driver by using the **Devices** option from the **Control Panel**. The Registry must contain a correct driver entry, and **Automatic** is the startup type you want to select. You can use **Windows NT Diagnostics** to check whether the driver is properly installed and to identify free memory locations. In Windows NT, memory locations used by drivers must not overlap, to avoid collisions.

To maintain a safe and stable system, use only nonreserved I/O locations for drivers to avoid the case that device drivers start in overlapping I/O areas. Note that the Registry should be edited only by the system administrators for good reasons!

`Start` is another important parameter, in addition to `IOPortAddress` and `IOPortCount`. The startup options available are listed in Table 4-1.

The correct value for `XOn_PortIO` is normally **2** (**Automatic**). If you want to use the driver to access addresses that are normally managed by the system (e.g., parallel port or SIO), select **1** (**System**). In this case, `XOn_PortIO` gets this memory location, and the subsequent system driver (e.g., `parport`) cannot install itself.

❏ 4.5.4   Port-IO VIs

Port-IO VIs are written in C code, which is implemented in the form of code interface nodes in the relevant virtual instruments. Listing 4-3 uses a header file called `gpioctl.h` as an example. This file contains important information for the CIN-C source text.

Listing 4-3

The gpioctl.h header file supplied by Microsoft

```
// gpioctl.h    Include file for Generic Port I/O Example Driver
//
// Define the IOCTL codes we will use.  The IOCTL code
   contains a command
// identifier, plus other information about the device, the
   type of access
// with which the file must have been opened, and the type
   of buffering.
//
// Robert B. Nelson (Microsoft  XE "Microsoft"  )    March 1, 1993
// Device type            -- in the "User Defined" range."
#define GPD_TYPE 40000
// The IOCTL function codes from 0x800 to 0xFFF are for
   customer use.
#define IOCTL_GPD_READ_PORT_UCHAR \
    CTL_CODE( GPD_TYPE, 0x900, METHOD_BUFFERED, FILE_READ_ACCESS )
#define IOCTL_GPD_READ_PORT_USHORT \
    CTL_CODE( GPD_TYPE, 0x901, METHOD_BUFFERED, FILE_READ_ACCESS )
#define IOCTL_GPD_READ_PORT_ULONG \
    CTL_CODE( GPD_TYPE, 0x902, METHOD_BUFFERED, FILE_READ_ACCESS )
#define IOCTL_GPD_WRITE_PORT_UCHAR \
    CTL_CODE(GPD_TYPE,  0x910, METHOD_BUFFERED,
    FILE_WRITE_ACCESS)
#define IOCTL_GPD_WRITE_PORT_USHORT \
    CTL_CODE(GPD_TYPE,  0x911, METHOD_BUFFERED,
    FILE_WRITE_ACCESS)
#define IOCTL_GPD_WRITE_PORT_ULONG \
    CTL_CODE(GPD_TYPE,  0x912, METHOD_BUFFERED,
    FILE_WRITE_ACCESS)
typedef struct _GENPORT_WRITE_INPUT {
    ULONG   PortNumber;     // Port # to write to
    union   {               // Data to be output to port
    ULONG   LongData;
    USHORT  ShortData;
    UCHAR   CharData;
    };
}   GENPORT_WRITE_INPUT;
```

Listing 4-4
CIN source code to read port information

```
/*
 * CIN source file
 */
#include "extcode.h"
#include <windows.h>
#include <stddef.h>
#include <stdio.h>
#include <stdlib.h>
#include <winioctl.h>
#include "gpioctl.h"        // This defines the IOCTL constants.
                            /* stubs for advanced CIN functions */
UseDefaultCINInit
UseDefaultCINDispose
UseDefaultCINAbort
UseDefaultCINLoad
UseDefaultCINUnload
UseDefaultCINSave
CIN MgErr CINRun(uInt32 *register_address, LVBoolean *read_a
_byte_or_a_word_F_byte_, uInt16 *value);
CIN MgErr CINRun(uInt32 *register_address, LVBoolean *read_a
_byte_or_a_word_F_byte_, uInt16 *value)
{
  // The following is returned by IOCTL.  It is true if the read
     succeeds.
  BOOL    IoctlResult;
  // The following parameters are used in the IOCTL call
  HANDLE hndFile;      // Handle  XE "Handle" to device, obtain from
  CreateFile
  ULONG   PortNumber; // Buffer  XE "Buffer" sent to the driver
                          (Port #).

  union
    { ULONG    LongData;
      USHORT  ShortData;
      UCHAR   CharData;
    } DataBuffer;          // Buffer  XE "Buffer" received from driver
                          (Data).

  LONG    IoctlCode;
  ULONG   DataLength;
  DWORD   ReturnedLength; // Number of bytes returned
  // The input buffer is a ULONG containing the port address.
  // It is specified as 0, 1, 2, ... relative to the base
  // address set in genport.h
  // or overridden by the registry.
  // The port data is returned in the output buffer DataBuffer;
  *value = 0xffff;                      // indicates an Error
  hndFile = CreateFile(
       "\\\\.\\GpdDev",         // Open the Device "file"
       GENERIC_READ,
```

```
              FILE_SHARE_READ,
              NULL,
              OPEN_EXISTING,
              0,
              NULL
              );
      if (hndFile != INVALID_HANDLE_VALUE)     // Was the device
                                                      opened?
      { if (*read_a_byte_or_a_word_F_byte_)
        { IoctlCode = IOCTL_GPD_READ_PORT_USHORT;
          DataLength = sizeof(DataBuffer.ShortData);
        }
        else
        { IoctlCode = IOCTL_GPD_READ_PORT_UCHAR;
          DataLength = sizeof(DataBuffer.CharData);
        }
        PortNumber = (ULONG) *register_address;
                            // Get the port number to be read
        IoctlResult = DeviceIoControl(
            hndFile,          // Handle  XE "Handle"   to device
            IoctlCode,        // IO Control code for Read
            &PortNumber,        // Buffer  XE "Buffer"   to driver.
            sizeof(PortNumber),  // Length of buffer in bytes.
            &DataBuffer,       // Buffer  XE "Buffer"   from driver.
            DataLength,        // Length of buffer in bytes.
            &ReturnedLength,     // Bytes placed in DataBuffer.
            NULL);           // NULL means wait till op.
                              completes.

        if (IoctlResult)      // Did the IOCTL succeed?
        { if (ReturnedLength == DataLength)
          { if (ReturnedLength == sizeof(UCHAR))
              *value = DataBuffer.CharData;
            else
              *value = DataBuffer.ShortData;
        } }
        CloseHandle(hndFile);       // Close the Device "file".
      }
    return noErr;
}
```

The `PortIO.11b` library contains VIs with a functionality exactly identical to the VIs of the previous LabVIEW versions and enhanced functions that support error handling and 4-byte words. The `ioport32.11b` library contains functions that allow a much better performance because their interface is fine-tuned for NT (open device, read/write, and close device are separate). Figures 4-17 and 4-18 illustrate the OutPort VI; Table 4-2 summarizes the `PortIO` libraries.

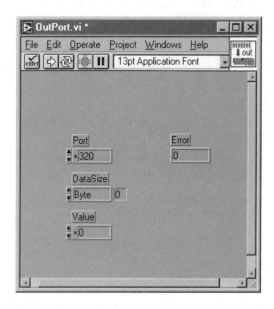

Figure 4-17

OutPort.vi panel

Table 4-2 *PortIO libraries*

Library	VI Name	Function
PortIO.llb	InPort.vi OutPort.vi	These VIs correspond exactly to the usual functions known from previous LabVIEW versions; they allow easy porting to existing applications.
PortIO.llb	InPort.vi OutPort.vi	Functions with enhanced interface (error handling, 4-byte words)
IOport32.llb	IOOpen.vi IOClose.vi IORead.vi IOWrite.vi	Functions for a new driver concept; when opening, a handle is generated and used for all read and write operations. Close releases the device.

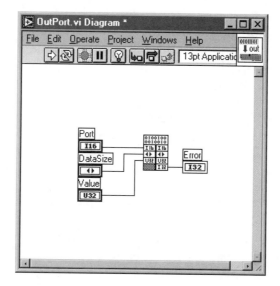

Figure 4-18

OutPort.vi diagram

☐ 4.5.5   Summary

The preceding sections provided an overview of the way LabVIEW handles networking and interapplication communications. To better understand the various concepts involved, these sections cover several underlying standards and techniques, such as file sharing, communication protocols, OPC, and Internet protocols. A separate section is dedicated to describing interapplication communication under Windows NT, including some NT-specific aspects. Finally, a specific example illustrates how you can access port-I/O resources in a safe and reliable microkernel multitasking/multiprocessing operating system like Windows NT. This example used the driver implemented by Xon-Software, which is based on Microsoft's generic port driver.

4.6   Database Access with LabVIEW

Measuring and testing generates a huge amount of data. In most cases, these results must be stored in such a way that the data can be retrieved at a given time. The most common way to facilitate retrieval is finding a file-naming convention that codes the most important parameters (e.g., Identifier, Date, Operator, Passed?) into the file name. This simple method is widely used, especially after the introduction of Windows 95, which allows longer naming conventions.

A real database system has a number of advantages: The number of search criteria is not limited, querying proceeds much faster, generating the search criteria is easier, mechanisms for backup of data are available, and so forth.

This section describes how LabVIEW can create, read from, and write to a database. A database is used in many applications to save acquired data, to manage the results, and to get results over a longer period of time. The complexity of the data structure (saving test results) is low, but the amount of data may be huge. Nearly all conventional database software meets this requirement. The DAO toolkit introduced in the following discussion provides the full functionality of the Microsoft Access database software. In addition, we illustrate how an object-oriented DLL can be used inside the LabVIEW environment.

☐ 4.6.1 Terminology

Before going into the details, we introduce some basic definitions that are frequently used in connection with databases.

- **ODBC** (Open Database Connectivity) — A standard protocol that enables applications to connect to a variety of external database servers or files. ODBC, first introduced by Microsoft, is a widely accepted standard, supported by the majority of database vendors. In general, the quality and reliability of ODBC drivers vary significantly. The ODBC drivers used by the Microsoft Jet database engine allow access to Microsoft SQL Server and several other external databases. The ODBC applications programming interface (API) may also be used to access ODBC drivers and the databases they connect to, without using the Jet engine.

- **Microsoft Jet Database Engine** — A database management system that retrieves data from and stores data in user and system databases. The Jet engine can be thought of as a data manager component by the help of which other data access systems, such as Microsoft Access and Visual Basic, are built. To work with other (non-MS) database systems, the Jet engine uses the ODBC interface. The Jet engine is included in MS Office 97 and can be distributed free of charge.

■ **DAO** (Data Access Objects) — The API that communicates with the Jet engine. DAO is written in C++, using an object-orientated design. You can use data access objects, such as the `Database`, `TableDef`, `Recordset`, and `QueryDef` objects, to represent objects that are used to organize and manipulate data, such as tables and queries, in C++ code.

■ **SQL** (Structured Query Language) — A language used to query, update, and manage relational databases. SQL can be used to retrieve, sort, and filter specific data in a database environment. SQL is supported by all the leading database vendors. Please note that there are a lot of different implementations of SQL dialects!

❑ 4.6.2   Comparing SQL Toolkit (ODBC) with the DAO Toolkit (Jet Engine)

Via the SQL toolkit, LabVIEW can communicate with every database system for which an ODBC driver is available. However, performance and reliability of this solution depend on the quality of the ODBC driver used. The DAO toolkit gives direct access to the Microsoft Access database without use of an ODBC driver. As shown in Figure 4-19, it is clear which is superior for a given application.

If the preferred database product is MS-Access, working with DAO is the better solution. If the employed database is Oracle, Paradox, or any other product that is not directly supported by the Jet engine, the SQL toolkit generates a lower software overhead.

The Jet engine as well as the DAO interface are royalty free. DAO can be installed either together with MS Office 97 or separately with the Jet engine. Xon Software, an alliance partner of National Instruments, developed a DAO toolkit (`www.xon.de`) that is shipped together with an installation program including the Jet engine and DAO. DAO is available only for the 32-bit Windows versions (Windows 95, Windows NT 4), not for Windows 3.x.

DAO is able to define and create a new database. This functionality is intentionally not implemented in most ODBC drivers because it enables users to work without buying the database product.

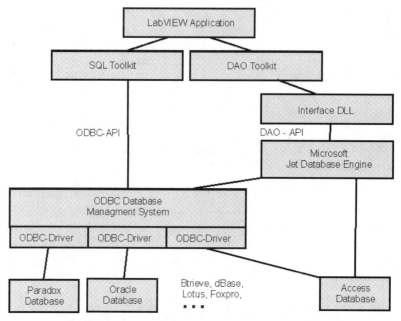

Figure 4-19

Structure of SQL and DAO database products

❑ 4.6.3   DAO Dynamic Link Library

The DAO API is provided as a DLL. As we have seen earlier, LabVIEW allows the integration of DLLs in a seamless way. However, the DAO DLL is written in C++, using an object-oriented design that exports classes instead of functions and thus cannot be used in LabVIEW. A class provides a lot of methods that are functions used by an application. Unfortunately, they use polymorphic parameters. Although LabVIEW can work with polymorphic parameters ("OLE variant") since Version 5.0, it is recommended that a separate function be written for every data type.

Figure 4-20 shows an "Interface DLL" between the DAO API and the DAO toolkit. This DLL can be called a "--DLL" because it returns C++ code back to ANSI C. The Interface DLL can be accessed by LabVIEW using the standard DLL Node. Every method of all classes is converted into at least one function. If there are polymorphic parameters, one method can be converted into a number of functions.

Writing the Interface DLL can be easily done by an experienced programmer. A C++ method has a self-pointer, which points to an instance of this class. A function passes this as an additional parameter, using a simple pointer type. To avoid programming errors, it is useful to define different data types for each kind of pointer. If this is done, the C compiler avoids passing a wrong pointer (e.g., a pointer to a `TableDef` instance to a function that expects a pointer to a `Recordset`). The next paragraph shows how to get the same security inside LabVIEW.

As an example, we will show how to build a function, `GetField-InfoByName`, using the method `GetFieldInfo` of a `TableDef`. Our new function points to the first parameter position of the `TableDef` structure. The next parameters are the name of the field and a pointer to get the result. A little error handling is built in. It is useful to first check if the pointer is `NULL`. Error handling inside the DAO DLL looks difficult. In case of an error, an exception is generated. Exception handling inside LabVIEW is not possible; the macros `ERRON` and `ERROFF` try to mask the exception if necessary and generate an error return value.

The code inside the Interface DLL looks like Listing 4-5.

Listing 4-5

Code in the Interface DLL

```
int TDFKT(GetFieldInfoByName) (TDHDL hdl, LPCSTR lpszName, FIHDL
    fihdl)
{ if (!hdl) return -1;
  ERRON
  TDCAST(hdl)->GetFieldInfo (lpszName, *FICAST(fihdl),
    AFX_DAO_ALL_INFO);
  return 1;
  ERROFF
  return -2;
}
```

☐ 4.6.4 Data Types Supported

A database system supports a number of different data types. It is not important for a user to know details of the internal representation, but of course it is helpful in estimating the amount of disk space needed for one record. Table 4-3 summarizes data types and their use; Table 4-4 summarizes number types and their use.

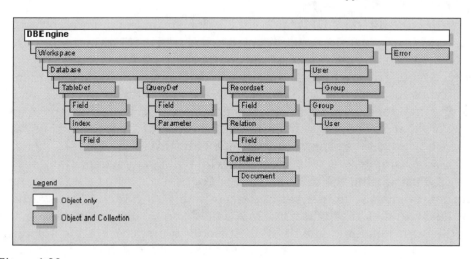

Figure 4-20

Data access object hierarchy chart

Table 4-3   *Database data types*

Data Type	Constant	Use for
Text	DbText	Text and numbers, such as names and addresses, phone numbers, and postal codes. A Text field can contain from 0 to 255 characters. You must set the size property to the maximum length of your text when you declare a field as type Text.
Memo	DbMemo	Lengthy text and numbers, such as comments or explanations. A Memo field's size is limited by the maximum size of the database.
Number	(various)	Numerical data on which you intend to perform mathematical calculations, except calculations involving money. See Table 4-4 for a further explanation of number fields.
Date/Time	DbDate	Dates and times. These values are stored in a combined field — the date is always stored with the time.

Table 4-3   (*Continued*)

Data Type	Constant	Use for
Currency	dbCurrency	Used where a field contains monetary values. Don't use the Number data type for currency values because numbers to the right of the decimal may be rounded during calculations. The Currency data type maintains a fixed number of digits to the right of the decimal. Set the field object Type property to dbCurrency for currency fields.
Counter	dbLong	Sequential numbers automatically inserted by the Microsoft Jet database engine. Set the dbAutoIncrement flag in the field object Attributes property. Numbering begins with 1. The Counter data type makes a good primary key field. If you want to establish a primary key/foreign key relationship between two tables by using a counter field, make sure that the foreign key column data type is also defined as Long. You can also set (or read) dbRandomIncr-Field in a new field object's Attributes property to indicate that a counter field is generated with random numbers. (This is valid only on fields that are counters.)
Yes/No	dbBoolean	Yes/No, True/False, On/Off, or fields that contain only one of two values.
OLE Object	dbLongBinary	Objects created in other programs using the OLE protocol.
Binary Image	dbLongBinary	Any value that can be expressed in binary up to 1.2 gigabytes in size. This type is used to store pictures, files, or other raw binary data.

Table 4-4   *Database number types*

Number Type	Use for
DbBoolean	Binary values of 0 or 1.
dbByte	Positive values from 0 to 255. Stored in Byte variables.
dbInteger	Values from –32,768 to +32,767. Stored in Integer variables.
DbLong	–2,147,483,648 to 2,147,483,647 (about 2.1 GB). Stored in Long variables.
DbSingle	–3.402823E38 to 1.401298E–45 for negative numbers and 1.401298E–45 to –3.402823E38. Stored in Single variables.
DbDouble	1.79769313486232E308 to 4.94065645841247E–324 for negative values and 4.94065645841247E–324 to 1.79769313486232E308 for positive values. Stored in Double variables.

The most interesting data types are OLE Object and Binary Image. Both of them consist of binary data that cannot be interpreted by the database system. An OLE Object always needs to have a Flatten method; this flat representation of the object is stored. The LabVIEW VIs Flatten to String and Unflatten from String provide an easy way to store complex data structures inside the database.

Of course, it does not make sense to flatten the header information (identifier) and store it into a binary image because in this case it is not possible to use this item for searching and sorting. Each part of the header information should be written into a separate field. The binary image is a good way to store curves, images, and other extensive data elements. Using the binary image saves disk space, increases speed, and allows the use of curves of different size.

What are the limitations of this approach? First of all, the records must be able to fit onto the 2-kilobyte pages supported by the Jet database engine. Any additional field object specified by the user, including binary and memo fields, have to fit into the 2K table space. The reserved field length for binary and memo fields, however, is only 14 bytes per non-null field and only 1 byte for null fields because the actual data for these fields is stored in additional pages. The size of data assigned to each text field is not set until the user actually saves data into the field.

☐ 4.6.5   Typified Handles

In Section 4.6.3, we recommended that you define different data types for all pointers. The same security mechanism can be used within LabVIEW. Although the mechanism is simple, users find it difficult to implement it:

1.   Take a **Data Log File RefNum** from **Path & RefNum** — Palette
2.   Take an **Enumerated Type** from the **List & Ring** – Palette
3.   Insert only one item into the **Enum** (the item name should be your type name)
4.   Drag the **Enum** into the **RefNum**

   This sequence results in a new data type that can be connected only with indicators and controls of the same type. It is advisable to save this control as a custom control. The internal representation, of course, is always an unsigned integer (U32). If typed handles are passed into or from a Code Interface Node (CIN) or DLL call, it is necessary to cast them explicitly (Figure 4-21).

   Figure 4-22 shows the function `TDGetFieldInfoByName`. In Section 4.6.3 we listed the C code for this function. Figure 4-22 shows how the parameters are passed into the DLL and back to the application. However, using such a programming style does not allow you to connect a Handle pointing to an element of a type different from the controls. This is important because wrong pointer types would cause a protection violation and at least, this task could fail.

☐ 4.6.6   *Creation and Release of DAO Objects in LabVIEW*

The C++ language provides a constructor method that is called automatically when an instance of a class is created. However, this feature is lost when going back to ANSI C. In this case, the application programmer has to take care when an instance is generated (including allocating memory and

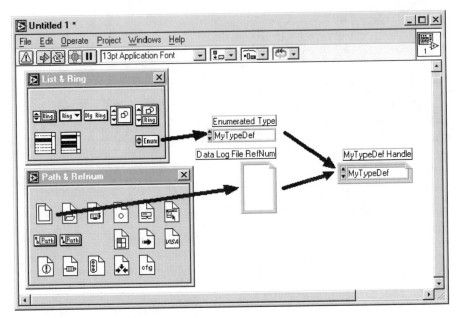

Figure 4-21

Creating your own Handle data types

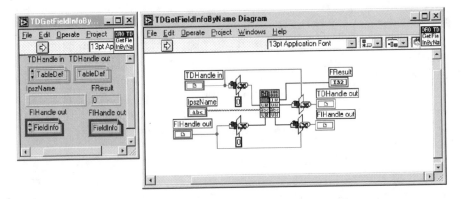

Figure 4-22

Using your own Handle data types together with a DLL call

presetting member variables). When the object is no longer needed, it must be released. This function call releases the memory. There is no need for a programmer to worry about memory consumption when using simple data types like numbers. LabVIEW automatically decides when to release the

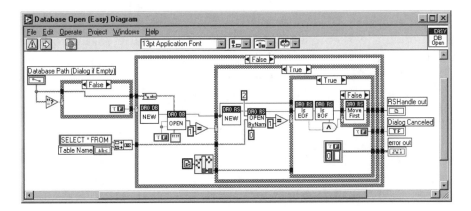

Figure 4-23

Diagram of Database Open (Easy).vi

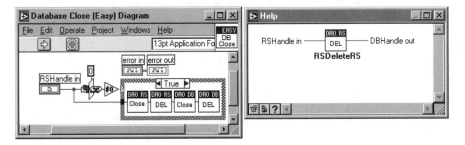

Figure 4-24

Diagram of Database Close (Easy).vi

memory and whether it is necessary to pass a copy to a subVI or a pointer/reference. The use of handles leads to a better performance.

Figure 4-23 shows the diagram of an "Easy" function. Because the number of functions is very high (184 VIs), there is a set of macro functions called Easy. These functions use default parameters to call the low level functions.

These functions generate two new handles, a Database Handle (DAO DB NEW) and a Recordset Handle (DAO RS NEW). The RSHandle is an output of this VI. The DBHandle seems to be lost. If the DBHandle needs to be released, it is important to first close/release all Recordsets. For easy wiring, the DBHandle is returned from the function RSDeleteRS, which releases the Recordset. Figure 4-24 illustrates the function that closes Recordset and Database and releases all memory used.

The VI `Database Open (Easy)` uses a simple SQL statement, `SELECT *` `FROM` to open the Recordset. This SQL term causes the engine to select all columns from the table. The SQL language can be used at this point to perform more sophisticated operations with the table (e.g., to select by using a filter or sorting criteria). Using the `ORDER BY` or `WHERE` clause to sort or filter a Recordset is more efficient than using the sort and filter properties.

Microsoft Jet database engine SQL is generally compliant with ANSI-89 Level 1. However, certain ANSI SQL features are not implemented in Jet database engine SQL. Conversely, Jet database engine SQL includes reserved words and features not supported in ANSI SQL. A full description of the Jet engine SQL language is part of the Microsoft Access documentation.

❏ 4.6.7   Reading and Writing Data

The best way to learn how to read and write data by using DAO is to take a closer look at Figure 4-25, the diagram of the `Easy` functions.

The VI expects a column list, which is an array of clusters. It is not possible to write VIs utilizing polymorphic inputs. In LabVIEW Version 5.0, the type OLE Variant type is available and can be used to do this. In this case, the programmer has to place the data into the OLE Variant variable. To avoid this, the VIs shown use a string as input for all data types. The data is converted inside the VI, then one OLE Variant (`DAO OV NEW`) is created, and the conversion is done according to the given data type (`Date` in the frame shown in the figure). The variant is written to its field (`DAO RS SetField-ValueByName`) and released once all entries have been completed.

If data need to be appended, `DAO RS AddNew` is used to create a new (empty) record. If the function is to work with the entry currently selected, it is necessary to use `DAO RS Edit`, which generates a temporary copy. If the record is filled or modified, the function `DAO RS Update` must be called. This VI physically writes to the database and releases the temporary copy.

❏ 4.6.8   Examples

Figure 4-26 shows a very small program. This VI uses a database containing a table. This table has (at least) two columns called "String" and "Double."

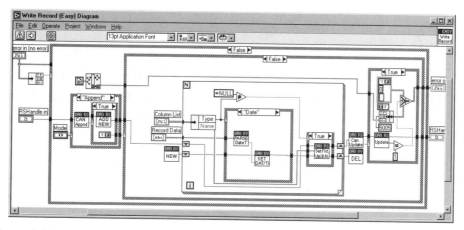

Figure 4-25

Diagram of the Write Record (Easy).vi

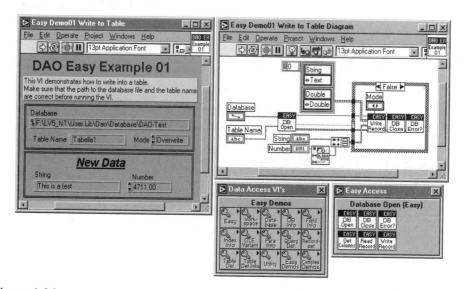

Figure 4-26

Writing to an existing table

The program is not designed to create the database or to modify the structure. VIs are available to read and change the structure of the database.

Looking at the VI hierarchy, it can easily be seen that the code looks simple only because of the easy VIs.

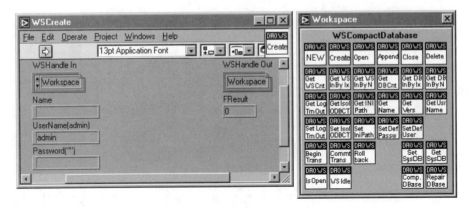

Figure 4-27

Workspace functions

Recall that on top of the hierarchy chart (Figure 4-20) there is an object called "Workspace" (Figure 4-27). This has not been used until now. DAO always opens a default workspace (called `Workspace(0)`) and uses this as long as the application does not create a different workspace. This may be necessary, for example, if the database is password protected.

Advanced users can use workspaces and transactions if the database is not used by only one LabVIEW application. Examples are systems consisting of many different computers that perform tests and save the results into a single database or systems where a big database is used to control a production process and the QS results have to be written to this system. With workspaces, a multiuser database, including all security options (password, record-locking, transactions, etc.), can be implemented.

☐ 4.6.9 Summary

The Jet engine is a powerful tool that works with database systems using the Microsoft Access data format. Because it is available royalty free, it is an elegant solution for all application programmers who do not require a more sophisticated database system. There are no functional restrictions in Access, but in practice, the maximum size of a database is limited.

The object-oriented design of the DAO API does not prevent its use in LabVIEW. Together with typified handles and high-level functions (`Easy`), the API inside LabVIEW is designed in a very user-friendly style.

LabVIEW and Automation Technology

5

If you want to build a ship, don't drum up people together to collect wood and don't assign them tasks and work, but rather teach them to long for the endless immensity of the sea.

— Antoine de Saint-Exupéry
The Wisdom of the Sands

Many applications in industrial automation pose requirements that often exceed the capabilities of conventional MMI (Man Machine Interface) and SCADA (Supervisory Control and Data Acquisition) products. This chapter shows how a graphical programming environment can close this gap. First, the general requirements in industrial automation technology are explained; then the BridgeVIEW environment is introduced.

5.1   From LabVIEW to BridgeVIEW

As mentioned in previous chapters, the term "virtual instrumentation" has become a catchword in measuring technology. Many use this term to generally describe computer-assisted systems used to acquire, evaluate, and visualize measuring data. In contrast to this, automation technology uses terms like *process visualization*, *MMI* (also referred to as *HMI — Human Machine Interface*) and *SCADA*. In practice, this wealth of terms can be reduced to a single core statement: The standard PC has been and will be an integral part of modern automation systems both in laboratories and in industrial production. In this connection, the question as to which software methodology can solve a process control task is much more important.

Traditionally, process visualization systems that allow the integration of add-on components exclusively on the basis of text-oriented standard languages are used. However, the complexity of such languages requires specialized programmers who generally do not have much knowledge of automation processes. On the other hand, for ten years the measuring technology has implemented the concept of graphical programming, which allows users without sound programming skills to create sophisticated programs quickly. Graphical programming offers a natural man-computer interaction, similar to the representation of complex processes in block diagrams derived from the process and control technology.

☐ 5.1.1   Motivation for BridgeVIEW

Thanks to its programming methodology, LabVIEW has been readily accepted in industrial automation applications despite its strong orientation toward measurement applications. This acceptance can also be accounted for by the increasing use of PC-based systems in this industrial sector and an ongoing removal of the rigid borders between data acquisition and process visualization. These circumstances led to the development of a wide range of solutions, such as for water economy, historical trend analysis in plastics extrusion plants, or distributed data acquisition systems in the pharmaceutical industry. With BridgeVIEW, National Instruments expanded LabVIEW to the visualization of process components. Wide acceptance led to the development of BridgeVIEW (see Figure 5-1), an application that is upward compatible to LabVIEW. By maintaining the LabVIEW concept, a graphical environment for MMI and SCADA applica-

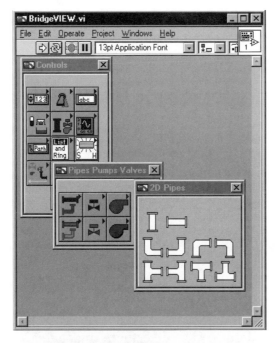

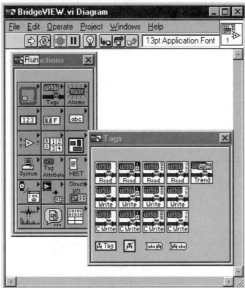

Figure 5-1

BridgeVIEW front panel and block diagram

tions was created with the goal to provide an integrated development environment for rapid prototyping and fast time-to-market applications.

❑ 5.1.2   Basic Requirements to MMI/SCADA Software

The basic structure of the available MMI/SCADA software for process visualization does not differ much from other modern visualization software. It generally offers the following properties:

- It connects processes via PLCs, field buses, data acquisition cards, and others.

- It provides a real-time database to use real-time data in a distributed database.

- It provides a visualization environment to map active real-life processes.

- It handles alarms by detecting, managing, and logging unexpected process states.

- It presents historical and current trends as a process progress over an optional period.

- It provides statistical process control (PLC) for process monitoring and quality assurance.

- It integrates external databases.

- It manages formulas.

Figure 5-2 illustrates these processes.

Although these key functions can potentially cover a major part of industrial applications, the software tool alone is normally not flexible enough to provide tailored solutions to real problems. It is often desirable or necessary to provide a way to adapt a given package by writing more or less extensive program code. In fact, most conventional packages offer a proprietary script language to let you write specific code, but it is generally too time-consuming. This is where BridgeVIEW comes in; its graphical programming environment sets new standards for customization.

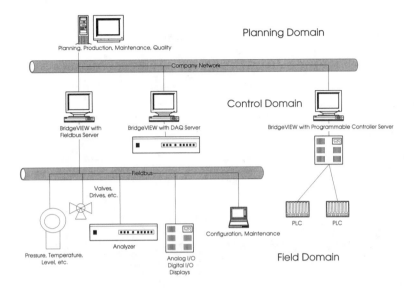

Figure 5-2

Process structure

☐ 5.1.3 The BridgeVIEW Architecture

BridgeVIEW consists of a modular, open system architecture. Software features include its real-time database, historical trend analysis, alarm and event logging, device connectivity, system configuration tools, and G, an object-oriented programming language, in which you graphically assemble software function blocks to develop MMI, measurement, analysis, and control applications.

The BridgeVIEW engine maintains up-to-date I/O values and status information it receives from device servers in a powerful real-time database, which is accessible to MMI and control applications written in G. The engine also performs other system functions such as engineering unit conversion, event monitoring, alarm processing and logging, historical logging and trending, and system security.

The basic structure, shown in Figure 5-3, consists of three functional units: At the top of the hierarchy is the MMI, allowing you to interact with the actual process. The BridgeVIEW engine with its real-time database is in the center. At the bottom of the hierarchy are device servers which act as

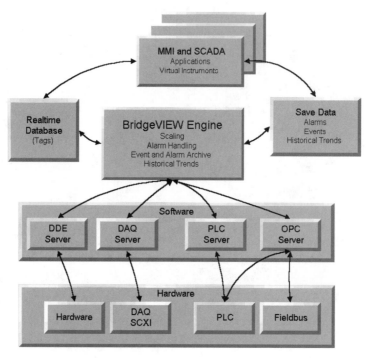

Figure 5-3

The BridgeVIEW architecture

connecting links to process hardware. They let the BridgeVIEW system communicate with I/O interfaces (e.g., PLCs, field bus controllers, data acquisition boards). These device servers operate in the background, similarly to stand-alone programs.

☐ 5.1.4   How Does BridgeVIEW Work?

BridgeVIEW uses a combination of tags, events, and data. Many system objects are created and maintained using BridgeVIEW's easy-to-use, fill-in-the-blanks configuration tools. The *Tag Configuration Editor* creates objects, called *tags*, which represent I/O values acquired from external devices and maintained in the real-time database. The tag editor also lets you configure tag parameters and system functions, such as name, description, device I/O connection, engineering unit and scaling, history logging, and alarming. Tag configuration information can be imported from existing device configuration files, thus maximizing your productivity. The following sections describe briefly how BridgeVIEW works.

First, tags are assigned to the real I/O points in the Configuration Editor. When the BridgeVIEW engine is started, the configuration file is read and the online monitoring of the tags and alarms starts; the I/O server runs in the background. Two methods are used to assign tag and alarm values to objects in the MMI. The conventional method uses a text-oriented configuration of visualization elements.

This approach is used in many visualization packages. Although it simplifies the assignment of real inputs/outputs to display elements, the configuration lists become less reproducible as the applications grow in volume. To increase transparency of all assignments, BridgeVIEW offers a graphical method in the form of a block diagram representing the entire process. MMI objects like trends or graphical maps of pumps, valves, and others, are placed on the front panel, then a menu option is selected to configure them and associate them with tags. The *G Wizard* (see Figure 5-4) creates a graphical source code according to the configuration of front panel objects and their functions. The result is a block diagram, representing the actual control program.

The part on the right in Figure 5-4 (BridgeVIEW diagram) is the source code generated automatically by BridgeVIEW. You can enhance or modify it according to your needs.

Because BridgeVIEW builds on LabVIEW and thus on G, it offers you control mechanisms that can be added directly to the block diagram. This additional capability of integrating control mechanisms in the graphical representation could be a total redefinition of development environments for MMI and SCADA packages.

❑ 5.1.5   Event Monitoring and Alarm Handling

Each tag has an associated set of alarms for high-high, high, low, and low-low conditions, with corresponding alarm set points that can be individually configured by the user. Each alarm condition also has a corresponding user-configurable priority and group designation, which is useful for filtering alarm conditions. For example, with the BridgeVIEW alarm summary display, you may elect to see only the highest priority alarms that have not been acknowledged. You can also list all alarms for certain groups of tags that may be associated with equipment or operating units being monitored and controlled. Events are associated with a change

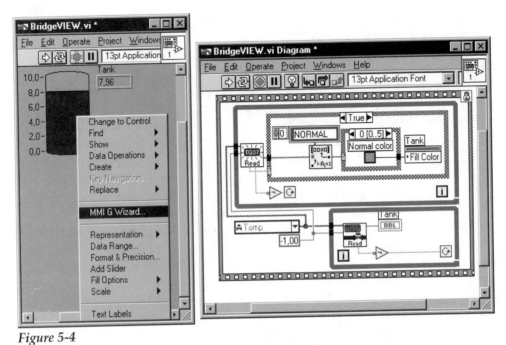

Figure 5-4

The G Wizard of BridgeVIEW generates block diagrams automatically

of state such as a motor starting, a high-level alarm, or an operator logging into the system. All alarms and events are logged to disk unless you choose to disable logging for a specific tag. Events logged to disk can be displayed with BridgeVIEW's *Event History Display*. Event information can also be accessed by user applications through graphical functions provided with G.

❑ 5.1.6   Security

BridgeVIEW provides a very flexible and configurable security system and access to the system. User accounts can be created and assigned a security level and privileges to fit that person's access requirements. Each user account can also be configured with a login ID and password. All user logins and operator actions are maintained in the event log.

❑ 5.1.7   I/O Options

BridgeVIEW offers a wide range of I/O connectivity options. Information from PC DAQ boards, FieldPoint modular distributed remote I/O, PLCs, motion controllers, analyzers, and industrial device networks is delivered to the real-time database through *device servers*. Device servers manage all device I/O and communication status information. Servers dramatically reduce software development costs because you do not have to spend valuable resources and time programming the low-level drivers required to communicate with your industrial devices. Device servers from National Instruments and alliance program partners provide a wide variety of popular PLCs and I/O devices. For instance, the software comes with a built-in OPC client (refer to Section 4.3 for a detailed description of OPC).

Several other communication options are included with the BridgeVIEW software. It includes, for example, DAQServe for National Instruments plug-in DAQ boards and remote SCXI. In addition, a series of other National Instruments servers are available to communicate with PLC, motion controllers and RTUs (Remote Test Units). BridgeVIEW lets you easily mix and match the communication components that best suit your industrial I/O needs.

❑ 5.1.8   Design Your Own Servers

You can customize your MMI further with the BridgeVIEW *Device Server Toolkit*, which contains specifications and programs to enable you to easily create your own device servers. The user interface tools embedded in BridgeVIEW will help you quickly create intuitive graphical displays, including process graphics with animation, trends, charts, and controls. To specify the functionality, you intuitively assemble block diagrams — a natural design notation for control and process engineers. The floating tool palettes in G contain many defined graphical objects that can be used at will without having to be created from scratch. The G graphical function library includes industrial automation symbols, I/O, analysis, historical, file, and networking functions. Other functions easily share information with other applications using standards such as OLE and DDE (see Section 4.2.3). With the BridgeVIEW open architecture, you can call any WIN16/32 DLL or VI library, thus leveraging your existing software development investment.

❑ 5.1.9   BridgeVIEW – Summary

The BridgeVIEW intuitive graphical user interface (GUI), combined with the powerful, graphical programming language, G, gives users powerful tools to perform data acquisition and analysis, create an MMI, and develop advanced PC-based control applications. G gives you the flexibility of a powerful application development language without the associated difficulty and complexity of conventional programming and scripting languages. BridgeVIEW applications span the manufacturing enterprise from research, pilot plant, production, quality control, environmental monitoring, and product testing. BridgeVIEW offers a cost-effective, scalable automation solution to help you solve your toughest industrial automation and control applications.

5.2   LabVIEW and CAN

The requirements of device networks depend on the devices connected and the applications for which they are used. For this reason, multiple networks are required to satisfy the full range of device networking needs. One of National Instruments' recent additions to its network interface offering includes products for connecting PCs to DeviceNet, an industrial networking standard based on the *CAN (Controller Area Network)* protocol.

CAN specifies how data can be transferred between multiple devices for communication and control. The specification defines how multiple devices access the network, including bus access when devices want to communicate at the same time. CAN also specifies how data packets are encoded, to include addressing and error checking, as well as actual process data from the device. Increasing consumer and commercial demands for CAN are the keys to decreasing prices and improving performance of CAN chips. As the deployment of CAN buses increases, it has been more frequently used to acquire measuring data. To avoid time- and cost-intensive "island" solutions, it is necessary to integrate the CAN bus transparently into the virtual instruments concept. This demand becomes clear when you consider that it offers both smooth exchange of data between the CAN bus and other measuring systems and a uniform software interface.

❑ 5.2.1   CAN Basics

Robert Bosch Corporation originally developed CAN to replace expensive wiring harnesses with a digital network in automobiles. CAN primarily targets the interconnection of in-vehicle controllers. Because of its design for automotive applications, CAN offers fast response and very good reliability under adverse environmental and electrical conditions. These attributes, along with cost-effective networking technology, also make CAN a good industrial network. In particular, CAN offers important advantages compared to other layer-2 protocols, which are mainly implemented in the software, because its protocol is fully implemented in hardware and it connects easily to application programming environments.

When implementing a CAN interface, you can use a number of available CAN chips. This means that all CAN-specific functions, such as bit arbitration, error handling, and exchanging messages, are implemented on the board. Such chips are offered, for example, by Intel, Intermetall, Motorola, NEC, Philips, Siemens, and NSC. However, the supported specification (CAN 2.0A or CAN 2.0B) differs from one vendor implementation to another and with regard to the host interface (full or basic CAN). To be able to switch output signals from the CAN chip to the bus line, you need an output driver, which is also available as an add-on component. Often, the CAN interface card features its own local processor to relieve the personal computer from preprocessing tasks to speed up response times.

❑ 5.2.2   National Instruments CAN Hardware

National Instruments CAN interface boards are available in three formats: The AT-CAN is a half-length ISA plug-and-play interface board, which offers complete software configuration with no jumper or switch settings under Windows 95. The PCI-CAN is a short PCI interface board. The PCMCIA-CAN is a PCMCIA type II card that is fully compliant with the PC Card standard. The PCMCIA-CAN can be used in a Notebook or other computer with either a type II or type III slot.

All National Instruments CAN boards use an Intel chip 82527 and an 80386EX microprocessor to handle communications directly on the interface board. The 82527 supports both CAN specifications (CAN 2.0A and CAN 2.0B), so that it can process 11-bit and 29-bit identifiers. An interesting feature of this controller is its combined full/basic CAN implementation. *Full CAN* means that the incoming message buffers accept only data with identifiers matching a mask which is preset in the control register. This relieves the card's processor from the need to filter messages. *Basic CAN* means that the bus accepts all messages and that the local processor has to filter them subsequently.

The Intel 82527 has 14 message buffers. The CAN board uses a masking mechanism to extract the desired information. In addition, a special message buffer is available to receive each message. This capability has a positive effect on the flexibility of the interface card: On the one hand, messages are received very quickly (e.g., for continuous monitoring); on the other hand, various devices can be polled by event.

The 80386EX provides a dedicated environment for reliable, high-performance execution of the CAN communications protocol stack. The physical layer of all National Instruments CAN boards fully conforms to the ISO 11898 physical-layer specification for CAN and is optically isolated to 500 V. Interfacing to the CAN bus is by means of a DB-9 connector or a connector defined by DeviceNet. AT and PCI-CAN interfaces can be configured to have their transceivers powered either internally or by the bus. PCMCIA-CAN boards can be ordered with either an internally powered or bus-powered transceiver cable. AT-CAN and PCI-CAN boards also include the National Instruments RTSI bus. With RTSI, your application can synchronize activities on the CAN bus with functions such as analog and digital I/O using a National Instruments plug-in DAQ board.

National Instruments CAN interfaces are designed to communicate with CAN devices. All CAN interfaces are available in 1- or 2-port configurations. Full Windows 95 plug-and-play compatibility gives you the benefits of automatic configuration for easier installation and maintenance. Windows 95 configuration is fully integrated into the Windows 95 Device Manager. All National Instruments CAN interfaces are designed to meet the physical and electrical requirements for in-vehicle (automotive) networks based on CAN.

☐ 5.2.3 Applications

The CAN boards described above are particularly interesting when combined with the LabVIEW development environment. Existing CAN applications can be divided roughly into two main fields of application: the *production control level* and the *process control level*. In most applications, the CAN bus does not appear to the outside, serving only for internal exchange of data between sensors, actors, and intelligent control units.

This design raises the important question as to how we can find the causes in case of errors. LabVIEW solves this issue by providing an excellent means to simplify and accelerate error diagnostics, thanks to its graphical user interface and error handling functions (see Figure 5-5).

The physical connection to the CAN bus can be implemented through a PC Card CAN bus card which is installed in a Notebook PC. The graphical LabVIEW development environment assists developers during the design

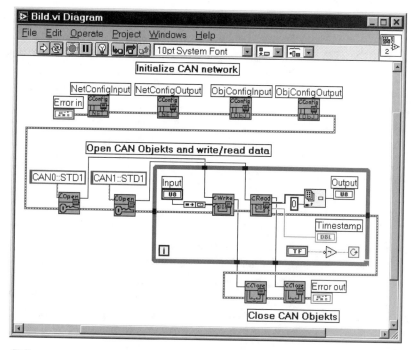

Figure 5-5

Programming a CAN bus in LabVIEW

and implementation phase, and it helps users by providing them this service tool.

On the process control level, the CAN bus appears initially to the outside because it networks with sensors, actors, and PLCs. In most cases, this application area requires you to create process diagrams and MMIs by using a process visualization software. The requirements of these software environments are data archiving in databases, alarms management, various security levels, real-time trends displays, formula management, and others, in addition to the graphical representation. This is the typical application area for BridgeVIEW, which meets all these requirements and ensures a high degree of flexibility.

❏ 5.2.4   CAN Tools

Each CAN interface is shipped with NI-CAN device driver software. This software provides a high-level API (Application Programming Interface) for reading and writing data frames. National Instruments CAN boards are shipped with NI-CAN software for Windows 95 (only AT-CAN boards) and Windows NT. NI-CAN software includes device drivers used for application development as well as firmware that runs on the embedded 80386EX microprocessor. The NI-CAN device drivers are full 32-bit drivers designed for Windows 95 and Windows NT. These device drivers are compatible with standard programming environments such as Visual C/C++ and Borland C/C++, as well as LabVIEW and LabWindows/CVI application software. The firmware is used to implement time-critical features provided by the NI-CAN software. NI-CAN software provides flexible yet easy-to-use functions for configuration and I/O on CAN networks.

The 80386EX processor on a National Instruments CAN interface provides the operating environment for execution of the CAN protocol stack. CAN specifies timing requirements to ensure reliable, deterministic operation of the bus. As the master in a typical system, a National Instruments CAN interface must provide the necessary system responsiveness. Because the majority of the CAN protocol executes on the embedded 80386EX on a National Instruments CAN interface, it achieves improved response to incoming messages. Embedded execution of the CAN protocol stack also results in more deterministic network performance because the on-board microprocessor is dedicated to CAN communication activities.

For program development, NI-CAN provides two different levels of access to a CAN network — the *CAN Network Interface Object* and *CAN Objects*. Both forms of access time-stamp incoming data and offer various forms of queuing.

The CAN Network Interface Object provides low-level access to a CAN network. Each CAN Network Interface Object maps to a specific CAN port, with no limitation on the maximum number of ports or cards (e.g., two AT-CAN/2 interfaces would provide CAN0 through CAN3). This object can be used to send and receive entire CAN frames. For example, to send a CAN frame, you would specify the outgoing arbitration ID, frame type (data or remote), data length, and data.

CAN Objects provide higher-level access to a CAN network. Each CAN Object maps to a specific data item (arbitration ID), and multiple CAN Objects can be used for a given port. When configuring a CAN Object for use, you specify the arbitration ID, direction of data transfer, data length, and also how you want the data to be accessed (e.g., periodically). For example, you could configure a CAN Object to send an outgoing data frame for a specific arbitration ID every 100 ms. After opening this CAN Object, you use the `write` function to provide data to send, and all periodic timing is handled by the NI-CAN embedded firmware.

☐ 5.2.5   LabVIEW and CAN – Summary

Considering the dominant position of the car-making industry in automation technology, a large potential is predicted for the CAN field bus.
Chances are good that market shares of CAN may grow rapidly during the
coming years. For this reason, high investments have been made to further
develop this interesting field bus technology, which will push standardization in this area. As far as software is concerned, CANOpen may become a
market leader as an application layer in Europe, and DeviceNet in the U.S.
market. Both application layers will be embedded into the LabVIEW CAN
libraries in the near future.

With a National Instruments CAN interface board and NI-CAN software, you can use a desktop, industrial, or Notebook PC running
Windows 95 or Windows NT for a variety of CAN applications,
including automotive testing and diagnostics, factory automation, and
machine control.

5.3   LabVIEW and HART

For many years, the field communication standard for process automation
equipment has been a milliamp (mA) analog current signal. The milliamp
current signal varies within a range of 4–20 mA in proportion to the process
variable being represented. In typical applications, a signal of 4 mA will
correspond to the lower limit (0%) of the calibrated range, and 20 mA will
correspond to the upper limit (100%) of the calibrated range. If the system is
calibrated for 0–100 psi, then an analog current signal of 12 mA (50% of
range) corresponds to a pressure of 50 psi. Virtually all installed systems use
this international standard for communicating process-variable information
between process automation equipment. The *HART Field Communications
Protocol* extends this 4–20 mA standard to enhance communication with
smart field devices.

The HART protocol was originally developed by Rosemount Inc. and is
maintained by the *HART Communication Foundation*. The HART Communication Foundation is an independent, nonprofit corporation, specifically
organized to coordinate and support the application of HART technology
worldwide. Educating the industry on the capabilities and value of this
important technology is one of the Foundation's key roles.

HART is an acronym for Highway Addressable Remote Transducer. The HART protocol makes use of the Bell 202 FSK (Frequency Shift Keying) standard to superimpose digital signals at a low level on top of the 4–20 mA. This enables two-way communication to take place and makes it possible for additional information beyond just the normal process variable to be communicated to/from a smart field instrument. The HART protocol communicates without interrupting the 4–20 mA signal and allows a host application (master) to get two or more digital updates per second from a field device. As the digital FSK signal is phase continuous, there is no interference with the 4–20 mA signal.

A wide choice of products supporting the HART protocol is available from major instrumentation suppliers, and the number of products and suppliers incorporating the technology continues to grow.

❑ 5.3.1   Sample Application

HART technology is being used in a wide variety of applications worldwide to obtain significant improvements in plant performance, provide solutions to regulatory compliance issues (ISO 9000, OSHA, EPA, DOT, etc.), and realize substantial cost savings in initial installation/commissioning and ongoing maintenance and operation.

The sample application described in this section concerns the development of a simple LabVIEW device driver that will accept the main measurement values from a single HART field device, delivering them to a personal computer (master/slave operation) (Figure 5-6). In this example, a current loop should close to accept or remove digital information. A power load within the 230–1100 ohms range is applied for this purpose. A HART modem is used to convert the information into signals the RS-232 interface can understand. This design establishes a relatively easy communication to the personal computer (master station).

A complete description of all commands and defined HART units would go beyond the scope of this section. Interested readers may obtain a complete specification and complete information regarding the HART protocol directly from the HART Communication Foundation (www.hartcomm.org).

Physical Master/Slave Connection

The HART modem we use in this sample application acquires digital information directly from the power transformer. In this case, the HART field

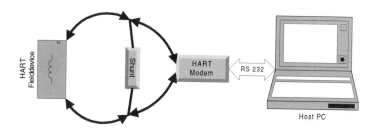

Figure 5-6

Connecting a HART field device to a personal computer

device does not need any auxiliary energy source to feed the current loops. The HART modem takes its electric energy directly from the computer's RS-232 interface.

HART Messages

Field devices and the personal computer communicate by exchanging messages. The HART protocol defines a series of message structures. Older HART devices know only the short telegram-style message format. Modern devices use a long message format, including an internationally unique 38-bit slave device ID. Table 5-1 shows the fields of both HART message formats.

To send commands to a field device, we first need to poll its address by sending a short message in the format shown in Table 5-2.

The preamble consists of several *FF* blocks. It defines the beginning of the message. The next field, 02, defines the message format. The address is 00h in this example because we address only one single field device. The 00h command requests the field device to send its ID. The device ID contained in this reply is needed to build long messages. The number of bytes field provides information on the data bytes that follow (excluding the checksum). The checksum field is determined by an XOR link of all bytes, starting from the start byte.

Initializing the Device Interface

Because LabVIEW is a true 32-bit application for Windows 95 and Windows NT, serial communication on this platform differs from the DOS platform in that the RTS (Request To Send) line is not set automatically during interface initialization. This applies also to initializing an interface in LabVIEW. This section describes briefly how you can initialize a field device interface in LabVIEW. Figure 5-7 is a schematic view of how to initialize the serial interface.

Table 5-1   *HART message format fields*

Short Format	Long Format
Preamble	Preamble
Preamble	Preamble
Preamble	Preamble
Start	Start
Address	Address
Address	Command
Address	Number of bytes
Address	Checksum
Address	
Command	
Number of bytes	
Checksum	

Table 5-2   *Message format to request a device ID*

FFFF FFFF FFFF	Preamble
02	Short message master/slave
00	Address
00	Read unique ID# command
00	Number of bytes
00	Checksum

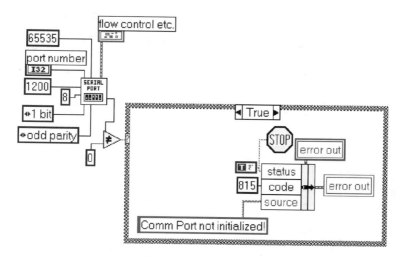

Figure 5-7

Initializing the serial interface

You need to initialize the interface according to the HART specification. The program should exit automatically in the event of a faulty initialization. To ensure correct communication, set your HART modem interface to the following values:

- 1200 baud speed

- 8 data bits

- 1 stop bit

- odd parity

Polling the Device ID

Before you send a message to request the device ID, you should run LabVIEW's `serial line ctrl.vi` to set the RTS line and to reset it immediately, as shown in Figure 5-8.

Note that there should be no data on the port from previous read/write cycles to make sure you obtain correct reply data. You can load the device reply into your computer in the way demonstrated in Figure 5-9.

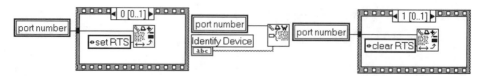

Figure 5-8

Requesting device identification

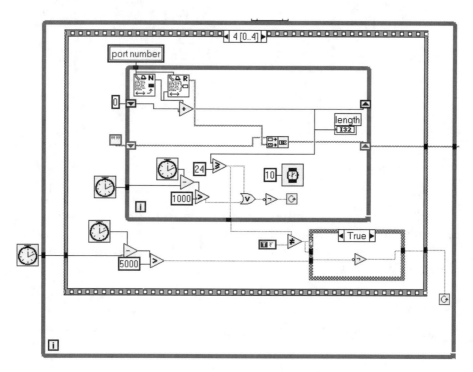

Figure 5-9

Reading a device ID into the computer

The inner while loop in Figure 5-9 continues reading bytes from the port until a string length of 24 bytes or the time-out is reached, whichever comes first. The outer while loop forces several retrials. It terminates communication if it does not receive successful results within five seconds. Once a valid string has been received, it is evaluated against the checksum and the device address is extracted from the string.

The program structure shown in Figure 5-10 shows how a preamble is filtered and how the check bit is determined by XOR, linking all data bytes.

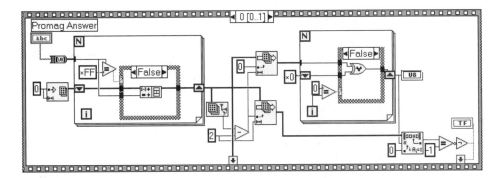

Figure 5-10

Determining the check bit

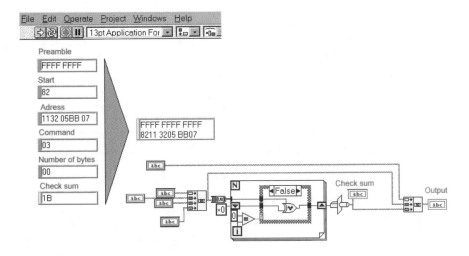

Figure 5-11

Messages for output to a HART field device

If the received check bit matches the calculated bit, the device ID is extracted from the byte array. It is now available for further communication.

Transferring Messages

One of the most important preparatory steps is to determine the device ID. Figure 5-11 shows the structure of valid HART messages.

This figure shows clearly that the checksum is recalculated each time a message is sent to the field device, depending on the command.

Deciphering Reply Messages

A HART field device responds to each command by sending a predefined string. Figure 5-12 shows how a reply message to a command 3 request is deciphered.

Command 3 requests the actual current on a 4...20 mA interface and four previously defined measuring values. This measuring data is contained in the reply message after the preamble (FFh), the device ID, the command retrial, and the number of data bytes.

The HART protocol specifies the structure for data bytes in Table 5-3.

Table 5-3   *Structure for data bytes specified in the HART protocol*

Data Bytes	Function	Comment
0–3	Current (mA)	Floating-point number (4 bytes according to IEEE 754)
4	Measuring unit code, measuring value 1	
5–8	Primary measuring value	Floating-point number (4 bytes according to IEEE 754)
9	Measuring unit code, measuring value 2	
10–13	Secondary measuring value	Floating-point number (4 bytes according to IEEE 754)
14	Measuring unit code, measuring value 3	
15–18	Tertiary measuring value	Floating-point number (4 bytes according to IEEE 754)
19	Measuring unit code, measuring value 4	
20–23	Measuring value 4	Floating-point number (4 bytes according to IEEE 754)

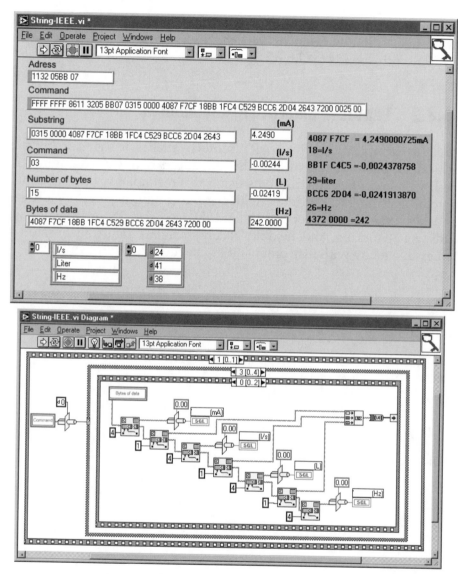

Figure 5-12

Deciphering a HART reply message

The `type cast` function converts a 4-byte measuring value into a floating-point number as specified in IEEE 754. The hex-coded string 4087 F7CF can be mapped directly to floating-point number 4.2490000725, corresponding to the actual current output, expressed in mA.

The coding of measuring units is based on international standards. The software takes the allocation of the read bytes to the corresponding unit (e.g., "29h" corresponds to "liter").

❑ 5.3.2   LabVIEW and HART – Summary

The relative simplicity of the HART protocol makes it easy for both end users and suppliers to benefit from the enhanced two-way communication capability of smart field instruments using this technology. Powerful multi-parameter instruments, efficiency with remote communication, field device diagnostics, cost-effective control in field devices, installation savings with multidrop networking, and flexible/accurate digital data transmission are all achievable today with instruments that use the HART protocol.

6

LabVIEW and Fuzzy Logic

What makes society turn is science, and the language of science is math, and the structure of math is logic, and the bedrock of logic is Aristotle, and that's what goes out with Fuzzy.

— Bart Kosko
Neural Networks and Fuzzy Systems

Fuzzy logic provides a precise mathematical theory to describe and process imprecise and uncertain information on a computer. Fuzzy systems model the linguistic handling of imprecise information, thus imitating human thinking by taking the imprecision inherent in all physical and technical systems into account. The theory of imprecise quantities (*fuzzy set theory*) was developed by Lotfi A. Zadeh of the University at Berkeley, California, in 1965.

6.1   Introduction

Humans, he argued, have the capability to reason under uncertainty and with imprecise data. Hard computing requires precision, certainty, and rigor. Zadeh recognized that precision and certainty carry a cost and that computation, reasoning, and decision making should exploit the tolerance for imprecision inherent in human reasoning.

Zadeh introduced his work at a time when the first microprocessors based on the digital technology with its dual Yes-No logic had been developed. Still, the fuzzy set theory has been experiencing a true application boom in various industries only since the mid-eighties. This delay holds true mainly for measuring and automation control systems. However, many of the standard tools in this area are still far from providing users with fuzzy solutions in an environment they are familiar with.

Today, users create fuzzy systems mainly in a predefined menu-driven development environment with a graphical user interface. Although this approach is much more comfortable for the user than creating a fuzzy system in a textual programming language, it usually lacks easy online connection for measured data or data output to the real world. In addition to its graphical programming capabilities for fuzzy systems in the sense of rapid prototyping, other essential advantages of LabVIEW are its capabilities to use A/D conversion to input analog values as control system parameters and to output the results from these calculations to processes or other interfaces by means of D/A conversion.

This chapter describes how fuzzy concepts can be implemented in LabVIEW and shows examples to explain how they are used in LabVIEW for online process control. Two sample applications are discussed. The first describes the implementation of a fuzzy controller that is used to detect and sort various small formed parts, where size and proportion of the parts may vary within a relatively wide range. The second sample application concerns the attenuation of undesired swinging movements of various loads transported on a portal crane model.

6.2   Basics – Fuzzy Logic and Fuzzy Control

One of the central terms in the world of fuzzy logic is the term *linguistic* (*lexical*) *imprecision*. In the strongly generalized realm of the classical dual logic with its two true or membership values, "Yes" or "1" and "No" or "0", fuzzy logic allows you to model imprecise colloquial terms and facts by means of so-called *membership functions* (fuzzy sets). Fuzzy sets allow arbitrary real intermediate values between the two membership values 1 and 0.

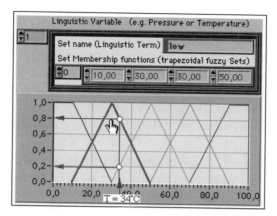

Figure 6-1
Linguistic variables and terms

For instance, the imprecise state "low flow temperature" can be modeled by a membership function "low." Each argument value, *T*, of the flow temperature — a physical parameter (linguistic variable) — is allocated to a membership degree μ, $0 \leq \mu(T) \leq 1$, which shows how well this argument value meets the state "low" (linguistic term). A linguistic variable is generally modeled by a set of overlapping linguistic terms (membership functions).

For instance, the "precise" value T = 34°C, shown in Figure 6-1, represents a "negative low" flow temperature. Thus the given value meets the state "low" at 80%, but it is also "medium" with a 20% compliance degree. Consideration of the neighboring terms in the calculation of conclusions from a given situation (value) accounts for the enormous robustness of fuzzy systems.

The second central term in fuzzy logic is control-based "closing," also called *fuzzy inference*. In practice, although expert knowledge of human engineers on optimum operation of an equipment or a process is available, it appears to be much more difficult to mathematically describe the physical relations in every detail. In general, time constants that have to be quantified as well as distributed, or time-dependent parameters and nonlinearities are generally the main reason why no appropriate exact mathematical process model is found within an acceptable time.

Fuzzy logic allows us to use expert knowledge in colloquial language in the form of if... then... control (e.g., IF position error = zero AND swinging angle = zero THEN drive = stop) as a basis for process control without having to develop a mathematical control members model in advance. Based on the input control basis, an algorithm called *fuzzy inference* is used to draw the conclusions needed in a given process situation and to output corresponding control actions. With regard to the suitability of fuzzy logic in control applications (*fuzzy control*), we obtain the basic structure shown in Figure 6-2.

The variables of the process to be controlled by sensor parameters (e.g., temperature T = 40°C, pressure p = 10 bars) are determined in three steps:

- **Fuzzification** — The technical parameters are transformed into their linguistic interpretation. To do this, it is necessary to determine how well the measurement values meet the various linguistic concepts (terms) of the respective linguistic variables.

- **Fuzzy inference** — This is the step where the control basis, that is, the derivation from the parameters described in linguistic form, is evaluated from the linguistically interpreted parameters by if...then... rules. The rule base contains the knowledge of the expert staff or the project engineer.

- **Defuzzification** — This step is applied to convert the linguistically described variables back into technical parameters to obtain a good representation of the control information contained in the total conclusion.

These steps are explained in the following example: Consider the situation of a car driver who, during a braking maneuver on a highway, has to determine a suitable value for output parameter, brake force K, in relation to the input parameters, distance A to the car driving ahead, and speed G of his own car.

The colloquial interpretation of the three parameters (distance, speed, brake force) is given in the form of the triangular fuzzy set shown in Figure 6-3. This example models distance A between 150 m and 350 m by the fuzzy set (term of the linguistic variable, distance) as "medium." The typical value of this fuzzy set is around 250 m.

Fuzzification of the input values is the process of determining the membership degrees of the current measuring values for the individual fuzzy sets.

For instance, the given distance value, 175 m, with membership degrees of 0.75 and 0.25, is assumed to be "low" and "medium," respectively. The current speed of 190 km/h with a membership degree of 1.0 is assumed to be absolutely "positive high."

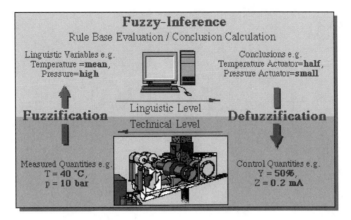

Figure 6-2
Basic structure of a fuzzy control

Let's assume our rule base evaluation proves that only the two given rules are met in this situation. If we AND-link the two input values using the minimum (min) operator, we obtain a compliance degree of 0.25 for the first rule and a compliance degree of 0.75 for the second rule. Accordingly, the conclusion from the first rule tells us to set the brake force to "3/4" and to evaluate it with a compliance degree of only 0.25, while the conclusion drawn from the second rule, that is, set the brake force to "full," is 0.75. This calculation of conclusions is called *fuzzy inference*.

To convert the two differently weighted conclusions into precise variable values — defuzzifying the inference result — we calculate a mean value from the two conclusion sets that are "truncated" at the level of the compliance degree for the respective rule. By applying this method, we apply the "best tradeoff" between the two differently weighted conclusions. Our example results in a brake force setting of 82.7%.

Fuzzy controllers have the static structure shown in Figure 6-4. They belong to the group of nonlinear characteristic diagram controllers. As shown in Figure 6-5, their dynamic behavior can be described in the form of a transfer characteristic (characteristic diagram).

6.3   Implementing Fuzzy Controllers in LabVIEW

In LabVIEW, a linguistic variable can be implemented quite easily as an array of linguistic terms. If you limit your model of linguistic terms to

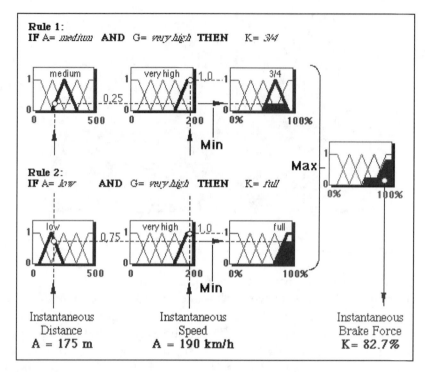

Figure 6-3
Schematic inference of a fuzzy controller

standardized trapezoidal and triangular membership functions (fuzzy sets), you can implement any term as a cluster consisting of a string (name of the term) and an array composed of four real values (the argument values for which the membership function takes the limit values 0 and 1), as shown in Figure 6-6.

A rule base can also be implemented easily as a cluster composed of one or several premise arrays (containing the premise terms) and a conclusion array (containing the conclusion terms), where the common index for all arrays describes the number of the individual rule.

With a fuzzy controller for two inputs and one output, we can implement a clear matrix form (similar to a state-event matrix) to represent the rule base, as shown in Figure 6-7. The row and column headings include the linguistic terms of the two input variables. The matrix elements, which are implemented as a menu ring, then allow the selection of the conclusion term desired for the term combination (AND-linking the two premise terms) given by the row and column coincidence.

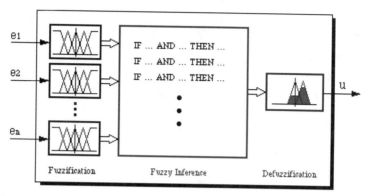

Figure 6-4
Internal structure of a fuzzy controller

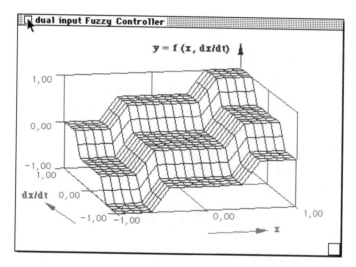

Figure 6-5
Characteristic diagram of a fuzzy controller

The situation shown in Figure 6-7 reflects the setting of the conclusion term "stop" for the "IF position fault = ZR AND swinging angle = ZR THEN output = stop" rule (where ZR means zero).

The structure of the fuzzy controller, CONTROL2, shown in Figure 6-7, corresponds to the three-step inference scheme described in the previous section. The measuring values of a real process (in this example, position deviation and excursion of loads conveyed on a portal crane) acquired over an input/output card and A/D-converted, serve directly as values for the two inputs (input value 1 and input value 2 in Figure 6-7). When called

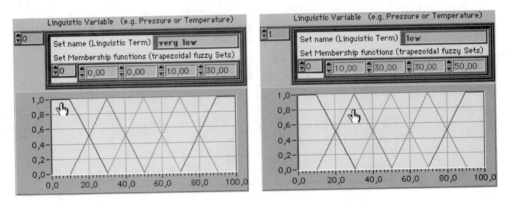

Figure 6-6
Implementing a linguistic variable and its terms in LabVIEW

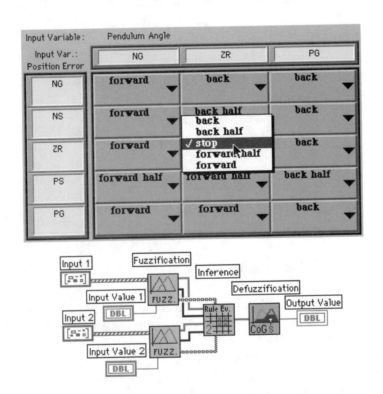

Figure 6-7
Rule base and structure of a fuzzy controller for two inputs

within a control loop, the fuzzy controller determines the current process state and supplies the variable required for control (`output value` in Figure 6-7). Finally, this value can be output to the process over a corresponding input/output card.

☐ 6.3.1   The Fuzzy Logic Toolkit for LabVIEW

Fuzzy logic can be used to accelerate the development of controllers for nonlinear or very complex systems. This is easy to comprehend because the control strategy is implemented with simple, intuitive, linguistic rules. Thus, the need for highly complex and difficult-to-understand control strategies is eliminated. In addition to control applications, the fuzzy logic software can be used for expert decision-making, such as pattern recognition or fault diagnostics.

The *Fuzzy Logic Control Toolkit* is application software from National Instruments to design fuzzy logic control systems for LabVIEW. In addition to helping you design your control system, this point-and-click software includes functions to implement your fuzzy control system in LabVIEW. The Fuzzy Logic Control Toolkit can be combined with NI-DAQ, PID Control Toolkit, Control and Simulation Toolkit, and SPC (Statistical Process Control) Toolkit for advanced control applications.

The Fuzzy Logic Toolkit for LabVIEW includes the following components:

- **Project Manager** — Used to edit fuzzy projects. In addition to online help topics explaining how to load and store fuzzy projects and to print project documentation, the Project Manager offers an input field that you can use to enter a project description. All other toolkit functions are called from the Project Manager. Figure 6-8 shows how a project is loaded into the Fuzzy Set Editor.

- **Fuzzy Set Editor** — Used to define and edit linguistic variables and terms. The Fuzzy Set Editor offers a series of functions to edit terms, in addition to the graphical display of linguistic terms for the linguistic variable currently selected. It shows the entire term arrangement. You can add, remove, or rename variables and terms. You can also change variable scopes and the forms of individual terms. Like all components of the Fuzzy Logic Control Toolkit, the Fuzzy Set Editor offers extensive online help. All actions are run through a plausibility check to cull those term arrangements that are not useful. Figure 6-9 shows the Fuzzy Set Editor panel with the **add term after** command from the **Define** menu currently selected.

- **Rule Base Editor** — Serves to define and modify the rule base. The Rule Base Editor offers functions to define and change individual linguistic rules. In addition to specifying the defuzzification method, you can define the output value to be suggested by the Fuzzy Controller in the event that an incomplete rule base does not match any of the available rules. The online help topics within the Rule Base Editor provide information on all functions available in this editor. Figure 6-10 shows the Rule Base Editor panel with the situation of a Fuzzy Controller with two input values during the selection of the defuzzification method.

- **Test Facilities** — Various facilities used to analyze the behavior of dynamic systems. This feature provides you with a series of test functions to analyze the behavior of dynamic systems. For example, you can determine and display the dependence of the behavior of dynamic systems on a single input value, with fixed values for the other input values. Figure 6-11 shows the panel for a simple controller characteristic. You can display both the current input situation and the relevant output values as well as the rules valid for the given situation.

Other facilities are:

- **Online Help** — This includes extensive information on all topics, explaining all functions of the toolkit.

- **Print Facilities** — You can use these facilities to output your work in various formats.

- **Ready-for-use samples** — Examples provided with the software include fuzzy control examples, a pattern recognition example, and an example of implementing a fuzzy controller with DAQ functions.

Each fuzzy controller can have up to four inputs and one output. For systems with large numbers of controller inputs, you can cascade multiple fuzzy controllers while implementing rules that are easy to understand.

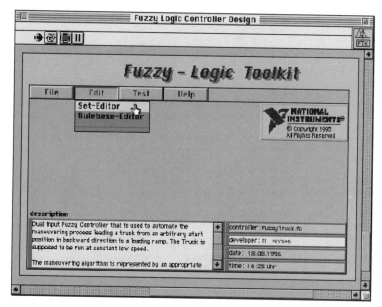

Figure 6-8
Project Manager panel (loading the Fuzzy Set Editor)

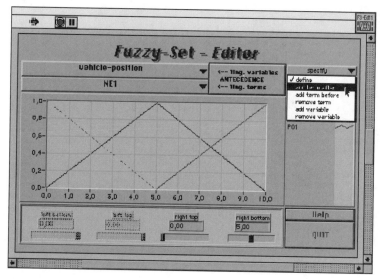

Figure 6-9
Fuzzy Set Editor panel (running the **add term after** command)

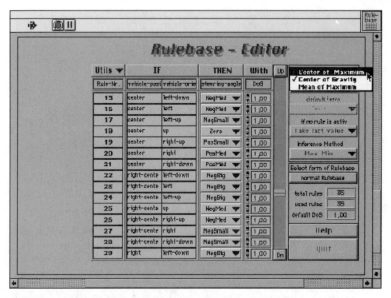

Figure 6-10
Rule Base Editor panel (while selecting a defuzzification method)

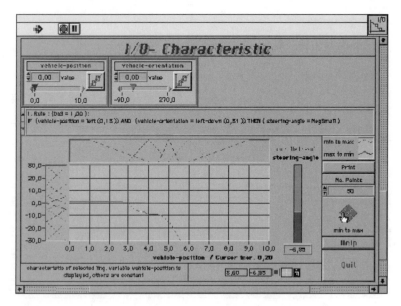

Figure 6-11
Function to determine the behavior of dynamic systems

Linguistic terms for rules are defined with membership functions in various shapes — triangular, trapezoidal, Z-shaped, S-shaped, or singleton. Rules are generated automatically to create a complete rule base of all possible input combinations, using the interactive Fuzzy Logic Control Toolkit. The output of each rule is manually selected, and weight values may be assigned for the purpose of controller tuning. Defuzzification methods include Center of Area, Center of Maximum, and Mean of Maximum.

The toolkit includes VIs to be used in your G application for control or decision-making. All controller information from the fuzzy logic control designer is saved to a data file, and retrieved and grouped into a single cluster with the Load Fuzzy Controller VI. This cluster of information is wired in your application to the Fuzzy Controller VI. Controller inputs are wired into this VI, and the controller output is calculated, based on your designed controller, and returned to your application. You can easily integrate fuzzy logic to NI-DAQ and data acquisition hardware.

☐ 6.3.2   Sample Applications

In the first sample application, we assume we need to detect and sort triangular, hexagonal, and rectangular or square parts. We further assume that both the size and the proportion of the parts may vary within a relatively wide range. The parts are placed on a conveyor belt and fed past a rather simple reflection light barrier, consisting merely of a series of LDR resistors (Figure 6-12). The strongly varying diffusion (shadow scattering), which is characteristic for many formed parts, is detected at the reflection light barrier and evaluated by a fuzzy controller. Signal acquisition, extraction of characteristics, and recognition are done online while the parts are conveyed.

Despite the large variance of the signal curves (allowed variation of component proportions and disturbing dependence on secondary light source), we can extract three distinct characteristics from the signal curves: the time to maximum coverage, `tup`, the time to maximum coverage, `thold`, and the time to release, `tdown`. These times can be determined relatively easily by preprocessing the signal (shown in Figure 6-13 for an ideal signal curve) in the fuzzy controller. If we add these three periods of time to the total time, `tSignal`, we obtain the values shown in Figure 6-13 for an extraction of characteristics that is largely independent of the component proportions.

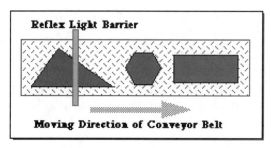

Figure 6-12
Sorting path model and real signal curves with a left triangle (TriLeft)

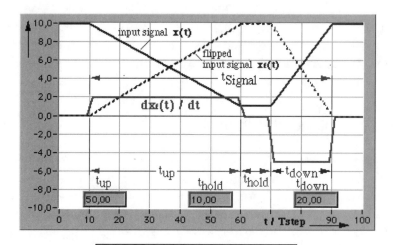

Figure 6-13
Signal preprocessing (ideal signals) and values for characteristics extraction

We can now use the typical signal curves measured for various parts to derive the linguistic terms for the two linguistic variables (input values), `Thold/Tsignal` (TH/TS) and (`Tup-Tdown/Tsignal` ((TU-TD)/TS), shown in Figures 6-14 and 6-15.

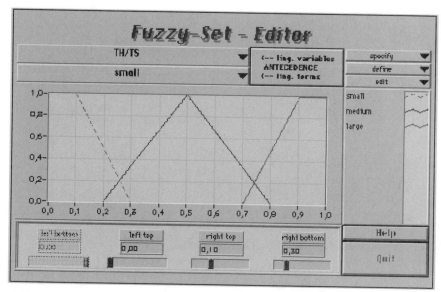

Figure 6-14
Term arrangement for the linguistic variable, TH/TS

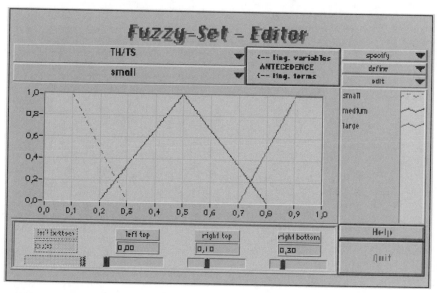

Figure 6-15
Term arrangement for the linguistic variable, (TU-TD)/TS

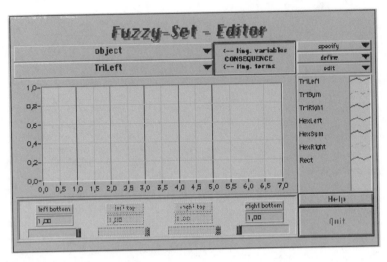

Figure 6-16
Term arrangement for the linguistic variable, object

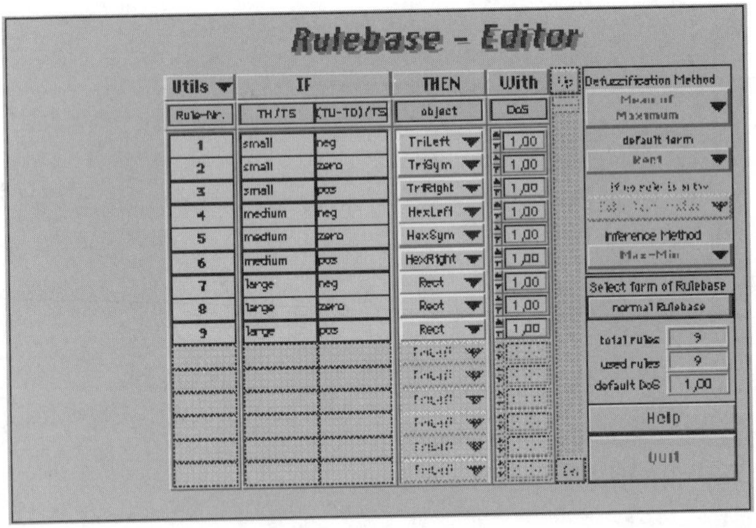

Figure 6-17
Rule base

The linguistic variable, object, shown in Figure 6-16, is used as output value. The individual object designations (terms) are modeled as singletons. The fuzzy controller used for pattern recognition uses the simple rule base shown in Figure 6-17.

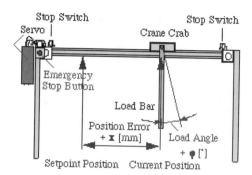

Figure 6-18

Portal crane model and determination of input and output values

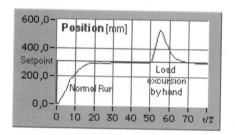
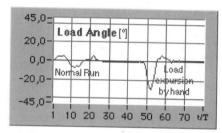

Figure 6-19

Portal crane model "position error and load angle curves"

The defuzzification method used in this example is *MoM* (Mean of Maximum). This method serves to determine the most plausible solution.

Our second sample application deals with the attenuation of undesired swinging movements during conveyance of various loads on a portal crane (see model in Figure 6-18). Our task is to drive a hanging load by means of a portal crane, within the shortest possible process time and without excessive swinging, to the destination where the load is dropped. The crane trolley's travel is controlled by a fuzzy controller, which is assumed to be appropriately preset. The position difference determined over an incremental sensor and the swinging angles acquired with a simple potentiometer will serve as measuring values.

We chose a rule strategy to avoid large swinging angles during the drive. We selected terms for output values so that relatively high process speeds for exceptional situations and rather moderate speeds for regular drive are obtained. "Center of gravity" is the defuzzification method used here.

The results are shown in Figures 6-19 and 6-20.

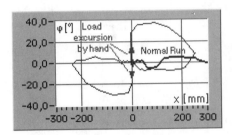

Figure 6-20
Portal crane model "position error and load angle in phase level"

☐ 6.3.3 Summary

The use of fuzzy methods in LabVIEW means that you can work in a single development system, from data acquisition over preprocessing and the use of fuzzy methods to presentation of results. This means an increase in productivity and efficiency because you do not have to leave your familiar LabVIEW programming environment while using fuzzy technology at the same time. In addition, the Fuzzy Logic Control Toolkit is written entirely in G, which means that it has cross-platform capability.

LabVIEW and Genetic Algorithms

The one remains, the many change and pass; Heaven's light forever shines, Earth's shadows fly; Life, like a dome of many-colored glass, Stains the white radiance of eternity...

— Shelley
Adonais

*T*here is hardly any production or service sector where people are not looking for optimum solutions to real problems. This is why optimization methods are very popular among developers and users in all disciplines. In general, most mathematical optimization methods provide satisfactory results, but only under very ideal assumptions (linearity of problem, differentiation of the functional approach to be optimized, only a few parameters to optimize, etc.). In practice, the use of mathematical optimization methods is rather limited. Due to the enormous rise of computer power, including personal computers, powerful optimization methods, oriented to the principles of evolution, have been developed recently.

The universal use of LabVIEW suggests that this type of optimization can be provided by LabVIEW concepts. Thanks to its graphical programming capability and the direct integration of analog and digital process values through appropriate data acquisition cards, users can optimize processes online.

This chapter describes how genetic algorithms can be implemented in LabVIEW and how they can be used directly to optimize fuzzy controllers. Practical examples illustrate this concept.

7.1   Basics – Evolution Strategies or Genetic Algorithms

Genetic algorithms are randomized parallel search algorithms that search from a population of points. Genetic algorithms are designed to search large, nonlinear search spaces where expert knowledge is lacking or difficult to write down in code, and where traditional optimization techniques fall short. They are flexible and robust, exhibiting the adaptiveness and graceful degradation of biological systems.

A genetic algorithm explores a subset of the search space during a problem-solving attempt, and its population serves as an implicit memory guiding the search. Every individual or object generated during a search defines a point in the search space, and the individual's evaluation provides a fitness. When stored, organized, and analyzed, this information can be used to explain the solution, that is, to tell which parts of the genotype are important and to allow a sensitivity analysis.

In practice, many optimization problems are normally defined by a more or less large number of n independent parameters searching for special parameter values, for which an evaluation or quality function (fitness) assumes a maximum. The problem can often be formulated so that the evaluation or quality function can be interpreted as (energy) cost, and special parameter values are searched for which the evaluation or quality function becomes minimal. In general, parameters are subject to certain restrictions, that is, they have to meet various marginal conditions so that a given optimization task always refers to the parameters that are valid within given restrictions in the entire search space.

The difficulty of a given optimization problem depends essentially on the characteristics of the quality function. In most real problems, the quality function can neither be differentiated steadily nor does it show a single minimum within the valid parameter sets. In systems with many inter-related parameters, the interesting search space generally includes a series of local optima (we then speak of a multimodal quality function), in addition to the global optimum (minimum). Evolution strategies or genetic algorithms have proven to be very suitable to find the global optimum because they do not require the quality function to have a special structure. They can be applied even if there are irregularities or "jagged" topologies with many local minima.

Genetic algorithms work according to the principle of evolution mechanisms applied in cycles — mutation, reproduction, and selection, to a population of parameter sets. A single cycle is normally called a *generation*. It should be observed that this type of evolution strategy can be paralleled to a high degree, in contrast to conventional optimization methods, so that it runs on parallel computers very efficiently.

Normally, the first step involves an initial population, which is selected depending on information available on the point of the optimum. In cases where no information is available (which is the regular case), a random, but equally distributed, initial population is selected within the valid search space. The three evolution mechanisms are subsequently applied to this initial population in cycles.

Mutation is the process needed to create new object variants (parameters sets). For this purpose, all individual parameters — $p_{i,n}$ — of a parameter set (also called *parent object*) of generation n are changed randomly, that is, by adding a normally distributed random number, ξ_i, without a mean value, and a diffusion of $\sigma_i(n)$ according to p_i, $n+1 = p_{i,n} + \xi_i$. In this way, we obtain a new parameter set (also called *child object*), the individual parameters of which diffuse around the parent object. Diffusion $\sigma_i(n)$, also called *incrementation* in analogy with deterministic optimization methods, corresponds to the mutation rate in nature. The higher the incrementation we select, the greater will be the probability for dramatic changes of a parameter during mutation. But even with small increments, we will always obtain a few "freak values," or runaways in the parameter variation, referred to the initial object (see Figure 7-1). These runaways make sure that a wider radius around the initial object is searched, which means that they contribute essentially to a high global safety of the optimization method. During the course of an optimization, the diffusion can be adapted as the cycle number n (number of generations) increases; however, it is normally reduced.

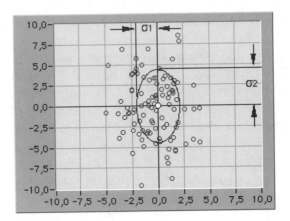

Figure 7-1
Various mutants of an initial object Q(0,0)

The changes of a parameter set generated by this method are random and undirected. This means that the new parameter set (child object) created by mutation may meet the quality function better or worse than the initial parameter set (parent object), in the sense of the quality criterion we want to optimize.

For each parameter set (parent object) of a population, k new parameter sets (child objects) are created in one cycle. These child objects scatter more or less densely around the parent objects, depending on the mutation process. This mechanism is called *reproduction*, and it is the process that gives well-adapted parameter sets a chance to spread out within a population.

Subsequent *selection* is needed to cull the population that has risen during reproduction. To achieve this in the simplest way, all parameter sets showing the least adaptation, that is, those with the highest quality function value, are sorted out successively until we obtain the original population density (survival of the fittest). The remaining parameter sets form the initial population for the next (generation) cycle.

This approach shows essentially the following degrees of freedom: population density, selection of an initial population, number of mutants in each generation, and mutation increments. The larger the number of parameters we want to optimize, the bigger must be our population density.

The following sections describe two examples to illustrate the mode of operation and the steps of an optimization process, using the multimodal quality functions shown in Figures 7-2 and 7-3.

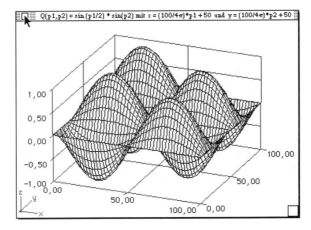

Figure 7-2

Multimodal quality function with four equal optima within a given search space

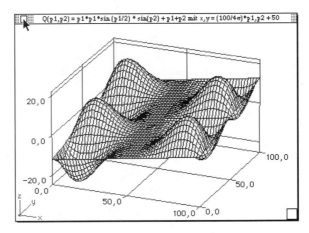

Figure 7-3

Multimodal quality function with one global optimum and three side optima

The bivalent multimodal quality function shown in Figure 7-3 has four equal optima within the given search space; all of them are found by the genetic algorithm implemented in LabVIEW. Each optimum contains almost equally strong populations. The diffusion of the individual objects reflects the topological form of each relatively flat optimum. Figure 7-4 shows the panel of a test application implemented in LabVIEW with the parameter sets selected for the genetic algorithm and the result obtained from running 15 generations.

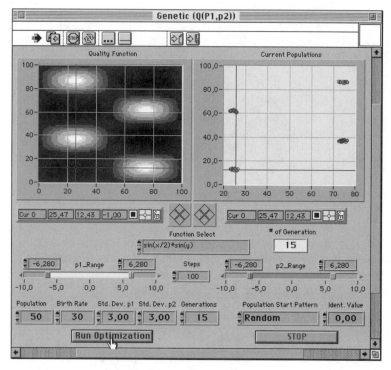

Figure 7-4

Optimization result for the quality function shown in Figure 7-2

The panel shows the given bivalent multimodal quality function as an intensity plot, where the maxima (mountains) are white and the minima (the optimum valleys in this example) are black. Note that the genetic algorithm implemented in LabVIEW will give us the same result even if we were to place the entire initial population in one of the minima.

The bivalent multimodal quality function shown in Figure 7-3 has three characteristic suboptima of various strengths and a global optimum within the given search space. Here, too, the genetic algorithm im-plemented in LabVIEW finds the global optimum when running only a few generations, regardless of how we arrange the initial population and how it is distributed within the valid search space. This holds true even if we were to place the entire initial population in one of the side minima.

Figure 7-5 shows the panel of the test application implemented in LabVIEW with the parameter sets selected for the genetic algorithm and the result obtained from running 15 generations for the quality function shown in Figure 7-3.

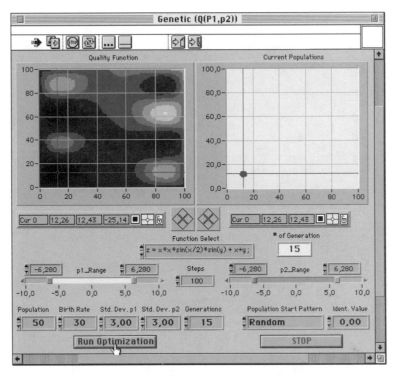

Figure 7-5

Optimization result for the quality function shown in Figure 7-3

Figure 7-6 shows the optimization curve in a series of interim "snap-shots" for an initial population that was randomly distributed within the valid search space, and Figure 7-7 shows the optimization curves after three and five generations.

7.2   Sample Application

This section describes a real-world application as an example. Our task is the online optimization of the parameter setting for a fuzzy PI controller to control a third-order path with a characteristic delay time. Our starting point is the (model) closed-loop control system represented in Figure 7-8.

We optimized the parameter setting of the fuzzy PI controller with regard to its command behavior, where the ITAE criterion was used as a quality criterion.

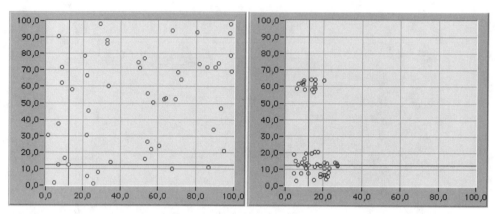

Figure 7-6

Optimization curve immediately after the start (left) and after one generation run (right)

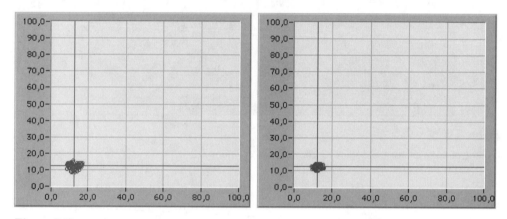

Figure 7-7

Optimization curve after three generations (left) and after five generations (right)

$$Q_{ITAE} = \int_0^\infty t \cdot |e(t)| \cdot dt \text{ ITAE criterion}$$

In general, fuzzy systems take many degrees of freedom with regard to their parameters, in particular, for the form and the position of membership functions for input and output values, rule base, operators, and defuzzification method. In this example, we optimized only the controller's rule base. For this purpose, we used the genetic algorithm described earlier to change the weighting factors (degree of support) assigned to the rules, then we determined the respective result (jump response of the closed-

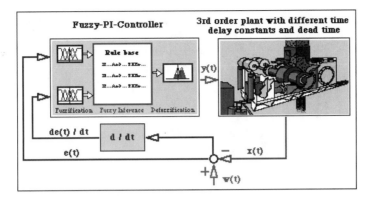

Figure 7-8

Fuzzy system consisting of fuzzy PI controller and controlled members

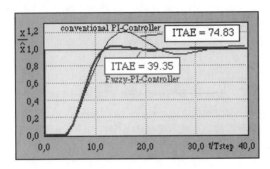

Figure 7-9

Optimized fuzzy PI controller versus an optimum conventional PI controller

loop control system) repeatedly until we obtained a minimum value for the quality criterion (ITAE criterion).

Our starting point for this rule base optimization was the standard arrangement included in LabVIEW's Fuzzy Logic Control Toolkit (see Chapter 6) for the two input values and the resulting rule base. This means that we completed this optimization without information or knowledge on how the controller should be preset. The extensive rule base (it includes all possible combinations of input sets as conditions and output sets as conclusions) embedded in this toolkit provides 125 rules, so that we obtain an equally extensive quality function. Figure 7-9 shows the result of this optimization compared to a conventional PI controller which had been optimally set according to the T sum method.

Note that, as a nice side effect, we were able to reformulate the valid rules that remained from this optimization process systematically with regard to

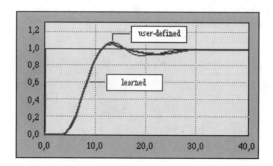

Figure 7-10

Optimized fuzzy PI controller "trained" to set itself to a predefined command behavior

their conclusion terms, thanks to the determined weighting factors, so that we obtained a complete consistent optimum rule base with only 25 rules.

Our second sample application involves the optimization of the desired command behavior (standard for the desired jump response) of the closed-loop control system represented in Figure 7-8, using the same fuzzy PI controller. We used the evolution strategy described earlier to determine the parameters for the fuzzy controller that will lead to a lowest possible deviation from the predefined jump response of the combined system. As in our first sample application and starting with the fuzzy controller's standard settings, we used the rule base itself as optimization object. Figure 7-10 shows our results. They demonstrate that evolution strategies or genetic algorithms can be used successfully to train a fuzzy controller for the setting of a predefined command behavior of the combined system.

7.3   Summary

This chapter and the sample applications demonstrate that evolution strategies or genetic algorithms can be used in many different ways to solve real-world problems. Particularly in combination with fuzzy systems and their large number of interrelated parameters, they are definitely superior to conventional methods.

8

Mathematics and Simulation in LabVIEW

The first man who noted the analogy between a group of seven fishes and a group of seven days made a notable advance in the history of thought. He was the first man who entertained a concept belonging to the science of pure mathematics.

— Alfred North Whitehead
Science and the Modern World

The full LabVIEW package features powerful mathematical functions that have mainly been used in supporting and supplementary operations for traditional measurement applications. However, this use does not, by far, exploit the capabilities of LabVIEW. Although the mathematical capabilities of LabVIEW are well known in these applications, it is less known that LabVIEW provides a suitable platform for numerically oriented mathematical routines and even symbolic manipulations. Its graphical user and programming interface and its dataflow architecture make LabVIEW suitable for describing numerical algorithms, allowing extensive experimentation with mathematical objects.

This chapter describes the mathematical features of the G Math Toolkit, a math-oriented add-on package for LabVIEW.

8.1   Introduction

The *G Math Toolkit* is a multipurpose LabVIEW add-on package for math, data analysis, and data visualization. It is the first toolkit that lets you merge measurement values and complex evaluation. G Math is a powerful collection of VIs written in G, which allows you to graphically assemble your mathematics program as a block diagram. You can use ordinary differential equations, optimization methods, root solving algorithms, and various mathematical functions. The toolkit's core is a parser, representing a kind of equation generator and manipulator, allowing you to enter mathematical equations on the front panel. The real capabilities of this package result from many different links among its components, shown schematically in Figure 8-1.

The real-world applications described briefly in later sections of this chapter illustrate the mathematical capabilities of LabVIEW and G Math, demonstrating how this package works. First, the fitting routines are introduced; they allow you to enter model equations, and they can be linked directly to a measuring process. Then, we demonstrate how you can visualize the Gibbs phenomenon that occurs in Fourier rows, using the G Math package. Finally, some practical wave equations are calculated and solved. The chapter closes with a real problem from applied physics — determining the movement behavior of a spinning object.

8.2   Adapting Measurement Values

Often in measurement applications, you need to adapt the acquired measurement data to predefined models. In theory, these models are often well known, but you normally adapt them by using known methods, like trigonometric functions or polynomials.

In general, we have pairs of measurement values (x_i, y_i), where y_i is interpreted as the actual measurement value at time x_i. Index i extends from 1 to the end value n. Assuming a model equation $y = f(a, x)$, parameter a can

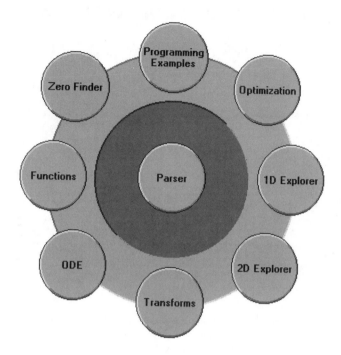

Figure 8-1
The parser plays a central role in the G Math toolkit; it handles equations. All components are interlinked.

consist of several individual components. We look for this type of *a* so that the following sum is as small as possible:

$$\sum_{i=1}^{n} \left| y_i - f(a, x_i) \right|^2$$

In practice, the Levenberg-Marquardt method is often used to solve this task, as shown in Figure 8-2. This method produces successful results even with nonlinear functions. The Advanced Analysis Library of LabVIEW offers the Levenberg-Marquardt method, but we have to fix the functional model *f(a,x)* before the program's run time. Using the G Math toolkit, we obtain unequally more degrees of freedom.

In the center of all this flexibility is the possibility to freely enter our model equation in the form of symbolic terms on the front panel. In particular, we can use known functions. You distinguish between model parameters and independent variables by the corresponding components of a cluster.

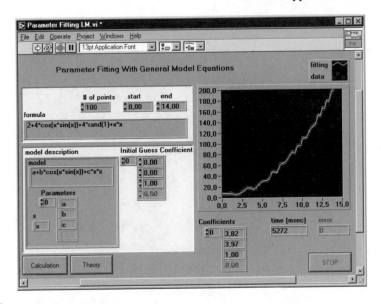

Figure 8-2

This screen illustrates how a well-defined fitting problem is dealt with. To better understand, we purposely did not use real measuring values, but rather another feature of the G Math Toolkit: the capability to generate arbitrary functional values with a given definition equation. When adding stochastic terms to the definition, we can define even very realistic fitting tasks. Based on a comparison of the original function versus the model equation, we can easily check whether or not the expected parameters match the real ones.

8.3 Solving Differential Equations

To solve differential equations, G Math provides Euler, Rung-Kutta, and Cash-Karp methods as well as numerical and symbolic routines for differential equation systems and *n*-order differential equations. The example in Figure 8-3 uses differential equations to describe the movement of a gyroscope. When fixing the gyroscope's center of gravity, we can describe this movement as three-dimensional space.

8.4 Solving Wave Equations

It is well known from physics and mathematics that you can use wave equations to describe physical values, for example, the strength of electric

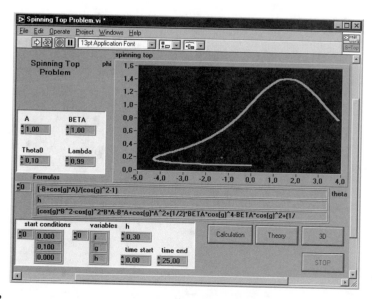

Figure 8-3

The top screen represents inputs of differential equations and parameters. The example in Figure 8-3 emphasizes the main benefits of G Math in LabVIEW. First, you can use measurement values as parameters for differential equation models. Second, the results return these calculations to the process (through DA conversion).

and magnetic fields, gas density and pressure, sound propagation in homogeneous media, and propagation of electromagnetic waves, involving some of the most important types of partial differential equations. In the simplest case of a one-dimensional homogeneous wave equation, there is a clear solution in closed form. The issue here is to find a suitable calculation tool to study such solutions. In addition, it is desirable to include space-distributed sources of the resulting field in the calculation. With most current applications, this could, for example, be sound sources installed in different positions. The basic analytical solution is no longer accessible directly in such cases. Figure 8-4 shows how this problem can be solved in G Math. Both the number of external sources (and their spatial arrangement and their direct time curve) and the time curve of the derivation can be optionally selected in the form of mathematical formulas. The resulting behavior can be studied in detail. For instance, you can set up an artificial source with special properties, which is implemented as the superposition of spatially distributed and simply structured sources. For example, sine-shaped sources can simulate any artificial source of interest if properly arranged.

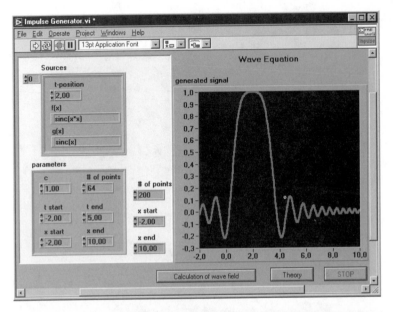

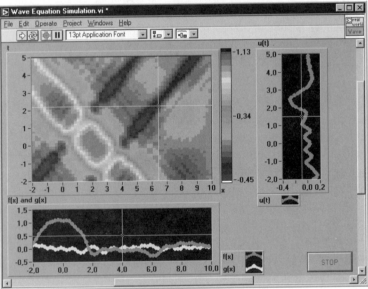

Figure 8-4

This example shows the setup of external sources (top) and how the wave equation solution is calculated (bottom). Interactive tools can be used to analyze the graphical solution in more detail.

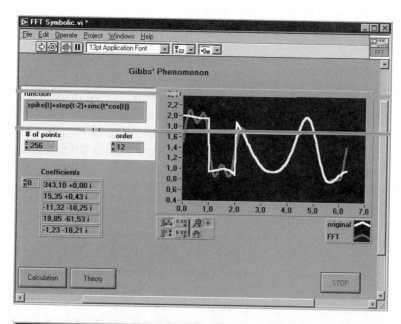

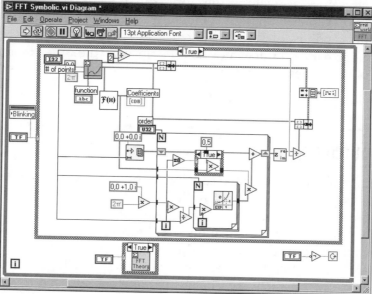

Figure 8-5

Visualizing the Gibbs Phenomenon

8.5   Visualizing the Gibbs Phenomenon

The didactic capabilities of LabVIEW, particularly in the field of education, play an important role. Of particular importance is the direct conversion of the concepts in the sense of rapid prototyping. The following example demonstrates the capabilities of the G Math Toolkit for educational purposes. The example discusses a signal processing problem, which has been discussed — to much controversy. It concerns the Fourier rows of $[0, 2\pi]$ periodic signals. It is known that each periodic function can be represented by a Fourier row because Fourier rows of many functions converge against these same functions. From the mathematical view, this convergence is still given, but it is often not equal for all points within the defined area. The consequence is that the Fourier row development produces overshooting near discontinuity points of given functions. This behavior is called the *Gibbs Phenomenon*. The G Math Toolkit allows you to visualize this effect quite easily (see Figure 8-5).

Both the functions to be studied and the number of parameters of the supporting values as well as the approximation order can be arbitrarily selected. We can see that overshooting occurs systematically near the discontinuity points of the function to be approximated. Increasing the order of approximation does not help because, as the order increases, only the area of the highest deviation narrows.

8.6   Summary

In general, we can state that the capabilities of LabVIEW in the field of education and simulation have been underestimated. The benefits of LabVIEW versus products like Mathcad, MATLAB, or Mathematica emerge particularly in the field of visualization and direct linking to the real world. In addition, many signal theory algorithms or applied mathematics algorithms can be represented clearly and efficiently in a graphically oriented language like G. You can interactively change parameters and see the results immediately on the front panel. The G Math Toolkit is a valuable add-on, making LabVIEW apt for new fields of application.

9

Fourier Transformation in LabVIEW

Entia non sunt multiplicanda praeter necessitatem. (One must not multiply entities without necessity.)

— William of Occam
The Razor

Digital signal processing represents the last member in the chain of PC-controlled measurement technology. The essential information contents are (in the ideal case) extracted from a digitized measuring signal. Fourier transformation plays a central role. This chapter illustrates the differences between the Discrete Fourier Transform (DFT) and the Fast Fourier Transform (FFT) versus the mathematical Fourier integral by using selected sample applications in LabVIEW. The chapter also presents methods that help reduce or avoid artifacts. For better readability, we do not use mathematical derivations, but instead refer to further reading. Finally, LabVIEW examples are used to discuss FFT applications.

9.1   Introduction

Digital signal processing has developed into a central issue within measurement technology during the past years. Both the underlying computers and the other hardware components required for signal processing and digitization have become much more powerful and affordable. In conventional analog measurement technology, the measuring process generally ends with the reading of values. First, PC-controlled measurement technology supplies a digitized measured signal, consisting of a large data field. Then, information has to be extracted from this set of data by applying digital signal processing. In addition to adapting curves based on the method of least squares and statistical signal analysis, the Fourier transform plays a central role in the extraction of essential information from a given signal. The significance of the Fourier transform reaches far beyond pure frequency analysis; it also simplifies complex mathematical operations (convolution, correlation, Hilbert transformation). The Discrete Fourier Transform (DFT) matches the mathematical result of the Fourier integral only as long as certain marginal conditions are met. These concern sampling conditions, measuring time constraints, and related clipping effects. Of course, the same constraints apply to the Fast Fourier Transform (FFT), which uses a very efficient DFT algorithm (e.g., the Cooley-Tuckey algorithm) if data number N complies with value 2^n (integer n). During digitization of analog signals, one has to take care that artifacts caused by sampling are reduced to a minimum. This reduction can be achieved with an analog-digital converter that supports a sufficient word length, and with a high-quality anti-aliasing filter.

9.2   Sampling Analog Signals

Analog signals are sampled by use of an analog/digital converter that acquires time-dependent voltage values $u(t)$ within time intervals T. The signal curve between two consecutive measurements is ignored. For this reason, one has to make sure that the signal between two consecutive measurements does not lead to undesired "surprises." This requirement corresponds to the *sampling theorem*, which defines the upper limit frequency f_{max} of an analog signal in relation to the sampling rate $f_T = 1/T$. As we generally do not know the frequencies contained in a measured signal, the analog/digital converter is equipped with an analog anti-aliasing filter with high edge contrast and a limit frequency below half the sampling rate.

❑ 9.2.1   Sampling Theorem

With a given sampling rate, $f_T = 1/T$, the sampling theorem supplies the maximum signal frequency f_{max}, which may contain the analog signal $x(t)$ to be sampled:

$$f_{max} < 2 \cdot f_T \ or \ f_{max} < 1 / (2T) \qquad (1)$$

The maximum frequency given by the sampling theorem is also called *Nyquist frequency* (f_{Ny}). When the sampling theorem is violated, low-frequency spectral shares appear in the discrete sampling sequence $x(n)$, which are not contained in the original analog signal $x(t)$. This effect is called *aliasing*. An anti-aliasing filter is normally used to avoid this effect. The undersampling of signals can be achieved only by means of analog filters because digital filters are not capable of separating spectral components caused by undersampling from "real" spectral components of the analog signal.

The following three examples illustrate the sampling theorem.

Example 1

When fine periodic image contents are acquired by a video camera, the image is undersampled by local quantization so that moiré structures will appear in the corresponding image areas. This effect can be prevented by flat tuning of the camera objective. Defocusing eliminates high-frequency image components and is therefore similar to optical anti-aliasing filter.

Limit frequency f_g of an anti-aliasing filter depends on its edge contrast and the required attenuation of the Nyquist frequency. For this reason, there are strict requirements for the quality of an anti-aliasing filter.

Example 2

Assume we want to use a simple RC anti-aliasing filter with a limit frequency of $f_g = 1 \ kHz$. The filter has an edge contrast of -20 dB/decade. For the Nyquist frequency f_{Ny}, the anti-aliasing filter should provide a -60 dB attenuation. In our case, three decades are required to reach this attenuation, i.e., at frequency $f = 1 \ MHz$. To meet the given requirements, we would have to select a sampling rate of $f_T = 2 \ MHz$! However, if we use a high-quality anti-aliasing filter with an edge contrast of -120 dB/decade, we are able to achieve the required attenuation after only 0.5 decades. In this case, the Nyquist frequency is at $f_{Ny} = 3.14\ kHz$, so that a sampling rate of $f_T = 6.4$ kHz is sufficient for the given requirements.

Of course, the sampling theorem is not limited to periodic signals; it also applies to transient and stochastic signals. These possess a continuous Fourier spectrum, with spectral components reaching to arbitrarily large frequencies. Although the undersampling effect does not appear initially, there will eventually be major artifacts if no anti-aliasing filter is used.

Example 3

Figure 9-1 represents a transient rectangular impulse with an amplitude of $A = 4\ V$ and a width of $T_p = 5\ ms$, which is being digitized at $f_T = 500\ Hz$. The range of the rectangular impulse comprises a total of three values with an amplitude of $A = 5\ V$. The width of the discrete impulse is $T_p = 4\ ms$ because sampling is done with an interval of $T = 2\ ms$. The sampling values $u(n)$ represent time-continuous rectangular impulses with a width that may fluctuate between $4\ ms \leq T_p < 5\ ms$. This means that the discrete Fourier spectrum, $X(k)$, is uncertain, and this uncertainty can be reduced only by increasing the sampling rate. Figure 9-2 shows various limit cases.

9.3   Fourier Transformation

Fourier transformation is a very powerful math tool that can be applied both to time-dependent and place-dependent signals. The Discrete Fourier Transform (DFT) plays a central role in digital signal processing. In digital image processing, it is applied to 2D signals. The 3D Fourier transform plays an important role in studying the structure of solid matter.

In the early 19th Century, the French mathematician *Jean Baptiste Joseph Fourier* (Figure 9-3) proved that any reasonably behaved periodic function $g(t)$ with period T can be constructed by summing a (possibly infinite) number of sines and cosines.

Fourier studied the mathematical theory of heat conduction. He established the partial differential equation governing heat diffusion and solved it by using an infinite series of trigonometric functions. He published his first major work in 1822 devoted to the mathematical theory of heat conduction. In this work he introduced the representation of a function as a series of sines or cosines, now known as *Fourier series*.

Fourier's work provided the impetus for later work on trigonometric series and the theory of functions of a real variable.

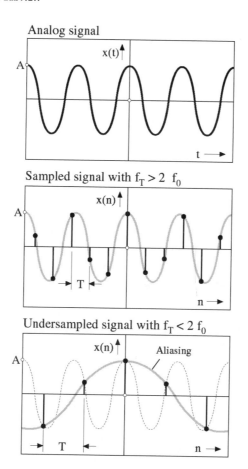

Figure 9-1

Time-continuous cosine signal with amplitude A and frequency f_n, which is acquired at various sampling rates f_T (top). The cosine signal curve can be reconstructed if sampling rate f_T exceeds the dual value of the Nyquist frequency (center). If the sampling rate matches exactly the Nyquist frequency, each measuring point can lie on the zero passage of the harmonic time function. Below this limit value, a low-frequency signal with a frequency depending on the sampling rate appears (bottom).

The allocation of time- or place-dependent signals to corresponding spectral functions can be illustrated in the easiest way with acoustic signals. An acoustic signal that humans can hear consists of pressure fluctuations within a frequency range between 30 Hz and 20 kHz. In contrast to a microphone, the human ear does not capture the time curve of pressure fluctuations, but captures their number per second. The spectral evaluation of an

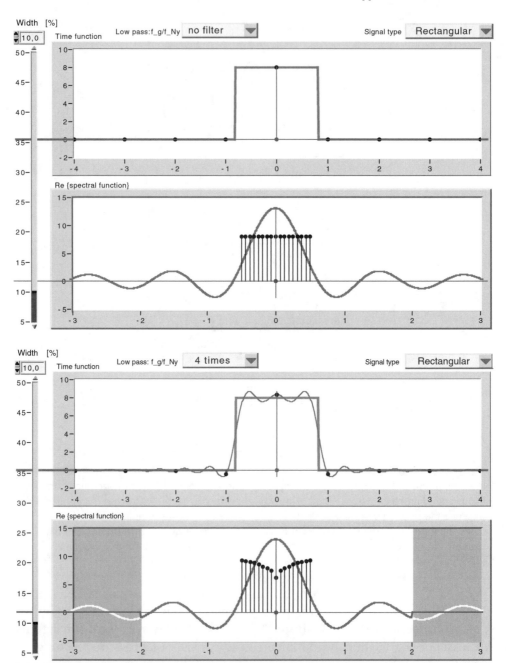

Figure 9-2

Violation of the sampling theorem with a transient signal (here: rectangular signal)

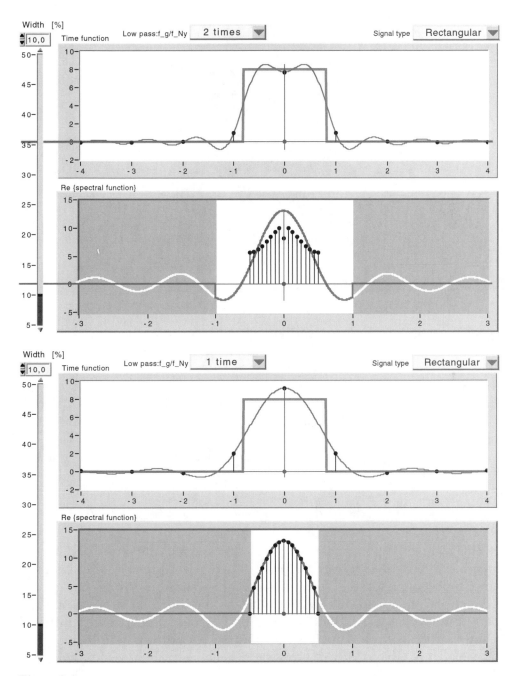

Figure 9-2

(Continued)

Figure 9-3

The French mathematician Jean Baptiste Joseph Fourier (1768–1830)

acoustic signal originates from the fact that hearing cells in the inner ear use resonance effects, working in a frequency-selective way. With a pure sine tone, this means that only hearing cells "calibrated" to the given frequency are excited. Although acoustic information is noticed only as spectral function, the acoustic perception is not limited to periodic signals because very short, totally inharmonic sound events that occur only once can be captured.

The Fourier integral over the continuous time function *f(t)* is defined as follows:

$$\underline{X}(f) = \int_{-\infty}^{+\infty} x(t) \cdot e^{-(j2\pi ft)} dt = \int_{-\infty}^{+\infty} x(t) \cdot \cos(2\pi ft) dt - j\int_{-\infty}^{+\infty} x(t) \cdot \sin(2\pi ft) dt \qquad (2)$$

This complex exponential function can be divided into a real part and an imaginary part so that we obtain two partial integrals. Multiplying the amplitude values of the "analog signal" *x(t)* with a cosine function of frequency *f*, the integral value from the product function supplies the real part of the complex-value spectral function *X(f)* with the given frequency *f*. Multiplication with the sine function supplies the corresponding amount of the imaginary part, which has to be multiplied by the factor −1. Note that the Fourier-transformed *X(f)* is a function of the real variable *f*, and that it is generally complex-value, even if *x(t)* is real!

Example

Figure 9-4 shows a sine-shaped analog signal with frequency f_0 = 3 Hz (left column). This signal is multiplied with various harmonic functions. The integral over the product function becomes non-zero if the phase *and* the frequency of both terms match (second line). In this case, only the imaginary part of $X(f)$ at frequency f_0 = 3 Hz has a value greater than zero. Supposing that we use frequency $-f_0$ for the sine function, then all parts of the product function lie below the time axis. This means that the Fourier integral supplies a value non-zero with both positive and negative frequencies, where both values are mutual complex conjugates.

9.4   Inverse Fourier Transform

The *Inverse Fourier Transform* (IFT) allows the calculation of the time function $x(t)$ from the corresponding spectral function $\underline{X}(f)$. The inverse Fourier integral is defined as follows:

$$x(t) = \int_{-\infty}^{+\infty} \underline{X}(f) \cdot e^{(j2\pi f t)} df \qquad (3)$$

When applying (3), $x(t)$ results clearly in $\underline{X}(f)$, and when applying (2), $\underline{X}(f)$ results clearly in $x(t)$. This means that both integrals form a transformation pair, which can be expressed symbolically as follows:

$$x(t) \leftrightarrow \underline{X}(f) \qquad (4)$$

The next section introduces the Discrete Fourier Transform and describes some basic properties of the Fourier transform in more detail. These properties originate directly from the Fourier integer and the inverse relation.

9.5   Discrete Fourier Transform

To calculate the *Discrete Fourier Transform* (DFT) of a continuous time-independent signal, $x(t)$, we need to form discrete sampling values $x(n)$ from an analog signal, $x(t)$, and limit the measuring time T_M to a finite length.

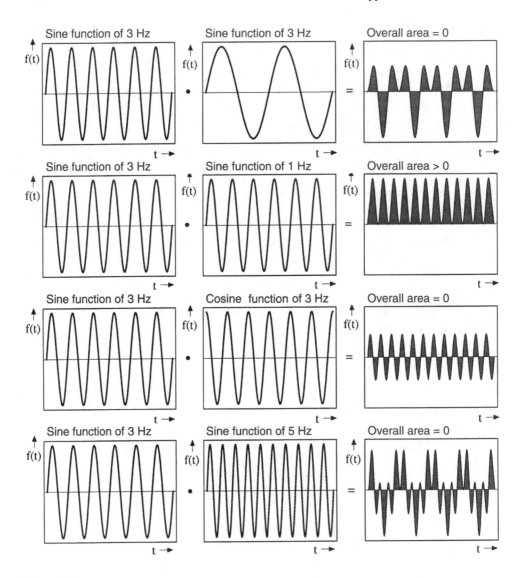

Figure 9-4

Schematic representation of the result of the Fourier integral for a sine-shaped measuring signal

If a total of N values are acquired from an analog signal, $x(t)$, during measuring time T_M, then the time distance between the individual sampling processes is $T = T_M/N$. After the sampling, the discrete function has only functional values non-zero in positions $t_n = nT = nT_M/N$ with $0 \le$

$n < N$. Integral variable n corresponds to a dimensionless discrete time variable. Continuous time variable t can thus be replaced by discrete values nT:

t	is replaced by	$t_n = nT = nT_M/N$
$x(t)$	is replaced by	$x(t_n) = x(nT) = x(nT_M/N)$
$exp(-j2\pi ft)$	is replaced by	$exp(-j2\pi f \cdot nT) = exp(-j2\pi f \cdot nT_M/N)$

After applying the DFT, we obtain the discrete spectral function $X(kF)$ with frequency interval $F = 1/T_M = (NT)^{-1}$ in the spectral range. The integral values k have the significance of a dimensionless discrete frequency variable. This means that, with DFT, continuous frequency variable f is replaced by the discrete values kF $(0 \le k < N)$:

f	is replaced by	$f_k = k/TM$
$exp(-j2\pi f\, nT)$	is replaced by	$exp(-j\, 2\pi f_k\, t_n) = exp(-j\, 2\pi nk/N)$

The two continuous variables t and f within the Fourier integral (2) can thus be replaced by integral variables n and k. To achieve compliance with the Fourier integral, the result of DFT has to be multiplied by a correction factor, A. This correction factor depends on the nature of the original measuring signal. We distinguish between two cases.

A time-limited signal, $x(t)$, with finite signal energy (e.g., impulse-shaped signal) is called an *energy signal*. In this case, the correction factor has the value $A = T$, where T corresponds to the sampling interval. In this case, the Fourier integral is replaced by the following Fourier sum:

$$X[k] \;=\; T\sum_{n=0}^{N-1} x[n]\, e^{-j\frac{2\pi}{N}nk} \tag{5a}$$

The value $exp(-j\omega)$ corresponds to a complex number with the amount *One* and angle ω to the positive real axis. Summation is done over N (complex) values. For readability, the following abbreviation is often used:

$$W_N^{nk} \;=\; e^{-j\frac{2\pi}{N}nk} \tag{5b}$$

We can now represent the Fourier sum in the following short notion:

$$X[k] \quad = \quad T \sum_{n=0}^{N-1} x[n] W_N^{nk} \tag{5c}$$

A time-unlimited signal (e.g., harmonic function or noise) has the character of performance signals. To obtain compliance with the corresponding Fourier integral (2), we have to weight DFT with factor $A = 1/N$:

$$\underline{X}[k] \quad = \quad \frac{1}{N} \sum_{n=0}^{N-1} x[n] W_N^{nk} \tag{5d}$$

Given that the Fourier sum now contains only index values k and n, this form of the Fourier transform is called Discrete Fourier Transform (DFT). With constant index value k, the (real) measuring values $x(n)$ are multiplied by the (complex) weighting factors $\underline{W}_N^{nk}$, and the value sequence

$$\left\{ x(0) \cdot \underline{W}_N^{0k}, x(1) \cdot \underline{W}_N^{1k}, \dots, x(N-1) \cdot \underline{W}_N^{(N-1)k} \right\} \tag{5e}$$

is summed. If the time function consists of $N = 2^a$ values, we can utilize the symmetric property of the discrete complex exponential function. This reduces the number of required complex multiplication operations in the DFT sum (5d).

By decomposing the discrete time function $x(n)$ into several shorter time functions, we can cut out more multiplication operations. This optimized calculation method is called Fast Fourier Transform (FFT). The continuous Fourier transform and DFT or FFT have the following identical properties:

- If the time function is real and even, then the spectral function is real and even.

- If the time function is real and uneven, then the spectral function is imaginary and uneven.

- If the time function is imaginary and even, then the spectral function is imaginary and even.

- If the time function is imaginary and uneven, then the spectral function is real and uneven.

- The negative part of the spectral function is a complex-conjugate to the positive part.

The Fourier transform (FT) differs from DFT or FFT in the following aspects:

- The spectral function $\underline{X}(k)$ is discrete and defined only in discrete frequency intervals $F = 1/T_M$. The discrete spectral values are repeated periodically after N values, thus forming a cyclic function, even if the Fourier spectrum $\underline{X}(f)$ is nonperiodic. The Nyquist frequency is in the center of the discrete spectral function $\underline{X}(k)$ and represents the upper limit of the spectral values that can be represented. The negative spectral values are in the second half of $\underline{X}(k)$. The negative Nyquist frequency coincides with the positive at index value $k = N/2$.

- The discrete time function $x(n)$ represents a cyclic function with period N, even if the corresponding analog signal $x(t)$ does not represent a periodic function.

- To achieve compliance with the result of the Fourier integral in DFT and FFT, we have to distinguish between performance and energy signals:

$$\underline{X}[k] \;=\; \frac{1}{N}\sum_{n=0}^{N-1} x[n]W_N^{nk} \qquad \text{(performance signals)} \qquad (6a)$$

$$\underline{X}[k] \;=\; T\sum_{n=0}^{N-1} x[n]W_N^{nk} \qquad \text{(energy signals)} \qquad (6b)$$

Example

Imagine we want to acquire $N = 8$ discrete values in time intervals of $T = 0.25\ s$ from an analog voltage signal, $x(t) = 4\ [V] \cdot sin(2\pi 4\ [Hz]\ t)$ (see Figure 9-5). The total measuring time is $T_M = NT = 2\ s$. As a result, from the Fourier integral over the analog signal, we obtain two Dirac impulses at frequency values $f_+ = 4\ Hz$ and $f_- = -4\ Hz$. The "part contents" of the two impulses are purely imaginary and have the values $-j2\ V$ and $+j2\ V$, respectively. In this case, the periodic continuation of the discrete sampling values $x(n)$ is identical to the sampling values that would result with an infinite measuring time. The spectral function of the discrete signal is repeated periodically in the $1/T = 4$ Hz frequency intervals (see Figure 9-5, bottom). Weighting the Fourier sum with factor $1/N = 1/8$, we obtain two imaginary spectral values non-zero, $k = 6$: $\underline{X}(2) = -2j$ and $\underline{X}(6) = +2j$, with index value, $k = 2$. All values except the two spectral values, $\underline{X}(2)$ and $\underline{X}(6)$, are zero.

We can conclude from the discrete time function $x(n)$ (Figure 9-5, top) that exactly two periods fit into the measuring window (gray area). With $T_M = 0.5$ s, we obtain a spectral resolution of $F = 2$ Hz. The spectral value at the index $k = 2$ position corresponds to frequency $f_2 = 4$ Hz. The second impulse at position $k = 6$ does not correspond to frequency $f_6 = 12$ Hz because it already lies in the area of the periodic continuation of the negative spectral parts. Using the periodicity condition, $a = -1$, results in the negative frequency $(6 - N) \cdot F = -2$ Hz. The actual "interface" of the discrete spectral function, $\underline{X}(k)$, is at index value $k = 4$. With the Nyquist frequency $f_{max} = N/(2T_M) = 8$ Hz, which is also defined by the sampling theorem (1), the positive and the negative spectral parts fall to the "interface" $k = 4$ in the center of the discrete spectral function $\underline{X}(k)$.

9.6   Spectral Functions

As mentioned earlier, the real time function $x(t)$ and the complex spectral function $\underline{X}(f)$ have a fixed allocation, which is clearly defined by the Fourier transformation. For this reason, the two functions are called a *Fourier pair*. The transition from the time area to the spectral area is symbolized by (4).

The various representation forms available for spectral functions can be best explained by a harmonic function, $x(t)$, (see Figure 9-5) because it has only one single spectral content. A harmonic function is defined by three parameters — amplitude A, frequency f, and phase angle φ. If it doesn't matter whether or not the phase angle is known to describe a signal, then it is sufficient to apply the amplitude amount as a function of the frequency.

The amplitude amount spectrum $|\underline{X}(f)|$ is real and defined only in the positive frequency domain. A real-value harmonic signal, $x(t)$, with fre-

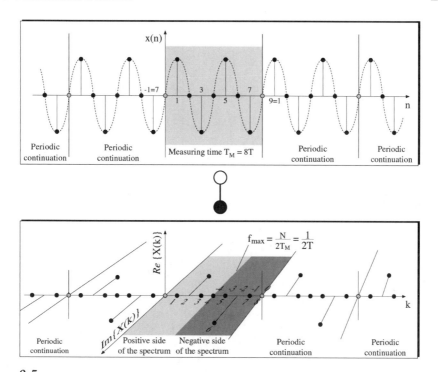

Figure 9-5

Illustration of the DFT properties

quency f_0 supplies one single spectral line at positive frequency $+f_0$, regardless of its phase position. The "surface" of this spectral line matches the amplitude of the harmonic signal. After retransformation of the amplitude amount spectrum, you will generally not obtain the original time function $x(t)$ but the complex exponential function $|\underline{x}(t)|$ (see Figure 9-6).

The phase spectrum $\omega(f)$ describes the phase position of a complex spectral component versus the real axis as a function of frequency, f. The phase spectrum $\omega(f)$ can be used to determine the shift of a signal, $x(t)$, versus a reference signal.

The complex one-sided amplitude spectrum $\underline{X}(f)$ also contains the phase information of the time signal $x(t)$ (Figure 9-7, top). The complex spectral line $\underline{X}(f)$ of a harmonic signal $x(t)$ corresponds to the pointer diagram, normally used in electrical engineering, which consists of one single rotating pointer with length A. The pointer turns counterclockwise at an angle speed of $\omega = 2\pi f$. The pointer diagram corresponds to the position of the rotating pointer to

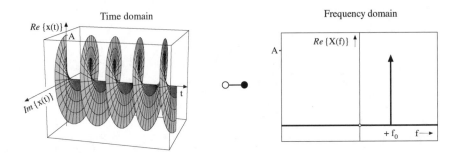

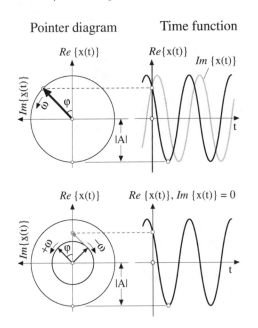

Figure 9-6

Harmonic signal in the time and spectral ranges

Figure 9-7

Harmonic time function and the pertaining pointer diagram

zero time. The frequency f of the harmonic signal $x(t)$ is not visible in this representation because, in this case, the frequency axis is vertical to the complex number level (= paper level). During its rotation, the pointer describes a sequence of complex number values, which correspond to the function curve of a complex exponential function (Figure 9-6, left). Now, if you turn the frequency axis to the paper level, you obtain the one-sided complex amplitude spectrum $\underline{X}(f)$. This representation contains all information required to recover the original time function from the spectral function. We can conclude from the pointer representation in Figure 9-7 (top) that the pertaining real harmonic time function $x(t)$ corresponds only to the real part of the complex pointer.

The two-sided Fourier spectrum $\underline{X}(f)$ is based on the complex exponential function contained in the Fourier integral (1). The relation between a cosine or a sine function and the complex exponential function is given by the Euler equations:

$$A\cos(\omega t) = \frac{A}{2}(e^{j\omega t} + e^{-j\omega t}), \quad A\sin(\omega t) = \frac{A}{2j}(e^{j\omega t} - e^{-j\omega t}) \tag{7}$$

On the right, each of these equations contain a sum of two complex exponential functions, the exponents of which have frequency f_0 as a parameter with a different sign. The corresponding pointer diagram consists of two pointers with length $A/2$, which rotate in opposite directions (Figure 9-7, bottom). As both pointers are always complex-conjugate in relation to each other, their sum always returns a pure real function.

The two-sided Fourier spectrum $\underline{X}(f)$ consists of two complex-conjugate, spectral parts at the positive and negative frequencies $\pm f$. The negative frequency corresponds formally to the pointer rotating clockwise within the pointer diagram.

In system theory, spectral functions are represented in a logarithmic scale. The **impulse response** $g(t)$ forms a Fourier pair with **frequency response** $\underline{G}(f)$ of the corresponding system:

$$g(t) \leftrightarrow \underline{G}(f) \tag{8}$$

Figure 9-8 represents the system functions $g(t)$ and $\underline{G}(f)$ for a first-order system (e.g., an RC element), the impulse response $g(t)$ of which corresponds to a causal exponential function.

This function forms a Fourier pair with the complex Debye function which, in this case, has the significance of a frequency response, $\underline{G}(f)$. The two system functions form a Fourier pair.

The zero point of the frequency response corresponds to the (real) static transmission factor, k. Figure 9-9 represents the decomposition of $\underline{G}(f)$ into a real part and an imaginary part as a projection to the corresponding levels.

The projection of $\underline{G}(f)$ to the complex number level results in a circle that reduces to a semicircle if we leave out the negative frequency domain. In system theory, this representation is normally called a *Nyquist diagram*.

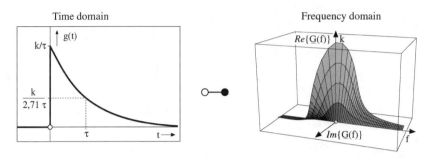

Figure 9-8

Impulse response, *g(t)*, within the time domain and frequency response, *G(f)*, in the spectral domain

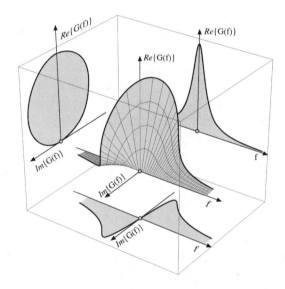

Figure 9-9

The projection of *G(f)* to the real level produces the real part of *G(f)*, while the projection to the imaginary level shows the imaginary part of *G(f)*. The lateral projection produces a circle on the complex-value level.

Another way to represent the frequency response is called *Bode plot*. This fractionates, logs, and multiplies the amount of the frequency response by a factor of 30 to the static transmission factor, *k*, to obtain manageable numerical values. These values have the unit [dB] and are eventually applied to a logarithmic frequency scale. Figure 9-10 shows the Bode diagram for a low-pass system of first order with limit frequency $f_g = 1$ *kHz*. The diagram on the right shows the phase curve in logarithmic scaling.

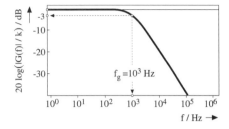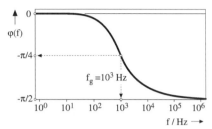

Figure 9-10

Bode diagram of a first-order, low-pass system

We obtain the **energy spectrum** W(f), or the **power spectrum** P(f), by squaring the amount values of the Fourier spectrum. In general, these spectral functions are represented only in the positive frequency domain. The energy or power part originating from the negative spectral domain has to be included in the one-sided representation. Therefore, apart from the equal share, the power values are multiplied by a factor of 2. A logarithmic scale for energy or power spectra is often used to make the entire value range visible.

The power spectrum of noise signals has a continuous curve. The amplitude of the power spectrum depends on the frequency resolution, which is determined by the measuring time in the case of DFT and FFT. To eliminate this impact, the power values are fractionated to the 1 Hz bandwidth. Consequently, the spectral distribution of the power of noise signals is represented as a power density spectrum with unit $[V^2/Hz]$. In semiconductor component data sheets you often find a type of representation called *voltage density spectrum* denoted with unit $[V^2/\sqrt{Hz}]$. The values of this spectral function result from the root of the values of the power density spectrum.

Example

Imagine we want to use the two-sided Fourier spectrum to show how to obtain various representation forms of the spectral function in LabVIEW. We use a voltage signal, u(t), with four harmonic components and one offset part for our representations. The parameters of the individual signal components are shown in Figure 9-11 in the data fields on the left. Each parameter remains unchanged in the image sequence.

The discrete signal x(n) shown in Figure 9-11 consists of $N = 32$ values. Selecting a sampling rate of $f_T = 32$ Hz, we obtain the measuring time $T_M = 1$ s. The number of cycles indicated in the middle data field matches the frequency under the given conditions. The phase refers to the cosine function, which shows a zero phase angle in the pointer diagram. By this definition, a

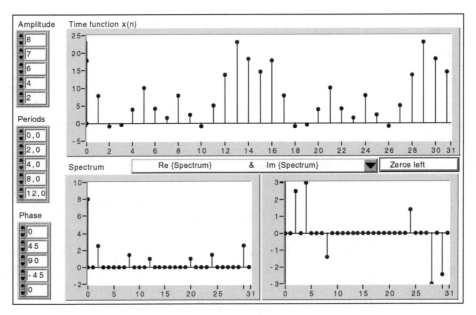

Figure 9-11

Representation of a spectral function in LabVIEW

sine function has a phase position of $-90°$. The quantity of $N = 2^5$ data values allows us now to apply FFT. The frequency resolution is $F = 1\ Hz$ under the given conditions. The Nyquist frequency ranges in the middle of the discrete spectral function $U(k)$ with an index value of $k = N/2 = 16$.

The equal share leads to a real value of $X(0) = 8\ V$ with an index value of $k = 0$. The first harmonic part with number of cycles 2 and amplitude 7 supplies both a real and an imaginary contribution at index value $k = 2$. The corresponding complex spectral value is $3.5\ V$ and has an angle of $45°$ to the positive real axis. With the index value, $k = 30$, we find a complex-conjugate spectral component, which has to be allocated to the corresponding negative frequency at $k = -2$.

The spectral function calculated by FFT has the properties of a cyclic signal that iterates in N cycles. The second half of the data block corresponds to the negative spectral part. The Nyquist frequency at $k = 16$ represents the interface of spectral function $X(k)$ at the position where the positive and the negative Nyquist frequencies meet. For this reason, cyclic signals are often described in a symmetric representation (see Figure 9-12).

From the complex Fourier spectrum $\underline{X}(k)$, we can calculate the amount spectrum $X(f)$ and the phase spectrum $\varphi(f)$ (see Figure 9-13):

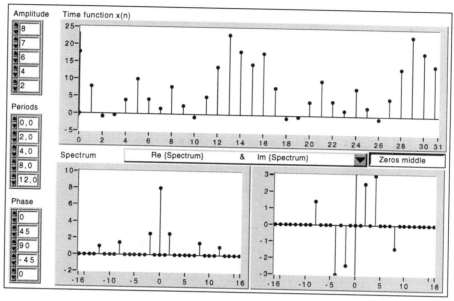

Figure 9-12
Symmetric representation of the Fourier spectrum

$$|x(k)| = \sqrt{\mathrm{Re}\{x(k)\}^2 + \mathrm{Im}\{x(k)\}^2}\ , \qquad \varphi(k) = \frac{\mathrm{Im}\{x(k)\}}{\mathrm{Re}\{x(k)\}} \qquad (9)$$

   The phase spectrum is often used to determine time shifts between two signals or to eliminate them numerically. When calculating the phase spectrum, we have to consider that the arctangent function is defined only in the range between $+\pi$ and $-\pi$, which means that discontinuities have to be expected if the phase angle exceeds the value $+\pi$. If this happens, we have to eliminate such discontinuities in either limit value. In LabVIEW, this can be achieved with the **Phase Unwrap.vi**. Another problem has to be dealt with when the amount of a spectral value falls below a minimum value. In this case, the noise share of the signal comes into the foreground, causing the phase values to take a random parameter. If, in the extreme case, a spectral component has a zero value, then the phase is not defined at all in this position. For this reason, when calculating the phase spectrum, we also have to check whether the corresponding spectral value exceeds a specified minimum value. If not, the phase value has to be set to zero in this position.

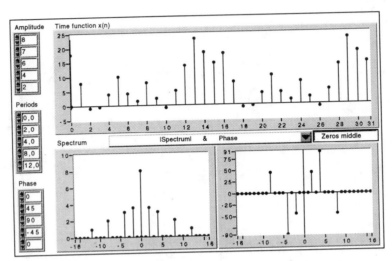

Figure 9-13
Symmetric representation using the amount and phase spectra

Now, when squaring the amount spectrum $|X(f)|$, we obtain the power spectrum $P(k)$ of the signal $x(n)$. This spectral function (Figure 9-14) serves for spectral evaluation of a signal, and it also offers a way to calculate the auto-correlation function of the signal $x(n)$ by applying the *Inverse Fast Fourier Transform (IFFT)*. The IFFT serves to separate periodic signal parts from stochastic parts.

The **amplitude spectrum** $A(k)$ is the most common representation of a spectral function. It supplies information on the amplitude a harmonic component has at frequency component, k. The amplitude spectrum $A(k)$ is calculated from the two-sided amount spectrum $|X(k)|$ as follows:

$$A(k) = |X(k)| \text{ for } k = 0 \text{ and } A(k) = |2 \cdot X(k)| \text{ for } 0 < k < N/2 \qquad (10)$$

The **Bode diagram** is used to represent the frequency response $G(f)$ of systems. This frequency response forms a Fourier pair with the corresponding impulse response $g(t)$. The zero point of the frequency response represents the static transmission factor k_0. The Bode diagram represents the frequency response in a dual-logarithmic scaling, where the amplitude values are fractionated to the static transmission factor:

$$\text{Bode diagram } (\log(kF)) = 20\log(G(k) / k_0) \text{ [dB] for } 0 < k < N/2 \qquad (11)$$

The unit of this scaling is dB (Decibel).

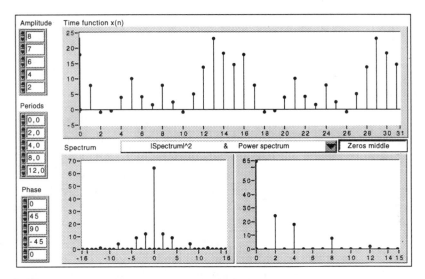

Figure 9-14

Symmetric representation using the power and amplitude spectra

Example

Figure 9-15 shows various spectral FFT representations of a discrete causal exponential function, $x(n)$. For comparison with the discrete spectral values, the Fourier spectrum $X(f)$ of the corresponding analog signal $x(t)$ is represented by a thick gray line. The causal exponential function forms a Fourier pair with the Debye function, shown in Figure 9-9 in a three-dimensional representation. The causal exponential function represents, for example, the impulse response $g(t)$ of a temperature sensor or an RC element. Thus, the Debye function describes the frequency response of this type of system.

The sample page shown in Figure 9-15 was created in LabVIEW to document and output test results in A4 paper size on a PostScript laser printer. These protocols were produced within a signal processing lab course at the University of Munich, Germany, Precision Mechanics Department, headed by Prof. Dr. Stockhausen.

Before applying FFT to the discrete values $g(n)$, the mean value from the first value $g(0)$ and from the last value $g(N-1)$ has to be calculated and used in the position of the first value. This action serves to reduce the discontinuity at the "interface" of the cyclic signal. Figure 9-15 shows various forms of representation for spectral functions. Notice that, even with a reduced discontinuity in the time domain, there is a clear deviation between the discrete frequency response and the corresponding theoretical curve. The Bode diagram is shown in the next-to-last line on the left side. To the right of it, you see the relevant phase response, also in logarithmic frequency scaling.

Explanation of the example given in Figure 9-15

The time function (top, left) represents a section of a causal exponential impulse, which was digitized with $N = 32$ values in this measuring time. The second line shows the real and the imaginary parts of the corresponding Fourier spectrum. The thick gray line corresponds to the theoretical curve of the spectral function. The spectral functions are shown in symmetric representation, where the zero point is in the center of the picture. The FFT leads initially to a causal sequence of spectral values, with index value $k = 0$ at the beginning of the cyclic spectral function. The symmetric representation results only after shifting the second half in front of the first. The discrete values show a clear deviation from the theoretical curve of the spectral function, particularly in the upper spectral domain. This deviation is caused by the time function, which is also cyclic. At the interface of the periodic continuation, the time function shows a clear jump that supplies a broadband component in the real part of the spectrum. The imaginary part represents an uneven function, which has a value of zero at index value $k = N/2$ in the Nyquist frequency. Therefore, the phase of the spectral function approximates closer to the zero value as the frequency increases (second row, left). Next to the Bode diagram (second row, right), the bottom row shows the two-sided and one-sided Nyquist plots. The deviation from the circular form is caused by different scaling of the axes.

The artifacts in Figure 9-15 can be reduced by calculating the mean value between the first value $x(0)$ and the last value $x(N–1)$ and inserting the result at the value's position. The signal manipulated in this way is shown in Figure 9-16 (top row). After this significant manipulation of the value sequence's curve, we obtain a much better compliance with the theoretical curve of the spectral function. The remaining deviations can be accounted for by the fact that the causal exponential function has not undergone the anti-aliasing filtering. The spectral function $X(f)$ of the causal exponential function corresponds to the Debye function shown in Figure 9-9, which does not have any spectral constraint. Although the mean value calculated at the beginning of the discrete time function causes a reduction of high-frequency signal parts, we cannot yet achieve full compliance with the theoretical signal curve by using this method. In practice, however, this tradeoff is normally accepted, particularly when maintaining the signal form is of prime importance.

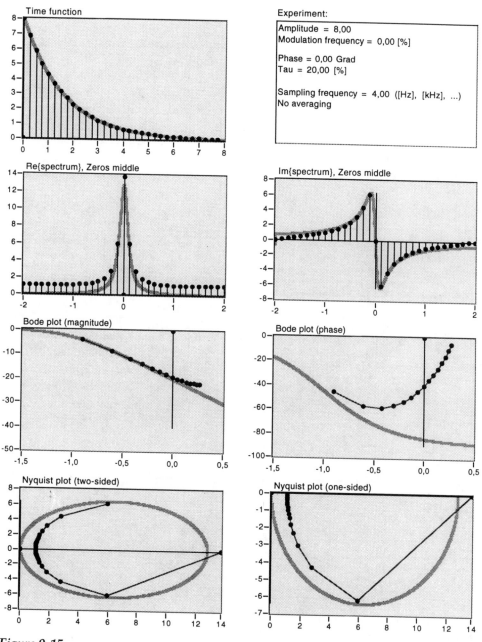

Figure 9-15

Various spectral representations of the FFT of a discrete causal exponential function in LabVIEW

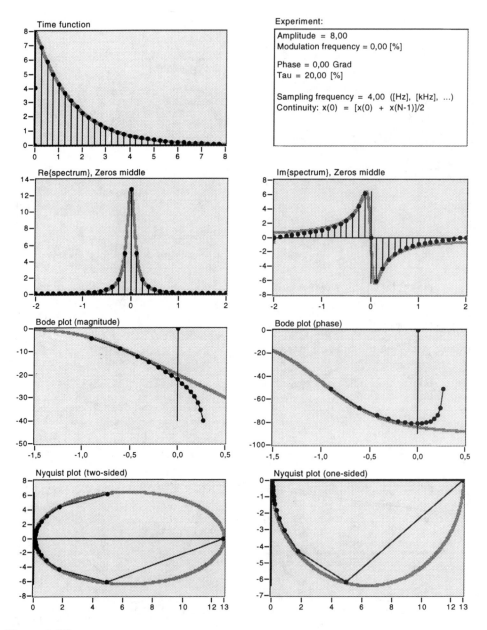

Figure 9-16

Reducing artifacts

For example, if we study the attenuation of an oscillating system, then the enveloping curve of the impulse response contains the information we look for. In this case, however, a theoretically exact digitization using an anti-aliasing filter would not produce any usable signal curve.

The Nyquist diagram corresponds to the projection of the frequency response to the complex number level. In the case of the Debye function shown in Figure 9-9, the Nyquist diagram produces a circle if the negative frequencies are included. The one-sided Nyquist diagram shows merely the positive frequency domain, thus forming only a semicircle. Both forms of these spectral representations are represented in the bottom row of Figure 9-15. In various physics disciplines, this form of representation is also called *Cole-Cole plot*. The representation of the frequency response projected to the Gaussian level allows us to quickly classify or define system properties. In physics, this form of representation is often selected to determine whether a process (e.g., relaxation) is composed of several subprocesses.

9.7 Leakage Effect

Limiting a continuous signal, $x(t)$, to a finite measuring time, T_M, leads to artifacts, as does discrete sampling, which can be minimized by suitable actions. From the mathematical view, the time constraint corresponds to multiplying a rectangular function with the time-unlimited analog signal. A rectangular function with width T_M and amplitude 1 is called a *rectangular window*. The rectangular window acts as a mathematical "mold" that sets all values outside the measuring interval to zero. Discrete sampling causes the conversion of this signal section into a cyclic signal. The curve of the discrete cyclical signal $x(n)$ matches the curve of a corresponding periodic analog signal, $x(t)$, as long as the measuring time T_M corresponds exactly to an integral multiple of the period time. It this is not true, a discontinuity will form at the "interface" of the cyclical signal, causing spectral lines in the spectral domain to blur. The amplitude of these lines is minimized, but at the cost of a broadband underground, caused by the irregularity at the interface of the cyclical signal. As a consequence of this *leakage effect*, it can be expected that the spectral lines of harmonic signal parts may disappear in the noise underground. Considering that the curve or period of an analog signal is normally not known in advance, leakage will generally occur.

The extent of this leakage is calculable when you consider the spectral function of the window function. An infinitely long window possesses an infinitely narrow spectral line in the zero point of a Fourier spectrum. In contrast to this, a rectangular impulse with infinite width leads to a split function in the spectral domain, the width of which increases as the measuring time decreases. When you multiply the rectangular window with a harmonic function, then two split functions will appear instead of two narrow lines in the spectral domain. The width of these split functions is determined by the measuring time. If the period of a harmonic signal exactly matches the measuring time, then the spectrum of the window function shifts exactly to the position of the two spectral lines — all other discrete spectral values fall into the zero transitions of the shifted split functions.

Example

Figure 9-17 shows a cosine-shaped signal, $x(t)$, from which $N = 16$ values are taken during measuring time T_M. The discrete values of the spectral function $X(k)$ follow one another at distance $F = 1/T_M$. The basic period of the discrete spectral function $X(k)$ reaches beyond $N = 16$ values. The positive frequencies' domain ends in the middle of the data block at $N = 8$. The negative frequency domain begins with the Nyquist frequency at $k = 8$ and ends with the −1st spectral component at $k = 15$. Thus, the second half of $\underline{X}(k)$ corresponds to the periodic continuation of the negative frequency domain. Index value $k = 8$ contains both the positive and the negative value of the Nyquist frequency. When you rearrange the (cyclic) discrete spectrum $X(k)$ in the following way,

$$\{ X(9), ..., X(15), X(0), X(1), ..., X(8) \},$$

then the discrete spectrum $\underline{X}(k)$ corresponds to the Fourier spectrum $X(f)$ of the time-unlimited analog signal $x(t)$, once the discrete values have been multiplied by a factor of $1/N$. The discrete spectrum $\underline{X}(k)$ possesses an enveloping curve, i.e., split functions at the positions of the positive and negative signal frequencies. These are caused by the time constraint of the analog signal imposed by the rectangular window $w(t)$. Except for the two sampling impulses $\underline{X}(4)$ and $\underline{X}(12)$, all other sampling impulses in Figure 9-17 fall to the zero points of the enveloping curve. The allocated time function (Figure 9-17, left) tells you that four entire periods fit into the measuring window. The cyclic iteration of the signal section is, under these conditions, identical with the time-unlimited cosine function.

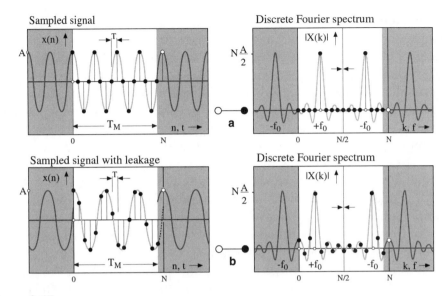

Figure 9-17

Sampling signal with constant continuation (top); sampling signal with leakage effect (bottom)

If, on the other hand, 3,6 periods are within the window, then a discontinuity forms at the "interface" of the periodic continuation (Figure 9-17, bottom). The periodically continued sampling signal now has a discontinuity at the interface so that it no longer corresponds to the curve of the original cosine function. The zero points of the continuous enveloping curve no longer match the positions of the sampling impulses in the spectral domain. Also, there is no sampling value at frequencies $\pm f_0$. The amplitudes of the sampling impulses near the "true" spectral lines do not reach the maximum value of the enveloping curve. Moreover, additional "smeared" sampling values appear over a broad spectral domain. You can think of this effect as a "leakage" of the Dirac impulses over the adjacent spectral domains, hence the name *leakage*.

9.8 Window Functions

This section describes some criteria for evaluating a window. Basically, there is no single ideal window function for all types of applications. Rather, you select a suitable window depending on the time signal to be analyzed and the requirements to the FFT spectrum, considering several criteria, described below.

Limiting the measuring time to a finite value, T_M, corresponds to multiplying the analog signal $x(t)$ by a rectangular window $w(t)$. Given the fact that you normally don't know the character of a measuring signal in advance, your discrete cyclic sampling function will probably produce discontinuities along the interface. The leakage caused by this discontinuity can be minimized by multiplying the discrete sampling values by a window function that gradually approximates zero along the margins. Although this changes the signal curve significantly, you achieve a clear reduction of the leakage effect in the spectral domain. The effect of a window function can be illustrated best when you study its spectral function $\underline{W(f)}$, which should ideally have a narrow peak and drop quickly towards zero in the environment. So, only the spectral properties of the most important window functions will be compared.

The continuous window functions $w(t)$ are denoted by a gray curve. The discrete flow of the window functions $w(n)$ is given by $N = 16$ sampling impulses at interval T. The corresponding log amount values of the discrete spectral function, $20 \cdot log(\mid W(k)/W_{max} \mid)$, are specified in dB units. The amplitude values are fractionated to the maximum W_{max} of the spectral amount of the window functions. At this point, the spectral function has a value of 0 dB. The vertical lines, following one another at interval $F = 1/T_M$, represent the case where entire periods fit into the measuring window of time, T_M. The points entered here correspond to the case where the discrete spectral values do not coincide with the maxima and the zero points of the spectrum $W(f)$ of the window function, i.e., when leakage occurs.

The Fourier pair of the rectangular window is represented in Figure 9-18 (top). The rectangular window is best suitable to detect harmonic signal parts when entire periods fall into the measuring window and when all spectral components existing in the measuring signal fall on the zero points of the split function. If leakage cannot be avoided by fine-tuning the selected measuring time T_M, then significant leakage will definitely occur.

To minimize the leakage caused by inconsistencies along the interfaces of the periodic continuation, the discrete data block $x(n)$ is multiplied by the values of a window function, $w(n)$, which takes the zero value at the interfaces $w(0)$ and $w(N)$. Initially, it appears logical to multiply the discrete measuring signal $x(n)$ with a triangular window so that the signal's amplitude values drop linearly to the interfaces. This window type, also known as the *Bartlett window* (Figure 9-18, center), is defined as follows:

$$w(n) = 1 - \frac{2 \mid n - N/2 \mid}{N} \qquad 0 \le n < N \qquad (12)$$

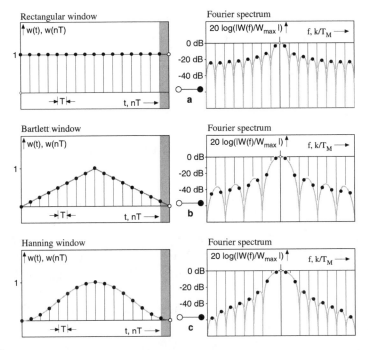

Figure 9-18

Three common window functions and their Fourier spectra: Rectangular window (top), Bartlett window (center), and Hanning window (bottom)

The side maxima have dropped noticeably compared to the spectral function of the rectangular window, but this benefit was achieved at the cost of a clear expansion of the main maximum.

One of the most frequently used window functions is the Hanning window:

$$w(n) = 0{,}5\left[1 - \cos(2\pi n / N)\right] \qquad 0 \le n < N \qquad (13)$$

When comparing the log spectrum of the Hanning window (Figure 9-18, bottom) to that of the rectangular window, you will notice that an additional attenuation of the side maxima can be achieved without considerable expansion of the main maximum. The attenuation of the first side maximum for the three window functions is:

$$\text{Rectangle: } -13\text{ dB} = 0.224 \quad \text{Triangle: } -27\text{ dB} = 0.045 \qquad (14a)$$
$$\text{Hanning: } -43\text{ dB} = 0.007$$

The width of the main maximum can be defined through the frequency interval in which the amplitude dropped from maximum value to –3 dB.

Referring this to the spectral resolution $\Delta f = 1/T_M$, the following main maxima result for the three window functions:

$$\text{Rectangle: } 0.45 \cdot \Delta f \quad \text{Triangle: } 0.64 \cdot \Delta f \quad \text{Hanning: } 0.65 \cdot \Delta f \tag{14b}$$

When comparing these values, you will notice that the Hanning window has the best properties with regard to selecting harmonic spectral components in a measuring signal. A digitized signal is normally weighted with a window function when the spectral contents of the analog signal changes relatively fast, when only short measuring times are possible, and when the signal within these measuring times behaves in a quasi-stationary way. The reason is that, when measuring times are short, the marginal effects play an important role because their contribution to the total signal energy increases as the measurement time decreases.

You should not use a window function in cases where the amplitude curve of a signal is of prime importance. Instead, you initially smooth only the direct neighborhood of inconsistencies to minimize leakage while maintaining the signal curve. The least impact on the amplitude curve is achieved by calculation of a mean value from the values around the interface (Figure 9-16). You can make a tradeoff between minimizing leakage and maintaining the original signal curve by using a cosine-tapered window. This window function limits the Hanning window to the marginal zones, while the mean value range remains unchanged.

Example

We want to demonstrate the leakage effect and its attenuation by applying window functions in a LabVIEW program. First, a harmonic signal is digitized with four entire periods within a measuring time of $N = 32$ values (Figure 9-19).

All representations are symmetric so that the zero point of each one is in the center. The wide gray lines correspond to the time-limited analog signal and its spectral function, respectively. There is no inconsistency at the interface of the time function, which means that the cyclic iteration of the signal section is identical to the time-limited harmonic signal. As expected, the real part of the spectral function (3rd row, left) also has only two spectral lines, while the imaginary part contains only zeros.

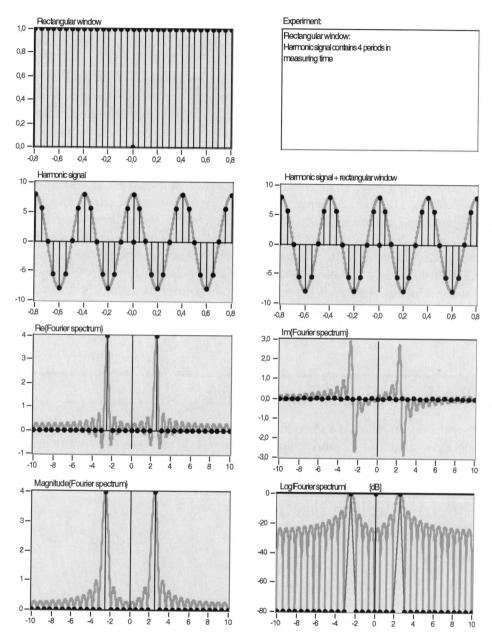

Figure 9-19

Applying the rectangular window to a harmonic signal with an integral number of cycles

When the measuring time does not match an integral multiple of the periodicity, then the spectral values no longer coincide with the zero points of the continuous spectral function (Figure 9-19). Also, none of the discrete sampling values fall within the maximum of the continuous spectral function. Discrete sampling values appear only along the edge of the spectral lines. When a cycle is terminated in the middle, in the extreme case, then the discrete spectral values fall on the maxima of the side lobes. This means that their height determines the maximum leakage impact.

We see in the log scaling (Figure 9-20, bottom left) that the side lobes drop only slowly as the distance from the spectral lines increases. If you multiply the discrete signal $x(n)$ with a Hanning window, $w(n)$, then the inconsistency along the interface of the periodic continuation disappears (Figure 9-21). However, this solution is at the cost of a significant change in the amplitude curve of the value sequence. As expected, the real and imaginary parts have clearly different curves, compared to the example shown in Figure 9-18. But the amplitude amount spectrum shows that the leakage was minimized considerably. This is visible particularly when you compare the amount spectra in the logarithmic scaling (bottom row, right).

9.9   Fourier Transform Properties

Several known basic laws are applied between time functions and spectral functions, which are useful to analyze signals. The most important properties are described in the following sections.

☐ 9.9.1   Addition Theorem

Two different signals, $x(t)$ and $y(t)$, have the spectral functions $X(f)$ and $Y(f)$, respectively:

$$x(t) \leftrightarrow \underline{X}(f) \quad \text{and} \quad y(t) \leftrightarrow \underline{Y}(f)$$

Applying the Fourier transformation to the sum of the two functions supplies the sum of the corresponding spectral functions:

$$x(t) + y(t) \quad \leftrightarrow \quad \underline{X}(f) + \underline{Y}(f) \tag{14c}$$

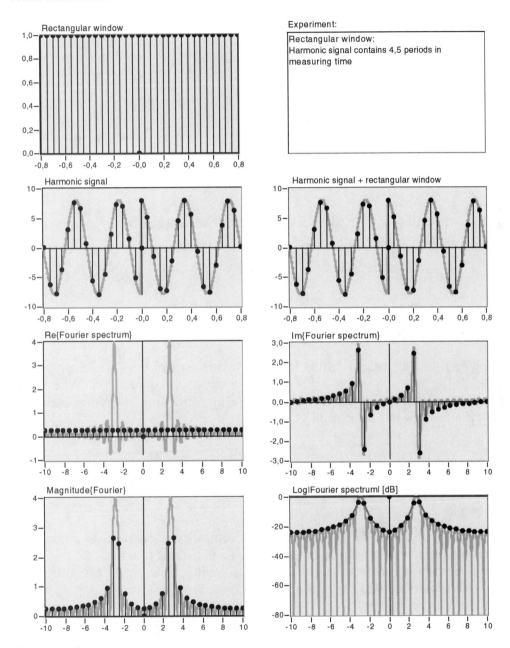

Figure 9-20

Applying the rectangular window to a harmonic signal with inconsistencies along the interface of the cyclic continuation

Example

$x(t) = A \cdot \cos(2\pi f_0 \cdot t)$ is added with an equal share of $y(t) = K$. The Fourier transformation of $x(t)$ consists of two Dirac impulses with surface $A/2$ at $f = \pm f_0$:

$$\underline{X}(f) = A/2 \cdot [\delta(f - f_0) + \delta(f + f_0)]$$

The spectral function of the equal share $y(t)$ corresponds to a Dirac impulse in the frequency zero point with surface K:

$$\underline{Y}(f) = K \cdot \delta(f)$$

The Fourier pair of the sum function from cosine function and equal share is then:

$$K + A \cdot \cos(2\pi ft) \quad \leftrightarrow \quad A/2 \cdot [\delta(f - f_0) + \delta(f + f_0)] + K \cdot \delta(f)$$

☐ 9.9.2 Symmetry

The Fourier integral (2) differs from the inverse operation (3) exclusively by the sign in the exponent of the complex exponential function. When you apply the Inverse Fourier Transform to a time function, $x(t)$, you obtain the complex-conjugate spectrum $\underline{X}^*(f)$:

$$x(t) \quad \leftrightarrow \quad \underline{X}^*(f) = \underline{X}(-f), \text{ if } x(t) \leftrightarrow \underline{X}(f) \tag{15}$$

Example

A Dirac impulse in the time domain with impulse surface A has a constant spectral function with the amplitude. A constant function in the time domain then has a Dirac impulse in the frequency zero point. Mirroring $x(f)$ -> $x(-f)$ plays no role in this case because both functions are even.

Example

A sine function in the time domain supplies an imaginary Dirac dual impulse in the spectral domain. Switching the two functions results in an inversion of the sign of the Dirac dual impulse. Only a conjugation in the spectral domain undoes this sign switching.

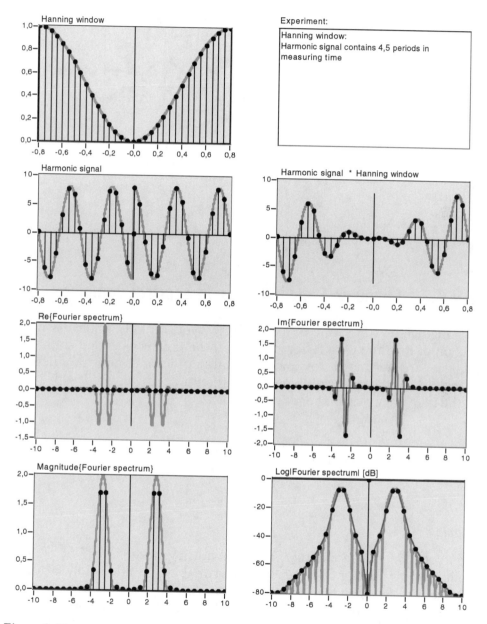

Figure 9-21

Applying the Hanning window to a harmonic signal with inconsistencies along the interface of the cyclic continuation

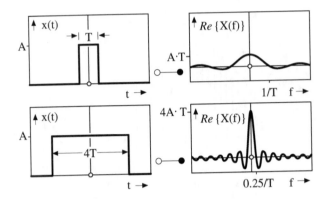

Figure 9-22

Changing scale in the time and spectral domains

☐ 9.9.3   Changing Scale in Time and Spectral Domains

By changing the time scale of a signal, $x(t)$, you actually achieve a multiplication of the time with factor k. At $a < 1$, the time is extended so that the
signal slows down, as in slow motion, while a quick motion is produced by
factor $a > 1$. In the spectral domain, the amplitude is weighted with $1/|a|$,
and the frequency axis is scaled with factor $1/a$:

$$x(at) \quad \leftrightarrow \quad \frac{1}{|a|} X\!\left(\frac{f}{a}\right) \qquad\qquad (16a)$$

Example

When you widen a rectangular impulse with impulse width T at constant
amplitude A by a factor of 4, then the time variable is multiplied by factor $a =$
0.25. The first zero points of the spectral function $X(f)$, shown in Figure 9-22
(top are around the frequencies $f \pm 1 = \pm1/T$. According to the Fourier pair of
the rectangular function shown in Figure 9-22, the amplitude of the split
function has the value $X(0) = AT$ in the frequency zero point. The Fourier pair
of the extended time signal is represented in Figure 9-22 (bottom). While the
first zero points of the split function approximate the zero point by a factor of
$1/a$, the amplitude in the frequency zero point rises to value $4AT$.

Example

When digital information is transmitted, rectangular impulses with time extension T, correlating to bandwidth B, are transmitted. Assuming that the main part of the signal energy of a rectangular impulse lies in the frequency domain within the first zero points of the spectral function ($|f| < 1/T$), you can determined its bandwidth at $B = 2/T$. A long impulse requires a low bandwidth for transmission, while a small impulse takes a large bandwidth.

Extending the frequency scale by a factor, a, upsets the time scale by factor $1/a$:

$$\frac{1}{|a|} x\left(\frac{t}{a}\right) \quad \leftrightarrow \quad \underline{X}(a \cdot f) \tag{16b}$$

☐ 9.9.4   Shifting Within the Time and Spectral Domains

Shifting the time function $x(t)$ by time t_0 produces a reverse phase angle rotation of the positive and negative spectral parts. The phase angle rotation grows linearly with frequency f. This relation is called *shift theorem*. From the mathematical view, you obtain a phase angle rotation by multiplying the spectral function $X(f)$ by the complex unit vector $exp(-j2\pi f t_0)$:

$$x(t - t_0) \quad \leftrightarrow \quad e^{(-j2\pi f t_0)} \underline{X}(f) \tag{17a}$$

Example

If you shift the function $x(t) = cos(2\pi f_0 t)$ with period $T = 1/f_0$ by one quarter period ($t_0 = T/4$), you obtain a multiplication factor to the right of (54a) of the complex unit vector $exp(-j\pi/2)$, which causes a rotation of the spectral line by $-\pi/2$ at frequency $+f_0$ and by $+\pi/2$ at frequency $-f_0$. The amount of the spectral function $\underline{X}(f)$ does not change because the rotating vector $exp(j\Delta\varphi)$ has the amount 1.

Example

The shift theorem can be used to determine the run time of an impulse. A simulation written in LabVIEW shows a Gaussian impulse (Figure 9-23, top); its symmetry axis is shifted to the right by eight index values against the time-zero point. The Gaussian impulse has a signal noise ratio of 30 dB. The amount spectrum (center) has also a Gaussian-shaped curve (flow). The spectral representations are symmetric to the frequency zero point.

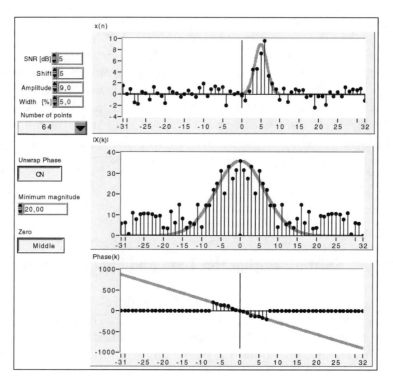

Figure 9-23

Applying the shift theorem to determine the time shift of a Gaussian impulse

The noise part dominates above the index range of $|k| > 20$. The rise of the linear curve of the phase spectrum (bottom) within index range $|k| \leq 6$ can be used to determine the impulse shift. A phase jump of $-180°$, caused by ambiguity of the arctangent function, appears between the index values $k = 6$ and $k = 7$. You can enable the **Phase Unwrap** function in LabVIEW to remove these inconsistencies (see Figure 9-24).

The phase spectrum now has an almost linear curve within the index range of $|k| < 20$. The spectral values above this index range should not be reused for adapting a straight to the phase spectrum. The reason is that the spectrum of the Gaussian impulse is dominated by noise outside this range, which supplies statistically distributed phase values. When reducing the signal-to-noise ratio to 5 dB, the noise part will dominate outside of index range of $|k| \leq 7$. This means that the phase spectrum can be evaluated only as long as you compare the amount spectrum against a specified minimum value and if invalid phase values are set to zero (Figure 9-25).

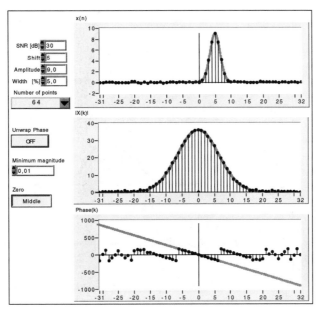

Figure 9-24
Using the Phase Unwrap function in LabVIEW to remove inconsistencies

Shifting a spectral function, $\underline{X}(f)$, by a frequency interval of Δf corresponds to multiplying the corresponding time function $x(t)$ by the (complex) factor $exp(j2\pi\Delta ft)$. A spectral function, $\underline{X}(f)$, which originates from the real-time function $x(t)$, forms a Fourier pair with the complex time function after a Δf shift.

$$e^{(j2\pi\Delta f \cdot t)} \, x(t) \quad \leftrightarrow \quad \underline{X}(f - \Delta f) \tag{17b}$$

Given that a time function normally has a real value, the corresponding (complex-conjugate) part of the spectral function has to be shifted to the opposite direction within the negative frequency domain. Under these conditions, the time function is multiplied by a cosine function:

$$x_{mod}(t) = 2 \cdot x(t)\cos(2\pi f_0 \cdot t) \quad \leftrightarrow \quad \begin{cases} X_{Mod}(f) = X(f - f_0) \textit{ for } f > 0Hz \\ X_{Mod}(f) = X(f + f_0) \textit{ for } f < 0Hz \end{cases} \tag{17c}$$

The function $2\cos(2\pi f_0 t)$ corresponds to a carrier frequency, to which the signal $x(t)$ was modulated. Shifting the spectrum of a band-limited signal toward the frequency zero point corresponds to a movement of the spectral parts into the baseband. This process is called *demodulation*.

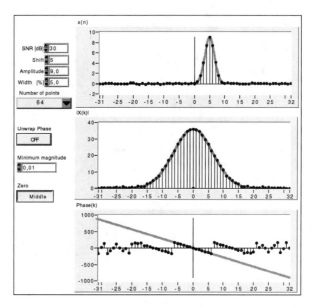

Figure 9-25
Determining the time shift of a Gaussian impulse becomes more difficult as the signal-to-noise ratio decreases

9.10   The Wiener-Khintchine Theorem

One of the most important properties of the Fourier transformation is that it simplifies convolution operations a great deal. *Convolution* plays an important role in filtering and restoring signals. The convolution operation with the two time-dependent signals, *x(t)* and *z(t)*, is defined as:

$$y(t) = \int_{-\infty}^{+\infty} x(\tau) \cdot z(t-\tau)d\tau = x(t) \otimes z(t) \tag{18}$$

Integral (18) represents the mutual signal coverage as a function of shift *t*. For this purpose, you replace function variable *t* in the two signals, *x(t)* and *z(t)*, by integration variable *t*. Initially, function *z(τ)* is mirrored around the zero point, which results in *z(−τ)*. This function, *z(−τ)*, is shifted by time interval $t = t_0$ so that we obtain function $z(t_0 - \tau)$. The product's integral, $x(\tau)z(t_0 - \tau)$, eventually supplies the value of convolution integral $y(t_0)$ at shift value $t = t_0$. Thus, the *y(t)* function is based on resolving an infinite number of integrals. The mathematically costly convolution operation in the time domain can be replaced by a simpler (complex) multiplication in the spectral domain. After retransformation of the (complex) product function, $\underline{Y}(f) = \underline{X}(f) \cdot \underline{Z}(f)$, we obtain result $y(t)$ of the convolution integral:

$$y(t) = \int_{-\infty}^{+\infty} x(\tau) \cdot z(t - \tau) d\tau = x(t) \otimes z(t) \quad \leftrightarrow \quad \underline{Y}(f) = \underline{X}(f) \cdot \underline{Z}(f) \qquad (19a)$$

The Fourier transformation is applicable in both directions so that the convolution theorem applies also in opposite direction. A convolution in the spectral domain corresponds to a multiplication in the time domain:

$$y(t) = x(t) \cdot z(t) \quad \leftrightarrow \quad \underline{Y}(f) = \int_{-\infty}^{+\infty} X(\gamma) \cdot Z(f - \gamma) d\gamma = \underline{X}(f) \otimes \underline{Z}(f) \qquad (19b)$$

Example

The convolution operation in two identical rectangular impulses, $x(t) = z(t)$, with amplitude A and width T (Figure 9-26, top) produces a triangular impulse with a maximum value, $A^2 \cdot T$ (center, top). This value is reached with a complete mutual coverage of the rectangular impulses. If we transform the impulses in the spectral domain, we obtain identical split functions (bottom). As a result of the multiplication with spectral functions, we obtain the squared function $\underline{X}^2(f) = \underline{X}(f) \cdot \underline{Z}(f)$ (center, bottom). This forms a Fourier pair with the triangle function. The shortcut using the Fourier transformation through the spectral domain and back to the time domain appears costly at first sight, but remember that the Fourier transformation supports very efficient algorithms (FFT) and special hardware (signal processors).

9.11   Discrete Convolution

When convoluting this time-discrete signal with a number sequence, $z(n)$, then the convolution integral $y(t)$ defined in equation (18) is replaced by the sum $y(n)$:

$$y(n) = \sum_{m=-\infty}^{+\infty} x(m) \cdot z(n - m) = x(n) \otimes z(n) \qquad (20a)$$

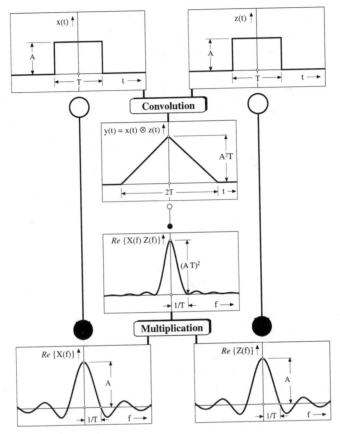

Figure 9-26
Convolution of two identical rectangular impulses using the Fourier transformation

Remember that the discrete time variable n has to be renamed to the discrete variable m before doing the summation to be able to apply the discrete time shift to the variable n. The time-discrete summation variable m acts as an integration variable τ in equation (18a). As a convolution results, we again obtain a discrete time function $y(n)$.

If we try to realize $x(n)$ and $y(n)$ with two number sequences in equation (20a), we need to deal with the problem that the summation can be done only through a finite set of numbers with N values:

$$y(n) = \sum_{m=0}^{N-1} x(m) \cdot z(n-m) \tag{20b}$$

If the lengths of the two data blocks are different, then not all terms of the sum (20b) are defined. We can solve this problem easily by padding the shorter of the two data blocks with zeros to full length. When you calculate the convolution sum using FFT, you need to consider that the discrete signals have a cyclic character. Therefore, some values of the convolution sum $y(n)$ are tainted with artifacts and have to be eliminated.

Example

A time-discrete step function, $x(n) = s(n - 4)$, with $N = 16$ values (Figure 9-27, top) is convoluted with the following impulse response: $g(n) = \{1, 0.7, 0.4, 0.2, 0.1\}$.

First, we have to append 11 zeros to $g(n)$ to produce data blocks with equal length (Figure 9-27, center). Second, the index values have to be renamed to m to reserve index n for the shift. When calculating the convolution sum based on the finite convolution sum (20b), we have the following problem: When mirroring the discrete signal $g(n)$, the values outside the zero point fall into the negative index range, while undefined signal values in the $n < 0$ range within index range $0 \leq n < N$ of the convolution sum emerge. If we use FFT to calculate the convolution operation, we have to treat the two signals $x(n)$ and $g(n)$ as a cyclic value sequence. In this case, the mirrored values (except $g(0)$) appear at the end of data block $g(-m)$. In the data block, the mirrored impulse response appears only in two spatially separated domains. This is why the convolution sum produces a faulty result. When applying the cyclic convolution, we have to check whether the mirrored impulse response $g(n - m)$ is preserved in compact form. Otherwise, the values of the convolution sum $y(n)$ have to be rejected in the corresponding shift values. Figure 9-27 (bottom) shows that the first four values of the convolution sum $y(n)$ are invalid. This "forbidden" index range is known as the *transient area* of the digital filter.

In most practical cases, the "transient process" of the filter does not pose any problem because normally long number sequences have to be filtered. In this case, the input signal $x(n)$ is decomposed into individual sequences and filtered separately. Subsequently, the individual parts of the convolution sums are put together again. When using overlapping number sequences for a segmented convolution, we can use the invalid values at the beginning of the data block for values of the previously calculated convolution sum in the overlap area. This method is called *overlap save segmentation*.

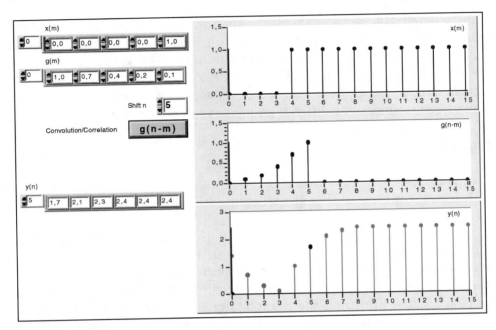

Figure 9-27
Example of a discrete convolution at a step function with an impulse response

Physically implemented systems can respond at the earliest at the time when an impulse is received at the system output. Therefore, the impulse response of causal systems is equal to zero in the index range $n < 0$. In contrast to real systems, digital filters are also able to operate convolutions with a noncausal impulse response. In this case, negative index values of the impulse response appear at the end of the data block.

9.12   Correlation Theorem

In addition to the application of statistical methods and the spectral analysis, the *correlation analysis* is one of the most important methods for the study of measuring signals. The correlation represents a measurement for the inner relationship of measuring signals and supplies information on the structure of signals. For example, the difference between a correlation analysis and a statistical analysis can be illustrated by a measuring signal that corresponds to the roughness of a surface. The statistical analysis supplies the mean surface roughness or the variance of the surface profile as parameters. A sandblast and a rotated surface can lead to identical results in the statistical analysis, although the surfaces differ distinctly. The statistical parameters supply information only on the distribution of incidence of measuring data, but they do not consider the structural properties of signals. In contrast, the correlation analysis can be used not only to determine the inner structure of signals but also to study the structural "relationship" between signals or to filter nonstochastic components from noisy signals. The *cross-correlation function (KKF)* $\Phi_{xz}(t)$ between a signal $x(t)$ and sample function $z(t)$ is defined as follows:

$$\Phi_{xz}(t) = \int_{-\infty}^{+\infty} x(\tau)z(t+\tau)\cdot d\tau \qquad (21a)$$

Also, resolving the correlation integral in the spectral domain can be replaced by a simple multiplication. Because correlation differs from convolution only by a lack of mirroring operation, the multiplication is done in complex-conjugate spectral functions:

$$\Phi_{xz}(t) = \int_{-\infty}^{+\infty} x(\tau)z(t+\tau)\cdot d\tau \quad \leftrightarrow \quad \underline{X}(f)\cdot \underline{Z}^*(f) \qquad (21b)$$

The spectral function $\underline{X}(f) \cdot \underline{Z}^*(f)$, which is normally complex, is called a *cross-power* or *cross-energy spectrum*. If one correlates a signal with its own copy, then $x(t) = z(t)$:

$$\Phi_{xx}(t) = \int_{-\infty}^{+\infty} x(\tau)x(t+\tau)\cdot d\tau \quad \leftrightarrow \quad \underline{X}(f)\cdot \underline{X}^*(f) = \left|X(f)\right|^2 \qquad (21c)$$

The *autocorrelation function (AKF)* $\Phi_{xx}(t)$ can be used, for instance, to extract periodical signal parts from noise. The spectral function of the auto-correlation function $\Phi_{xx}(t)$ is real and identical with the squared amount of the spectral function of $x(t)$. The function $|X(f)|^2$ is called *autospectrum* or *energy density spectrum* — $W_{xx}(f) = |X(f)|^2$, and the squared amount of the spectral function of power signals is called *power density spectrum* — P_{xx}.

Example

In radar and ultrasound applications, the distance of an object from a sender/receiver system is determined from the run time of an impulse. However, the reflected impulse is often so weak that it cannot be extracted from the noise. The impulse strength of the sender is limited by physical boundaries; on the other hand, increasing the impulse time reduces the resolution capacity. This is the reason for sending a defined signal pattern, which can be extracted from the noisy receiver signal by means of the correlation method. The signal pattern $x(t)$ shown in Figure 9-28 (top) is called *Barker code*. This signal extends over 13 s. Assuming that the signal amplitude of $x(t)$ has unit V, then we obtain a signal energy value of $W_{xx} = 13\ V^2s$.

The AKF pertaining to this signal is shown at the bottom of Figure 9-28. This AKF has a needle-shaped amplitude curve in the zero point and otherwise oscillates only by small amplitude values. The AKF's maximum corresponds to the value of the signal energy and, in this case, exceeds the amplitude of the Barker code by a factor of 13. The impulse in the zero point has only a half-width value of 1 s, which corresponds to a shortening of the signal time by a factor of 13. In addition to an amplitude increase, we also achieve a shortening of the signal length. This method is called *impulse compression*.

9.13   Discrete Correlation

To calculate the discrete cross-correlation function $\Phi_{xz}(n)$, we replace the integral (21a) by the following sum:

$$\Phi_{xz}(n) = \sum_{m=0}^{N-1} x(m) \cdot z(n+m) \quad \leftrightarrow \quad \underline{X}(k) \cdot \underline{Z}^*(k) \tag{22a}$$

The time-discrete signal $x(n)$, which is renamed to do the sum on $x(m)$, plays the role of a "reference function," while the discrete function $z(n)$ is considered a "pattern function" used to find its structure in $x(n)$. Index n now has the significance of a shift value. The correlation sum corresponds to a complex multiplication in the spectral domain. After retransformation of the cross-power spectrum in the time or space domain, we obtain the (cyclic) cross-correlation function. Correlating the data sequence $x(n)$ over itself, $z(n) = x(n)$ results in the autocorrelation function:

$$\Phi_{xx}(n) = \sum_{m=0}^{N-1} x(m) \cdot x(n+m) \quad \leftrightarrow \quad \underline{X}(k) \cdot \underline{X}^*(k) = |X(k)|^2 \tag{22b}$$

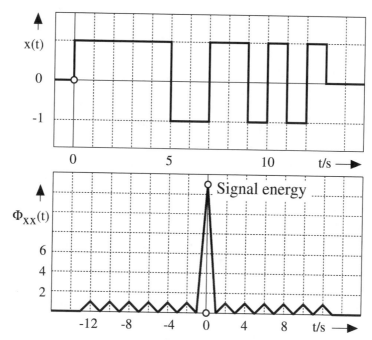

Figure 9-28
Determining the distance of an object by means of the autocorrelation function

If we use FFT to calculate the discrete correlation, then the two signals, $x(n)$ and $z(n)$, are cyclic. With a cyclic correlation, the correlation sum $\Phi_{xz}(n)$ or $\Phi_{xx}(n)$ cannot be calculated easily because the pattern function emerges along with a shift at the end of the data block so that it decomposes into two separate parts. Unless we take additional actions, we can expect a faulty result from calculating the correlation sum.

Example

Figure 9-29 (top) shows a discrete harmonic voltage signal $u(n)$ with amplitude $U_0 = 8$ *V*. The measuring time comprises 5.5 cycles, which means that there are inconsistencies at the interface. The gray line indicates the curve of the corresponding time-unlimited analog signal $u(t)$. The copies of signals $u(n)$ and $u(t)$ are shown in the center of Figure 9-29.

The result of the correlation sum is depicted in the bottom diagram. Harmonic signals have a cosine function as autocorrelation function, regardless of the phase position, with an amplitude corresponding to the signal power. With a signal amplitude of 8 V, the amplitude of the corresponding AKF is *32 V^2*.

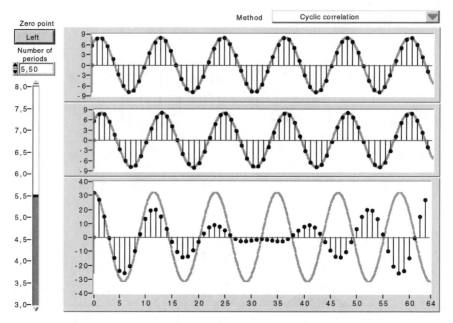

Figure 9-29
Effect of the zero point and number of cycles on the discrete AKT

If we now use the correlation theorem (22b) to calculate the discrete AKF, we obtain a clearly deviating signal curve, compared to the "true" AKF. An approximate match between the correlation integral (gray line) and the discrete cyclic AKF can be found only around the zero point. With larger shift values, the inconsistencies at the interface of the cyclic continuation lead to an increasing deviation from the theoretical result. If the measuring time includes integral cycles, then the cyclic KKF matches the "analog" KKF over the entire range.

Figure 9-30 represents all functions symmetrically to the zero point; the second half of the data blocks is placed in front of the first. This representation illustrates the different curve (flow) of the cyclic signals versus the corresponding analog signals. A signal match is achieved only provided that the measuring time includes an integral number of cycles. A reduction of the amplitude of the discrete AKF in line with an increasing shift of the pattern function can be accounted for by the fact that more and more values, showing a phase jump versus the second signal, flow into the (cyclic) correlation sum.

In practice, one has to expect that a discontinuity at the interface of the cyclic continuation occurs, so that it is necessary to make room by applying zero padding to be able to shift the pattern function versus the signal function without producing artifacts. Several methods are available to accomplish this.

If N zeros are appended to each of the two data blocks, the signal cycle doubles to $2N$ values. Within one cycle, the two signals appear as a product with a rectangular window function $w_x(n)$ or $w_z(n)$. The pattern function $z(m)$ can now be shifted by up to N steps along the reference function $x(m)$ in either direction:

$$\Phi_{xz}(n) = \frac{N}{N - |n|} \sum_{m=0}^{2N-1} x(m) \cdot w_x(m) \, z(m+n) \, w_x(m) \quad \text{for} \quad 0 \le |n| < N - 1 \qquad (22c)$$

A shift by n values reduces the length of the product function $x(m) \cdot z(n+m)$. This means that increasingly fewer terms contribute to the correlation sum. If the pattern function is shifted by N positions versus the reference function $x(n)$, the correlation sum results in a value of zero. Therefore, the correlation function is evaluated with a triangle window, $w_\Phi(n)$. This is referred to as the *Bartlett method*. The "true" correlation function results only after an inverse weighting, which undoes the impact of the triangle window. In the case of stochastic signals, as the shift of the pattern function increases, the variance of the correlation function also increases because it is additionally increased by the inverse weighting. In this case, only the region around the zero point of AKF can be used.

Example

The harmonic signal shown in Figure 9-29 is processed according to the Bartlett method (see Figure 9-31). A total of 64 zeros is appended to the 64 discrete values (top), then the result is used to produce a copy which serves as pattern function. Applying the correlation theorem after re-transformation, we obtain the AKF autopower spectrum shown at the bottom of Figure 9-31. Placing the second half of the value sequence before the first, we can see that AKF has a symmetrical curve toward the zero point and a constant flow in the zero point. The AKF shown in Figure 9-31 has its interface at index value $n = 32$. The correlation function calculated by means of the Bartlett method corresponds to a cosine function, weighted by using a triangle window. The basic width of the triangle window that is symmetrical to the zero point is $2N \cdot T$, where T is a sampling interval.

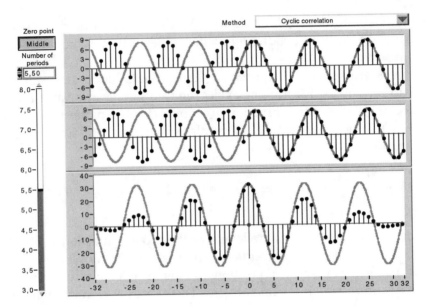

Figure 9-30
A signal match is achieved only provided that the measuring time contains an integral number of cycles

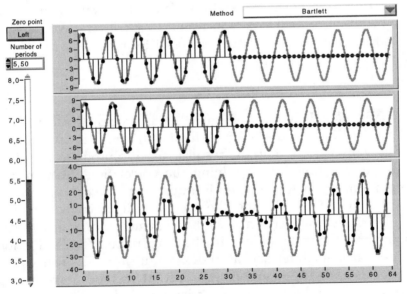

Figure 9-31
The harmonic signal shown in Figure 9-29 is processed with the Bartlett method

Another way of calculating the discrete correlation function consists in shortening the length of the pattern function $z(n)$ to K values ($K < N$). The remainder is padded with zeros. The cyclic correlation is valid only in the region in which zeros exclusively pass the interface. So, we can calculate a total of $N - K$ values, which are now weighted by a constant:

$$\Phi_{xz}(n) = \sum_{m=0}^{N-K-1} x(m) \cdot w_x(m)\, z(m+n)\, w_z(m) \quad for\ -(N-K-1) < n \leq 0 \tag{22d}$$

The process of shortening the pattern function $z(n)$ with subsequent zero padding to length N of reference function $x(n)$ is known as the *Rader method*. If one reduces the pattern function $z(n)$ to very few values, the variance of the correlation sum increases. In turn, one achieves large shift values. In the opposite case, one achieves a low variance and a high security to find textures or signal structure, but only small shift values. In general, the length of the pattern function $z(n)$ is cut in half and the remainder is padded with zeros. With a total number of N values, $N/2 -1$ signal shifts can be realized. The correlation sum contains only $N/2$ terms that are non-zero. This corresponds to a constant weighting with factor $1/2$. In the case of stochastic signals, the variance of the correlation function now remains constant over the entire (valid) shift range.

Example

The signal represented in Figure 9-29 is processed according to the Rader method (see Figure 9-32). In the copy of the discrete signal, the values in the second half of the data block are replaced by zeros (center). Although the actual objective of the calculation is the AKF of the harmonic signal, first the KKF of both signals (22a) is calculated. FFT is applied to transform both signals into the spectral domain, and both (different) spectral functions are used to calculate the (complex value!) cross-line spectrum. When subsequently operating a retransformation, the AKF so obtained is not even. The reason is that the causal sequence of the pattern function is maintained only within a shift range of $-32 < n \leq 0$. In the causal representation form (zero point at the left margin), the cyclic AKF is in the index range $32 < n \leq 64$ and valid at $n = 0$.

This representation does not yet match with the theoretical flow of the AKF (gray line in Figure 9-32, bottom). Only the symmetric representation

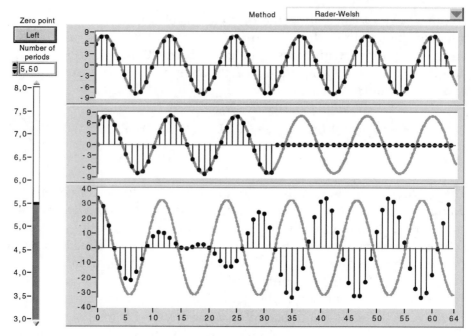

Figure 9-32
Example of signal processing by the Rader method

(zero point in the center) will match the discrete AKF with the theoretical AKF flow. The positive region can be completed by mirroring the negative region because autocorrelation functions always represent an even function. If one calculates the discrete cross-correlation function by using the Rader method, it is more useful to replace the first and last amplitude values of the pattern function by zeros. If zero sequences extend over $N/4$ values, then the correlation sum will also be weighted by a factor of $1/2$.

In this case, the valid region of the cyclic correlation function covers the shift values $-N/4 \leq n \, \delta \leq N/4$. This method is applied, for example, to locate noise sources based on two microphone signals that are built into an artificial head.

9.14   Summary

Computer-controlled measurement and analysis of signals require not only the right combination of computer hardware and software but also the appropriate use of available analytical functions. The user should not blindly rely on computer output but be able to estimate the correctness of results. This is why it is necessary to understand common analytical methods. This chapter described basics and applications from the field of FFT-based signal analysis. In particular, the basic properties of window functions were explained and window-based sampling of real signals was introduced. Further reading is suggested in the Bibliography chapter. In addition, LabVIEW builds in a series of examples that deal with different aspects of signal analysis. We recommend that you experiment with these programs. Various add-on toolkits for LabVIEW, e.g., the Filter Design Tool, the Joint Time Frequency package to analyze nonstatic signals, or the Octave Analyzer, are ideal tools for special applications.

10

Time-Frequency Analysis of Signals

I compared these various impressions with one another and found that they had this in common, namely, that I felt them as though they were occurring simultaneously in the present moment and in some distant past, which the sound of the spoon against the plate, or the unevenness of the flagstones, or the peculiar flavor of the madeleine even went so far as to make coincide with the present, leaving me uncertain in which period I was. In truth, the person within me who was at this moment enjoying this impression enjoyed in it the qualities which it possessed that were common to both an earlier day and the present moment, and this person came into play only when, by this process of identifying past with present, he could find himself in the only environment in which he could live, that is to say, entirely outside of time.

— Marcel Proust
Remembrance of Things Past

Signals with a frequency spectrum that changes over time represent a known phenomenon. Both the analysis and the evaluation of such signals pose some problems that have not been solved entirely. The first part of this chapter describes the concepts underlying the time-frequency analysis. The second part introduces LabVIEW tools for time-frequency analysis that allow the calculation of various measuring data distribution and that can be integrated in applications.

10.1   Introduction

Conventional methods applied to analyze signals that were acquired from physical measurements normally handle either the time domain *or* the frequency domain. The tool normally used to determine the frequencies contained in a time signal is the traditional Fourier transform, representing a kind of connecting link between the time signal and a physically inter-pretable spectrum. The two forms of representation are fully equivalent and contain exactly the same information. However, certain characteristics of a signal become more evident in one representation than in the other. The visualization of a signal in the time domain is particularly useful, for example, when you want to identify characteristic signal forms. For instance, if you want to detect an aperiodic anomaly in a biomedical signal, you will want to search the signal's time flow for certain typical patterns. If you want to detect periodic signals, e.g., certain sounds, you will want to pay particular attention to their frequency contents. This approach is common in practice, but it is often insufficient. How does all this link up?

10.2   What Is Time-Frequency Analysis?

Physical signals found in the real world usually have frequency contents that vary in time. Two examples will illustrate this. When we see the sky is turning gradually red while we are watching a sunset, we observe the time change of the light spectrum. Voice and music are examples of acoustic signals. It is known that a voice signal consists of a series of blended sounds, which differ in their characteristic frequencies (formants). In many tech-nological fields, the joint *time-frequency analysis* is widely used. Typical applications are voice recognition, sonar applications, acoustic quality control, and musical or biomedical signal processing. The classical frequency spectrum of such signals can merely show the characteristic frequencies that are contained in a sample, but it provides no information on the time the individual frequencies occur. This is the reason why the properties of these and many other nonstationary signals are studied by a joint analysis of the time and frequency domains.

Let us look at this concrete example. Figure 10-1 shows the analysis of a bird's voice, specifically, that of a Canadian wren. The first part represents the signal as a function of time as acquired by a recording device. This shows us the time flow of the voice's volume, but it is impossible to predict anything about the frequency content. The top right part of the figure shows the spectrum calculated by the classical Fourier transform. Although it shows the frequency content for the time signal, it contains no information on the current spectrum and its change over time. This information is supplied by the spectrogram represented in the center. It shows the subtle up and down movements of intensity and pitch that characterize the singing of this bird. Although the spectrogram is the most popular time-frequency distribution, it is only one of many. The bottom part of Figure 10-1 shows an alternative distribution. This is a *smoothed Wigner distribution* (cone-shaped distribution) of the same signal. The difference between the two distributions is striking: The time-frequency pattern of the bird's voice becomes much more evident in the cone-shaped distribution. Which method provides optimum decomposition in practical applications? Which one is better suited for real-time applications? How costly is the development or computing work? The following sections attempt to answer these questions.

10.3   Benefits and Drawbacks of Spectrograms

The main objective of a time-frequency analysis is to determine the energy density of the signal, which indicates how much energy is contained within a certain time interval in a given frequency band. Normally, one uses the classical spectrogram, which is calculated by using the *short-time Fourier transform* (*STFT*). It is based on the simple concept illustrated in Figure 10-2. A time-shifted window is used to extract a short sequence of measuring values from a lengthy measuring signal and then it is Fourier transformed.

The result is a frequency spectrum that belongs to the current window position. The spectrogram is created when shifting the window position by one or several measuring values and by repeating the Fourier transform for

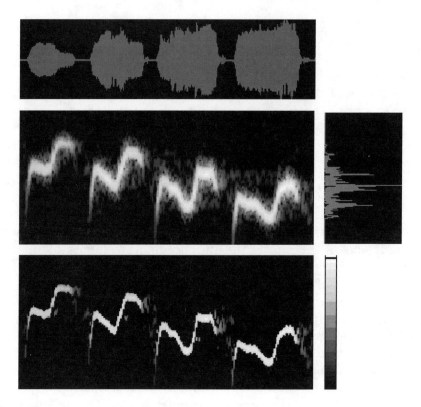

Figure 10-1

Time-frequency analysis of a wren's song. The top and right-hand parts show the time signal and the frequency spectrum, respectively, of the bird's voice. The center part shows the spectrogram, and the bottom part represents a smoothed Wigner distribution, i.e., a "cone-shaped distribution." Time is horizontal and frequency is vertical. Notice the striking difference between the two distributions.

each window position. In mathematical terms, we obtain the following definition for the short-time Fourier transform of a signal *s(t)*:

$$S(\omega) = \frac{1}{\sqrt{2\pi}} \int s(\tau)h(\tau - t)e^{-j\omega t}d\tau, \tag{1}$$

where *h(t)* is an arbitrary window function. In the center of the window is *h* ≈ 1. As the distance from the center decreases, *h* drops and approximates *h* ≈ 0 at large distance. In practice, an entire series of different window functions

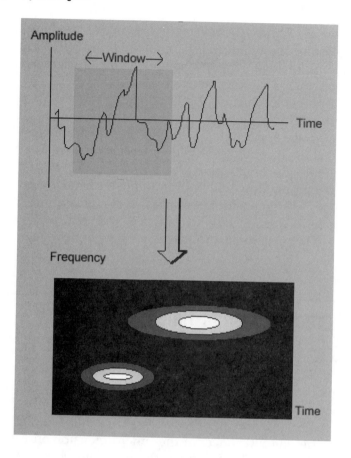

Figure 10-2

The classical spectrogram is calculated by the short-time Fourier transform. The selection of a window width has a decisive influence on time and frequency decomposition.

are used: Blackman, Hanning, Hamming, Gaussian, and others. Which one to choose depends on the application (see Chapter 9). From the Fourier transform $S(\omega)$ of the signal to time t, we obtain the energy density, i.e., the classical spectrogram by

$$P_{STFT}(t,\omega) = \left| S(\omega) \right|^2$$

(2)

The STFT method is an easy tool with which to study the properties of a signal in relation to time and frequency. In practice, the method produces good results, but it is not always suitable, as we can conclude from the example given in Figure 10-1. One of the main problems of the STFT method is selecting the window type and the window length. How is the user to decide in advance whether he should use a Hanning or Blackman window or whether a simple rectangular window will do for a specific application? Another problem is optimizing the window length. It can be easily understood that the window length has a major impact on the decomposition in the spectrogram. While a short window results in a good time decomposition, it contains only few measurement values, which means that the frequency decomposition (*broadband spectrogram*) is poor. In the reverse case, a long window will poorly define the time, but it will supply a better frequency decomposition because it contains more measurement points (*narrowband spectrogram*). In practice, one normally has to find a suitable tradeoff between time and frequency decomposition.

10.4   Characterization of Time and Frequency

How in the world does this work? A time signal *s(t)* is a function of time and time *alone*, at least in the form of representation we know. This signal is to be transformed into a function, *P(t,ω)*, of time *and* frequency, as shown in Figure 10-2. This may appear strange at first sight. How can one single variable become two all of a sudden? In other words: Which of the signal's properties should reflect in the time dimension and which ones in the frequency dimension? Of course, there are good reasons to study a signal as a function of both time and frequency. It is clear that, for a signal to form, various physical phenomena contribute. On the other hand, physical processes often depend on frequency, and this dependence on the frequency is controlled by parameters that change over time. So our motivation to study the time flow of a signal's frequency content is primarily based on our intuition and our understanding of the underlying physical processes.

Roughly, the joint time-frequency analysis separates fast processes from slow ones. Fast processes running within a window width are projected as a momentary spectrum to the frequency dimension. Slow processes, which take much longer than a window width, determine the time dependence. When STFT is used, the window width determines the position of the separating line. This is why it is so important to select an appropriate window width.

10.5   The Wigner Distribution

The spectrogram yielded from STFT is by far not the only option to implement a time-frequency distribution. The *Wigner distribution* is representative for an entire class of so-called *bilinear distributions*, which differ from the classical spectrogram in terms of quality. This distribution was originally developed by the American quantum physicist Eugene P. Wigner for quantum theory in 1932. About 15 years later, J. Ville used it for the first time to analyze signals. The developments of Wigner and Ville inspired a series of activities that continue to today. In the course of time, both benefits and drawbacks of this distribution have come to light.

The Wigner distribution is defined on the basis of a time signal, $s(t)$, as follows:

$$P_{Wigner}(t,\omega) = \frac{1}{2\pi} \int s^*(t - \frac{\tau}{2})s(t + \frac{\tau}{2})e^{-j\omega t}d\tau, \tag{3}$$

The signal part that lies on the negative side at time t is transferred to the positive side, then multiplied and Fourier-transformed. In contrast to STFT, where integration takes place within a limited window width, integration in the Wigner distribution extends over the entire signal duration. This means that each point in the Wigner distribution depends on the global signal. This phenomenon, which is also referred to as *nonlocality* in the literature, has the contradictory consequence that the distribution at time t can reflect such properties of the signal as are far away from the studied point in time. For example, if we study a signal that is very noisy in a certain time interval, the Wigner distribution can also contain noise in time intervals where there is no noise in the signal itself.

Apart from this contradiction, the Wigner distribution offers a series of important benefits. First, it is capable of sharply mapping a pure sine oscillation. In the spectrogram, a pure sine appears as a smeared peak. The reason for this behavior is that the short-time Fourier transform modifies the original signal in line with the window function, so that the spectrogram reflects the properties of the modified signal, not that of the original signal. On the other hand, the Wigner distribution can do without such a window function, which means that the problem of selecting an optimum window type and window width does not arise. Figure 10-3 compares a spectrogram to a Wigner distribution for a chirp, i.e., a signal with a momentary frequency that increases or decreases linearly with time. Notice the dramatic

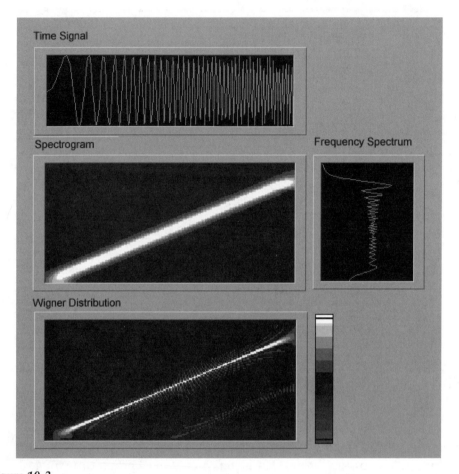

Figure 10-3

Comparing the spectrogram to the Wigner distribution of a chirp signal. The momentary frequency can be read clearly from the Wigner distribution, but not from the spectrogram.

improvement of the decomposition: The momentary frequency of the chirp signal can be determined exactly from the Wigner distribution, but not from the spectrogram.

Another nice property of the Wigner distribution concerns the projection of two-dimensional distributions to the time and the frequency axes. It should be reasonably expected that the projection to the frequency axis is identical to the Fourier spectrum. In fact, this is a fundamental requirement with important consequences. Notice that this requirement is met by the Wigner distribution and its smoothed variants, but not by the spectrogram!

Its behavior is entirely different when analyzing a signal that is composed of two additive components:

$$s(t) = s_1(t) + s_2(t) \tag{4a}$$

If we subject this sum to the Wigner distribution definition, we obtain four terms:

$$P_{Wigner}(t,\omega) = P_{11}(t,\omega) + P_{22}(t,\omega) + P_{12}(t,\omega) + P_{21}(t,\omega) \tag{4b}$$

Each of these terms has the following form:

$$P_{ik}(t,\omega) = \frac{1}{2\pi} \int s_i^*(t - \frac{\tau}{2}) s_k(t + \frac{\tau}{2}) e^{-j\omega t} d\tau, \tag{4c}$$

In addition to the two main terms P_{11} and P_{22}, two interference terms P_{12} and P_{21} appear. The general rule says that the Wigner distribution of a signal composed of several components does not equal the sum of the Wigner distributions of the individual components. Each component pair produces additional interference terms.

Figure 10-4 (center) illustrates this strange property, showing the Wigner distribution of a sum drawn from two Gaussian-modulated oscillations. The interference term lies between the two components. It has both positive and negative fields and — worse — it oscillates up and down wildly. Figure 10-5 shows a closeup of these oscillations. How can such interferences be interpreted? Obviously, we cannot easily interpret them as energy density because negative values and oscillations do not make any physical sense. Nevertheless, this distribution is usable for practical purposes! One obvious concept is to smooth the Wigner distribution before drawing any physical conclusion. This approach makes most of the interference terms with their quick oscillations (see Figure 10-4) disappear, but it leaves the main terms virtually unchanged, so that it is acceptable to interpret the smoothed Wigner distribution as energy density.

10.6   Cohen's Class

In 1966, Leon Cohen developed a general method to generate an entire class of smoothed time-frequency distributions. He intended to construct distri-

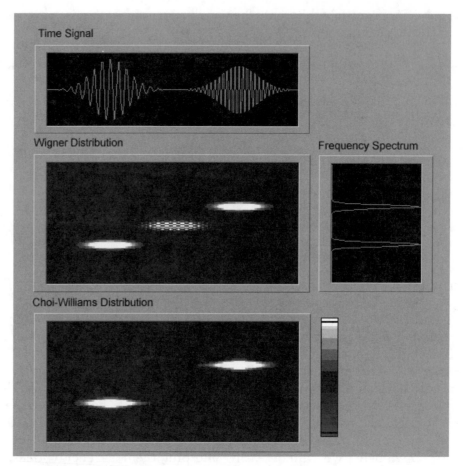

Figure 10-4

Comparison of the Wigner distribution with the Choi-Williams distribution of a signal consisting of two components. The Wigner distribution shows interferences that oscillate quickly between negative and positive values. In the Choi-Williams distribution, a smoothed variant of the Wigner distribution, such interferences disappear almost entirely.

butions that were to have all the good properties from the Wigner distribution but would concurrently be positive over the entire domain so that they could be interpreted as true energy density. Cohen's approach is based on a convolution of the integrand in the Wigner distribution with a complete kernel function. This convolution smooths the distribution, that is, it removes strange and physically meaningless oscillations. Depending on what kernel function you select, you obtain another variant from *Cohen's*

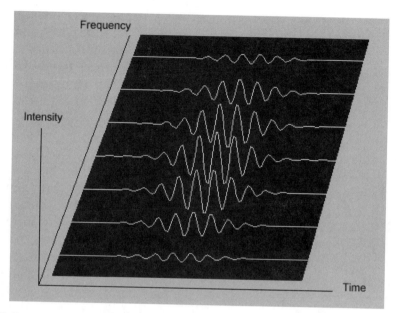

Figure 10-5

Closeup of the interference term from the Wigner distribution shown in Figure 10-4

class of distributions. Of course, the kernel has to fulfill a series of require-ments if the resulting distribution is to have all desired properties. However, this limitation leaves enough margin. In the simplest case, if you select the kernel function to be equal *1*, then you obtain the old Wigner distribution. Also, the spectrogram turns out to be an interesting special case from Cohen's general class. Excellent further reading is Leon Cohen's book, which discusses the concepts and mathematics underlying this method, illustrated with many examples.

During the past years, those time-frequency distributions that are signifi-cant for practical application have excelled because of the wide range of feasible and admissible time-frequency distributions. One representative example is the *Choi-Williams distribution*. Its mathematical formulation is:

$$P_{Choi-Williams}(t,\omega) = \frac{1}{4\pi^{3/2}} \int\int \frac{\sqrt{\sigma}}{\tau} e^{-\left(\frac{\sigma(u-t)^2}{4\tau^2}+j\tau\omega\right)} s_i^*\left(t-\frac{\tau}{2}\right)s_k\left(t+\frac{\tau}{2}\right)e^{-j\omega t}d\tau \quad (5)$$

Although this expression is not particularly clear, one can see that it weights the Wigner integrand with a Gaussian distribution. This helps achieve the desired smoothing effect. Weighting depends on a parameter, *s*. For an infinitely large *s*, the Choi-Williams distribution turns into a Wigner distribution, i.e., there is no smoothing. For finite values of *s*, you obtain the smoothing effect, which becomes stronger as you select a smaller *s*. This means that this parameter determines by how much the Choi-Williams distribution will deviate from the Wigner distribution.

The bottom part of Figure 10-4 shows the Choi-Williams distribution for a signal composed of two Gaussian-modulated oscillations. The interference term, which gave rise to quick oscillations with high amplitude in the Wigner distribution, is totally gone in this example. In general, the Choi-Williams distribution and other distributions from Cohen's class can still have negative fields. However, the interference terms are distributed over a certain surface so that they no longer oscillate. Notice that the main terms remain almost totally unaffected by the smoothing process, and all other good properties of the Wigner distribution are maintained. In fact, the two components in our sample signal appear as sharp in the Choi-Williams distribution as in the Wigner distribution. But this improvement has its cost: The calculation of the dual integral for each single point on the time-frequency level requires massive computing power, which means that real-time application is most likely excluded.

10.7   Time-Frequency Analysis in LabVIEW

Signals with a frequency spectrum that fluctuates over time are the rule, not an exception, in routine laboratory applications. The previous sections described several methods that can be applied to yield time-frequency distributions from such measurement data. Of course, the joint time-frequency analysis can be applied in practical applications as long as there are fast algorithms to run calculations from discrete measurements. In recent years, good work has been performed in this specific field. For example, the *JTFA Toolkit* (*Joint Time-Frequency Analysis*) for LabVIEW can be used to calculate various time-frequency distributions and to integrate them into more complex applications at little cost. The JTFA Toolkit is part of the *Signal Processing Suite*, a comprehensive package for signal processing applications.

10.8   Selecting a Spectrogram

Table 10-1 gives an overview of the distributions contained in the JTFA Toolkit, including some important properties. The table adds the *Gabor spectrogram* and the *adaptive distribution* to the distribution types discussed in the previous sections. In addition, most distribution types allow you to vary parameters, for instance, the degree of smoothing in Choi-Williams and cone-shaped distributions. Some distributions can also be calculated from the analytical signal. (Remember: The complex analytical signal is derived from the measured real signal in such a way that its spectrum for positive frequencies is identical to that of the real signal but equals zero for negative frequencies.) This manipulation of measurement data makes certain interferences disappear, but it may lead to distortions. All these options add up to a confusing multiplicity, making it more difficult to determine the optimum distribution type for a specific application. So it is all the more welcome to know that you can use the JTFA Toolkit to calculate all kinds of variants interactively and compare them.

In short, the Gabor spectrogram is created by developing a one-dimensional time signal within a basis composed of two-dimensional time and frequency functions. The time-frequency level is singled out by means of a grid. The basic functions (sine oscillations with Gaussian envelopes) "sit" inside the grid cells, and their contribution is limited to the near neighborhood of the respective grid cell. The coefficients of this development are then a measure for the intensity of the original time signal in the corresponding time-frequency cells. This shows that the Gabor spectrogram interpolates between the classical STFT spectrogram and the Wigner distribution. Depending on the spectrogram's order, approximation is to one or the other extreme. Thus, the Gabor spectrogram offers a better decomposition than the STFT spectrogram, but it contains interferences, although they are less significant than in the Wigner distribution. In addition, by being able to select the spectrogram order, you have a certain control over the extent of interferences.

The adaptive distribution is also based on developing basic functions. In contrast to the Gabor spectrogram, the width of the Gaussian envelopes and the grid cells are not defined in advance. A more sophisticated algorithm determines these parameters to ensure optimum results for a given time signal. Practical tests have shown that the algorithm works best for quasi-stationary signals, the spectrum of which changes only slowly over time. The adaptive distribution is always positive, and it offers the best decomposition of all distributions included in the JTFA Toolkit.

Table 10-1   *Time-frequency distributions offered by the JTFA Toolkit for LabVIEW*

Distribution	Method	Parameter	Interference	Decomposition	Computing Cost
STFT spectro-gram	Local window-based Fourier transform	Window type, window width	none	Depending on window; good time *or* good frequency decomposition	low
Wigner-Ville	Global bilinear transform-ation	Analytical signal?	strong	Good *joint* time-frequency decomposition	average
Choi-Williams	Smoothed Wigner-Ville distri-bution	Smoothing degree; analytical signal?	less	Good *joint* time-frequency decomposition	massive
Cone-shaped	Smoothed Wigner-Ville distri-bution	Smoothing degree; analytical signal?	less	Good *joint* time-frequency decomposition	massive
Gabor	Develop-ment in localized basic functions	Order	minor	Better than spectrogram	average
Adaptive	Adapts basic func-tions auto-matically	none	none	Best *joint* time-frequency decomposition	average

10.9 Time-Frequency Analyzer

Among the many tools included in the JTFA Toolkit is an executable *time-frequency analyzer*. This is a stand-alone application to analyze data stored on disk. All distributions listed in Table 10-1 can be calculated and represented as two-dimensional intensity graphs, including color coding. In addition, you can output the analyzed time signal and optionally the global or the momentary power spectrum. The resulting distributions can be stored in a text file. Figures 10-6 and 10-7 show a few examples. Figure 10-6 shows the front panel of the time-frequency analyzer and, as an example, the analysis of a short section from a longer EEG measurement by means of the Choi-Williams distribution. Figure 10-7 compares the STFT spectrogram (center) of a seismic signal with the cone-shaped distribution (bottom). The latter appears to be able to decompose in much finer details. Are these real, or could they be artifacts?

The time-frequency analyzer proves to be extremely useful for quick comparison of various analytical methods and for determination of the relevant parameters. Again, there is no absolute "best" method for all cases. Which method you select depends on the information you want to extract from a signal. Before you choose a specific method, there is a series of questions you need to answer, such as: Do you need to calculate global or local mean values? Do you need to localize peaks? How quickly does the spectrum change? Are time decomposition and frequency decomposition equally important in your application? Another important factor when selecting an analytical method is the computing cost. While the Gabor distribution can be calculated on a powerful workstation in real time, smoothed Wigner distributions, i.e., Choi-Williams and cone-shaped, are not suitable for real-time application unless you have a supercomputer.

In addition, the JTFA Toolkit contains a small library of subVIs for time-frequency analysis. "Small" refers only to the number of VIs it contains, but not to its performance: You will find exactly one VI for each of the six time-frequency distributions listed in Table 10-1.

These VIs require surprisingly few links. In fact, you need to set only a few parameters before you input the data array to be analyzed. This easy and intuitive operation is one of LabVIEW's strengths. You do not have to bother with complex algorithms; instead, you fully concentrate on evaluating the resulting distributions. The complexity of numerical calculations remains hidden. Figure 10-8 illustrates how easy it can be to program in LabVIEW.

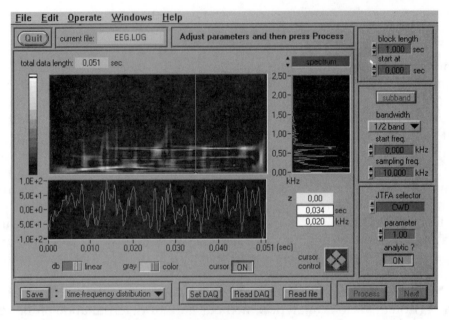

Figure 10-6

Front panel of the time-frequency analyzer. It shows the analysis of a short section from a longer EEG measurement by means of the Choi-Williams distribution.

This mini-application reads a data array from disk, calculates an STFT spectrogram, generates an intensity graph, and stores the results on disk. The entire process requires exactly three subVIs! Calculations using the other distributions listed in Table 10-1 are equally straightforward.

The Toolkit contains an additional VI for the professional programmer to calculate arbitrary distributions from Cohen's class. This class includes the Wigner distribution and its smoothed variants, e.g., Choi-Williams and cone-shaped distributions. The *Cohen's Class* VI can be used to calculate further variants. To do this, you need to specify the kernel function, which determines the smoothing type, in the form of a two-dimensional array with function values. Of course, this requires a sound knowledge of the underlying theory.

10.10   Real-World Applications

Acoustic signals are particularly suitable for time-frequency analyses. Potential applications include music editing, acoustic quality control, vibra-

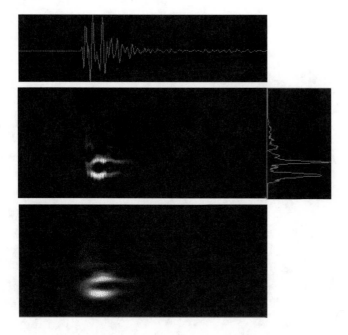

Figure 10-7

Analysis of a seismic signal by means of the STFT spectrogram (center) and the cone-shaped distribution (bottom), respectively. The top and right sections show the time signal and the power spectrum, respectively. Time is horizontal; frequency is vertical.

tion analysis, sonar identification of objects, and voice recognition. A spoken word can be identified in the time-frequency representation with a relatively high reliability. We know that spoken language is a time sequence of voiced and unvoiced sounds. Each voiced sound manifests itself in the frequency spectrum by characteristic peaks, i.e., formants, which correspond to the combined resonance of throat, mouth, and nasal cavity. As form and size of the resonance spaces change constantly during speech, a complex pattern forms in the time-frequency space, which serves as a key for recognition.

In industrial quality assurance, ultrasound is used for nondestructive identification of hidden material flaws. A test system consists basically of a pulse generator/receiver, an activator, a sensor, and a time-frequency analyzer. The pulse generator supplies the electric impulses. These are subsequently taken by the activator and converted into mechanical impulses in the ultrasonic domain, and then linked to the material sample.

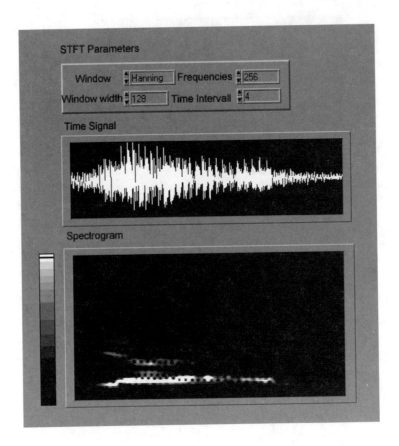

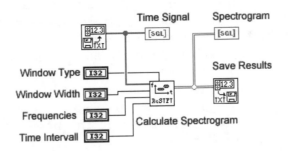

Figure 10-8

Programming a spectrogram in LabVIEW. The top section shows the front panel, and the bottom section shows the block diagram of a simplified analyzer.

Nonhomogeneity in the material, particularly structural faults that are of interest in this case, cause echoes. The sensor detects these echoes and reconverts them into electric signals. Subsequently, these electric signals are amplified and analyzed. By nature, echo signals are nonstationary, and it has been a long-known problem that time signals or frequency spectra themselves are not sensitive enough to respond to structural flaws one wants to find. A group of researchers at the University of Massachusetts in Dartmouth use the Gabor spectrogram as a more sensitive indicator. They estimate that this approach has a great potential for nondestructive material testing.

Interesting applications are found not only in technical fields but also in commercial fields. Economic data is normally treated as noisy time signals and analyzed with the help of statistical models. Although the models supply indicators such as mean values, variances, and other parameters, they do not provide information on such regularities as may be present in time-slot patterns. Practical experience has proven that business cycles exist, although they do not always last equally long. A group of researchers at the University of Texas in Austin uses time-frequency distributions to analyze economic indicators, such as the S&P 500 stock exchange index, or the unemployment index, over very long periods of time. They use this tool not only to determine the cycles within various time-slot patterns but also to find out when they occur and how long they last. For example, when developing the S&P 500 stock exchange index, they determined indications of the existence of three cycles at different time patterns.

10.11   Summary

Signal processing is generally a demanding subject, so that it doesn't hurt to have some basic knowledge. Some of the software tools included in the Signal Processing Suite are based on complex algorithms. Fortunately, their use is straightforward. You can run quick comparisons to find the most suitable among the available methods. You can easily experiment with various options and optimize parameters.

In summary, the Signal Processing Suite is a software package that gives users ready-to-run signal processing capabilities and developers high-level digital signal processing tools and utilities for a wide range of applications in the technical, commercial, and educational fields.

11

Designing Digital Filters in LabVIEW

To see a world in a grain of sand
And a heaven in a wild flower,
Hold infinity in the palm of your hand
And eternity in an hour.

— William Blake
Auguries of Innocence

The importance of digital filters is well established. Digital filters, and more generally digital signal processing algorithms, are classified as discrete-time systems. They are commonly implemented on a general-purpose computer or on a dedicated digital signal processing (DSP) chip. Due to their well-known advantages, many classical, analog filters are more and more being replaced by their digital counterparts. In this chapter, we introduce a new digital filter design and analysis tool implemented in LabVIEW. The tool allows developers to graphically design classical IIR and FIR filters, interactively review filter responses, and save filter coefficients. In addition, a real-world filter testing can be performed within the digital filter design application by means of a plug-in data acquisition board.

11.1 Digital Filter Design Process

Digital filters are used in a wide variety of signal processing applications such as spectrum analysis, digital image processing, pattern recognition, etc. Digital filters eliminate a number of problems associated with their classical, analog counterparts and thus are preferably used in place of analog filters. Digital filters belong to the class of discrete-time LTI (linear time invariant) systems, which are characterized by the properties of causality, recursibility, and stability and may be characterized in the time domain by their unit-impulse response and in the transform domain by their transfer function. Obviously, the unit-impulse response sequence of a causal LTI system could be of either finite or infinite duration, and this property determines their classification into either a finite impulse response (FIR) or an infinite impulse response (IIR) type of system.

To illustrate this, we consider the most general case of a discrete-time LTI system with the input sequence denoted by $x(kT)$ and the resulting output sequence $y(kT)$. As can be seen from (1), if for at least one v, $a_v \neq 0$, the corresponding system is recursive; its impulse response is of infinite duration (IIR system). If $a_\mu = 0$, the corresponding system is nonrecursive (FIR system); its impulse response is of finite duration, and the transfer function $H(z)$ is a polynomial in z^{-1}. Commonly, b_μ is called the μth forward filter coefficient, and a_v, the vth feedback or reverse filter coefficient. For a detailed discussion, refer to standard signal processing textbooks.

$$y(kT) = \sum_{\mu=0}^{m} b_\mu x(kT-\mu T) - \sum_{v=1}^{n} a_v y(kT-vT) \Leftrightarrow H(z) = \frac{Y(z)}{X(z)} = \frac{\sum_{\mu=0}^{m} b_\mu z^{-\mu}}{1 + \sum_{v=1}^{n} a_v z^{-v}} \qquad (1)$$

The design of digital filters involves the following basic steps:

1. Determine the desired response. The desired response is normally specified in the frequency domain in terms of the desired magnitude response and/or the desired phase response. Select a class of filters (e.g., linear-phase FIR filters or IIR filters) to approximate the desired response.

2. Select the best member in the filter class.

3. Implement the best filter, using a general-purpose computer, a DSP, or a custom hardware chip.

4. Analyze the filter performance to determine whether the filter satisfies all the given criteria.

11.2 Digital Filter Design Application

The Digital Filter Design (DFD) application is a complete filter design and analysis tool implemented in LabVIEW. It can be used to design digital filters to meet required filter specifications. The toolkit allows users to graphically design IIR and FIR filters, interactively review filter responses, save filter design work, and load design work from previous sessions. In addition, DFD DLLs can be accessed from other Windows applications, or other applications can load the filter coefficient files directly. Lastly, real-world filter testing can be performed with the DFD application, using a standard National Instruments data acquisition device. The time waveforms or the spectra of both the input signal and the filtered output signal can be viewed, showing how the present filter performs on real-world signals.

❑ 11.2.1 Classical IIR Filter Design

The Classical IIR Filter Design (Figure 11-1) panel is used to design classical IIR digital filters. These filters include the classic types such as lowpass, highpass, bandpass, and bandstop and the classic designs such as Butterworth, Chebyshev, Inverse Chebyshev, and Elliptic. The IIR Design panel consists of a graphical interface with the Magnitude vs. Frequency cursors and plot on the left side and a text-based interface with digital controls on the right side. Classical IIR filters can be designed by adjusting the filter specifications on the panel. The bandpass and bandstop requirements can be defined by using either the text entry or the cursors in the Magnitude versus Frequency graph. As the cursors are moved, the text entries are updated accordingly.

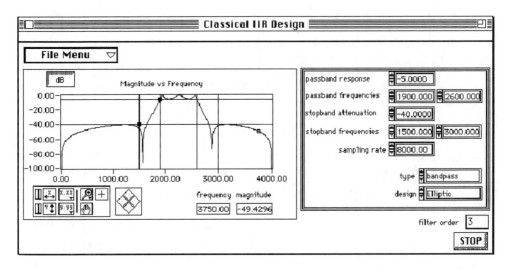

Figure 11-1

Classical IIR design panel

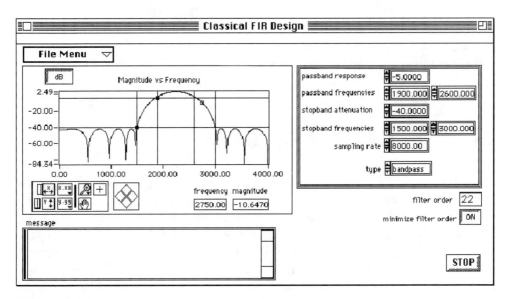

Figure 11-2

Classical FIR design panel

❑ 11.2.2 Classical FIR Filter Design

The Classical FIR Design panel (Figure 11-2) is used to design FIR filters. These filters include the classic types such as lowpass, highpass, bandpass, and bandstop and employ the Parks-McLellan equiripple FIR filter design algorithm. This panel is very similar to the classical IIR Design panel and operates in much the same way. The panel includes a graphical interface with the Magnitude vs. Frequency cursors and plot on the left side and a text-based interface with digital controls on the right side. Classical FIR filters can be designed by adjusting the desired filter specifications. The desired bandpass and bandstop requirements define a filter specification. The filter requirements are defined by using either text entry or the cursors in the Magnitude versus Frequency graph. As the cursors are moved, the text entries are updated accordingly.

❑ 11.2.3 Pole-Zero Placement Filter Design

Figure 11-3 shows the Pole-Zero Placement filter design panel. The panel includes a graphical interface with the z-plane pole and zero cursors on the left side and a text-based interface with digital controls on the right side. This panel can be used to design IIR digital filters by manipulating the filter poles and zeros in the z plane. The poles and zeros initially may have come from classical IIR designs. This panel can also be used to move or delete existing poles and zeros directly on the z-plane plot, allowing an accurate control of their important characteristics. One may describe the poles and zeros by using either the text entry or the cursors in the z-plane plot. As the cursors are changed, the text entries update automatically. Likewise, as the text entries are modified, the pole/zero cursors update automatically. The pole and zero locations in the z-plane, the characteristics of each pole and zero, the gain, and the sampling rate fully describe pole-zero filter designs. Any change in these parameters corresponds to a change in the filter coefficients. The DFD application matches the poles and zeros and creates stable second-order stages for IIR filter coefficients. It then uses these coefficients to compute the filter magnitude response. The Magnitude versus Frequency plot updates automatically whenever the poles or zeros are changed, giving an immediate graphical feedback to the user's pole-zero filter designs.

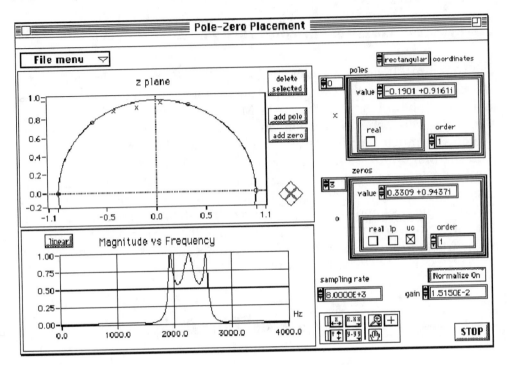

Figure 11-3

Pole-zero placement panel

11.3   A Design Example

As an example, we consider the design of an IIR lowpass filter with the following specifications: minimum bandpass magnitude response of 0.9 at frequencies at and below 400 Hz, and a maximum allowed bandstop magnitude response of 0.1 at and above 500 Hz, with the sampling rate at 2000 Hz. With the DFD application, these values are simply entered into the text entries in the classical IIR design panel, as shown in Figure 11-4.

We have selected a Butterworth lowpass filter. The filter order and internal design frequencies are computed, and the magnitude response graph is automatically updated (Figure 11-5).

We can now select other designs and get an immediate feel for the required filter order as well as the actual filter shape. With the above specifications, the Butterworth filter requires tenth order, the Chebyshev and

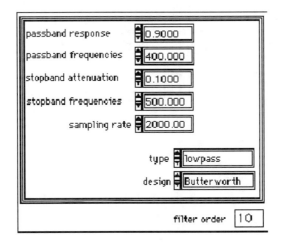

Figure 11-4

Filter specification

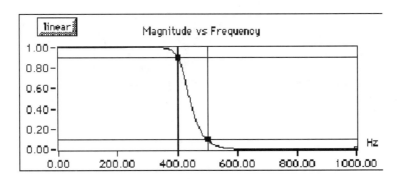

Figure 11-5

Magnitude response

Inverse Chebyshev require fifth order, and the Elliptic filter requires only a third-order IIR filter.

Figure 11-6 is the plot of the third-order Elliptic filter that meets our filter specifications and its pole-zero plot (Figure 11-7).

We can save the coefficients to a text file (see Table 11-1) containing all necessary information for implementing the designed digital filter.

The corresponding transfer function H(z) is given by:

$$H(z) = \frac{0,1888360(1+z^{-1})(1+0,1137z^{-1}+z^{-2})}{(1-0,3504z^{-1})(1-0,4992z^{-1}+0,7281z^{-2})} \tag{2}$$

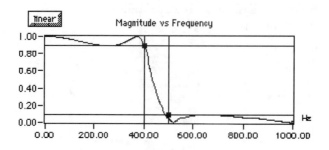

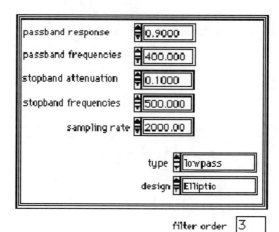

Figure 11-6

Third-order Elliptic filter

Table 11-1   *IIR filter cluster*

Coefficient	Value
Sampling rate	2 kHz
Stage order	2
Number of stages	2
4a coefficients	-3.503960E-1; 0; -4.991879E-1; 7.280681E-1
6b coefficients	1.888360E-1; 1.888360E-1; 0; 1; 1.137000E-1; 1

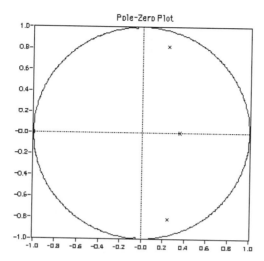

Figure 11-7

Pole-zero plot of the third-order Elliptic filter

In this example, the IIR filter is composed of two second-order stages. Because the underlying filter is really only third order, the second *a* coefficient and the third *b* coefficient are identically zero and can be ignored for efficient implementation. To see how the present filter design performs on real-world signals, you can set up a National Instruments data acquisition device within the DFD environment to acquire real signals. The acquired data then passes through the currently designed filter, and the DFD application plots the input and output waveforms and spectra.

11.4   Summary

Digital filter design and implementation consist of several interacting steps and call for efficient design and simulation tools. To facilitate this, we presented a new graphical filter design and analysis tool, called Digital Filter Design application, developed in LabVIEW. It was shown that a graphical user interface greatly eases the burden associated with the specification-design-test cycle at the heart of digital filter design. The result of repeated interactive graphical design sessions is that the designer can acquire a feel for how design parameters affect filter performance.

12

Image Processing in LabVIEW

Yesterday evening, I was busy painting a rather sloping ground in the wood, covered with moldered and dry beach leaves. You cannot imagine any carpet so splendid as the deep brownish-red, in the glow of an autumn sun, tempered by the trees. The question was — and I found it very difficult — how to get depth of color the enormous force and solidness of that ground. And while painting it I perceived for the first time how much light there was in that darkness, how was one to keep that light and at the same time retain the glow and depth of that rich color.

— Vincent van Gogh
Diaries

In view of dropping prices, increasing computing power, and rapid developments in semiconductor technology, industrial image processing has been experiencing a true upturn. This chapter describes the basics of PC-based image processing and how LabVIEW can be used to integrate image processing into the uniform concept of graphical and easy-to-use virtual instrumentation.

12.1   Introduction

Recent market studies showed that industrial image processing, as a cross-section technology for various industries, has been booming during the past years. Experts believe that this swift trend will continue in the future. Hardly any other sector in industrial measurement technology can come up with similar growth rates. Image processing can be used in many fields of application, from car manufacturing, electrical engineering, to food and tobacco processing industries and medical technologies. In the age of quality management, its enormous potential lies mainly in the acquisition of quality parameters that had been acquired manually. The popularity of PC-based image processing systems increases particularly in view of a low-priced and powerful combination of hardware and software components. The main benefit of PC-based systems is that they offer a wide choice of configuration and expansion options. This also facilitates the transition from image processing systems to virtual instrumentation. In particular, state-of-the-art development environments like LabVIEW and BridgeVIEW are not only tailored to interests pertaining to the measurement and control technology, but they also offer image processing options so that all you need to do basically to get started is to install a suitable image acquisition card in your computer.

12.2   Setting Up Your Image Processing Equipment

Analogous to the concept of virtual instrumentation, digital image processing can also be divided into three basic blocks: image acquisition, image processing, and image analysis. A generalized structure of a PC-based image processing system includes the following components:

- Capturing system, consisting of camera, lens, and lighting equipment

- Image acquisition card

- Personal computer and suitable image processing software

For each element, you will find a wide range of products from different vendors. The products you eventually select depend on your intended application. For example, camera, lens, and lighting equipment has to be suitable for the information you want to define. An image acquisition card has to be equipped with an appropriate interface to connect a camera. It should offer sufficient image acquisition frequency and data transfer rate to handle high-volume data. The underlying computer has to be sufficiently powerful. The performance of a computer is determined by its processor, data bus, and operating system. Finally, you have to install acquisition and processing software, which should support the required functionality, performance, and scalability to integrate your image processing system into existing and future virtual instrument applications. All these components are described in the following sections.

12.3   Camera and Lens

Each field of application uses an appropriate type of camera. Normally, *CCD cameras* are used for PC-based image acquisition because as they use a field composed of light-sensitive semiconductor sensors to capture images. Instead of acquiring the image one field at a time (two fields make up a frame) and then interlacing the fields for display, the CCD array in a progressive scan camera acquires the entire scene (one frame) at once. In this interlacing process, lines are synchronized with the HSYNC signal, and frames are synchronized with the VSYNC signal.

Linescan cameras are often used in industrial applications to make simple height and width measurements. Unlike an area scan camera, a linescan camera acquires an image that is a few pixels or one pixel wide, a *line image*. Linescan cameras use a single line of CCD sensors instead of a rectangular array. The camera is focused on a narrow line as a part moves past; it acquires lines at a very fast rate.

Digital cameras offer high-speed image output. Some digital cameras can output data at a rate greater than 100 Mbytes/s.

For most applications it is easier to configure and acquire images from a standard video camera. The standard video formats are EIA RS-170, CCIR for monochrome video, and NTSC (National Television Systems Committee) and PAL (Phase Alternate Line) for color video. For example, camcorders most often output color composite videos in NTSC or PAL formats which combine the luminance (brightness) and chrominance (color) components into a single analog signal.

Nonstandard cameras often use specialized video and a large image size to improve performance for a specific application.

12.4 Lighting Conditions

Selecting appropriate lighting equipment is of primary importance in image processing applications because it simplifies image acquisition and subsequent processing. You may have to select different lighting equipment, depending on the optical information you intend to evaluate from an object.

There are several reasons for arranging lighting for your image processing. First, it is important to provide sufficient illumination for the camera. Even the best camera makes a sharper image in better lighting conditions. More light also provides greater depth of field, that is, the more light you provide, the more of the foreground and background of the picture will be in focus.

There is a variety of lighting equipment to accomplish those purposes. There are spotlights to direct light on a specific area and floodlights to provide broad, even illumination. Reflectors are used to counter intense light coming from a single direction.

For example, to capture a flat image, you normally use indirect lighting because a direct light beam could cause undesirable reflections on the object, which could cause problems when you analyze the image. To prevent this effect, you can install a diffuser between the light source and the object.

Representing edges and structures requires direct lighting, which means that bundled light is radiated onto the object from one direction. This radiation causes shadows on edges and elevations, which improves the detection of structures.

Background lighting is used to make contours of objects visible. For this purpose, a light source is used underneath the object. Notice that it may be difficult to install such a light source, for example, when objects on conveyor belts have to be captured.

Moving objects can lead to blurred images. To "freeze" a moving object while it is captured, stroboscope lamps are normally used. These lamps radiate briefly (a few milliseconds). During this short time, the object moves only slightly so that the image is not blurred. However, this type of timing light is disturbing for people, so it should be shielded. In addition, the light impulse and the image acquisition have to be synchronized exactly.

In addition to a series of factors that have to be observed during the installation, there are factors such as environmental light sources and the object itself. A good source of reference for lighting plans is the Lighting Advisor of the University of Wales, Cardiff, United Kingdom. You can consult the Lighting Advisor at the University's web site at www.cm.cf.ac.uk.

12.5   Image Acquisition Cards and Drivers

Image acquisition cards are among the most important components of a PC-based image processing system. The electric signal yielded by the acquisition system is digitized in the image acquisition card — the *frame grabber* — and can then be processed by the computer.

NI-IMAQ is a complete, full-function IMAQ driver that includes capabilities for IMAQ I/O, buffer and data management, and resource management. The IMAQ hardware and NI-IMAQ driver software from National Instruments are designed to acquire data from progressive scan cameras, linescan cameras, digital cameras, and nonstandard cameras.

NI-IMAQ controls the National Instruments scatter-gather *MITE DMA* (*Direct Memory Access*) chip for interlacing video lines and copying image data to PC memory instead of the CPU. The MITE DMA chip is used for fast data transfer directly to noncontiguous memory blocks. This means that there is no need to lock down a large block of memory (i.e., 20 Mbytes) in the driver for image acquisition. In addition, the MITE DMA controller works in parallel with the host processor. The MITE chip moves the image to PC memory and handles the interlacing while the host (i.e., MMX Pentium or PowerPC) is performing image processing calculations in parallel.

The IMAQ boards are designed to work with the National Instruments DAQ products (see Chapter 3). The IMAQ boards have the RTSI bus, which shares timing signals among other IMAQ and DAQ boards. For the PCI series boards, the RTSI bus sits on the top of the board, and a ribbon cable is used to send triggering and timing information from one board to another. For the PXI-1408, the RTSI bus is part of the PXI backplane. A DAQ board can trigger the IMAQ PCI-1408 to begin acquiring data and synchronize image capture with DAQ board acquisitions. You can use this feature to correlate images with transducer data, for direct comparison of incoming data acquisition measurements and images.

The IMAQ PCI-1424 and NI-IMAQ driver software are designed to acquire images from digital cameras. The board and driver software are designed for fast digital image acquisition, large images, and high resolution. The PCI-1424 board has a 50 MHz pixel clock and can acquire images from a digital camera at a data rate of 200 Mbytes/s. With NI-IMAQ control of the PCI-1424's onboard memory buffer, you can acquire an image to one memory buffer while offloading another buffer to PC memory. This double-buffer image acquisition technique delivers sustained high-speed throughput and greater overall system performance.

In machine vision applications that require control, you may need a digital I/O signal to control a solid state relay or solenoid, for example, if your application is an inspection on a conveyor belt. Once you image the object, if a flaw is detected, then a digital control line can be used to control a solid state relay in order to remove the object from the assembly line. All the IMAQ series boards have dedicated digital I/O lines. Each line can be programmed to be either an input or an output.

12.6   Image Acquisition and Configuration Software

NI-IMAQ driver software is included with the National Instruments IMAQ hardware products. NI-IMAQ is an extensive library of functions that you can call from your application programming environment. These functions include routines for video configuration, image acquisition (continuous and single shot), memory buffer allocation, trigger control, and board configuration. NI-IMAQ performs all functionality necessary to acquire and save images.

NI-IMAQ has both high-level and low-level functions for maximum flexibility and performance. High-level functions include single-shot and continuous-mode image acquisition. Low-level functions include imaging sequence setup.

NI-IMAQ internally resolves many of the complex issues between the computer and IMAQ hardware, such as programming interrupts and DMA controllers. NI-IMAQ provides the interface path between LabVIEW, LabWindows/CVI, or a conventional programming environment and the hardware product. It is available on the Windows 95/NT and MacOS operating systems.

Overall, you can easily configure standard cameras using the NI-IMAQ configuration utility. You can use the NI-IMAQ configuration utility to set up the IMAQ hardware to acquire gray-scale images from a color video signal (NTSC or PAL). In hardware, an antichrominance filter is applied that removes the color information. When this filter is selected, the incoming color video is translated to an 8-bit gray scale.

Also, you can use NI-IMAQ's *StillColor* to acquire color images. StillColor eliminates cross-color and cross-luminance distortion, resulting in a clearer and higher-quality color image. You should use StillColor in applications where there is little or no motion in the image field because multiple frame acquisition will cause blurring in the image.

Using the NI-IMAQ configuration utility, you can easily configure the hardware to acquire images that are not the standard sizes or to work with progressive scan cameras. Using the IMAQ-1408 hardware, you can acquire up to one million pixels in one image. You can configure the IMAQ-1408 to acquire video from nonstandard cameras that have large image sizes, transmit noninterlaced images (progressive scan), and use cameras that transmit images with a variable frame rate (variable scan).

You also use the NI-IMAQ configuration utility to easily configure acquisition from a digital camera. With NI-IMAQ, you can acquire 4 channels of 8-bit data, or 2 channels of 10-bit, 14-bit or 16-bit data, or 1 channel of 32-bit data. In addition, two IMAQ-1424 boards can be used together to acquire 1 channel of 64-bit data.

You can configure standard and nonstandard video capture easily with the NI-IMAQ configuration utility. This utility lets you set up the camera type (RS-170, CCIR, NTSC, or PAL), region of interest, autoexposure, and antichrominance filter interactively. You can use this tool to set up acquisition from noninterlaced progressive scan cameras and to create your own camera configurations for nonstandard video by setting the VSYNC, HSYNC, and other timing information. The external lock feature is used to set variable scan acquisition for microscopes and other sources that generate their own PCLK, HSYNC, and VSYNC. NI-IMAQ saves you time because these settings are automatically used in your application development environment. Use the utility to snap and grab images and then save the images to file. And, calculate the histogram to verify the contrast, dynamic range, and lighting conditions.

12.7   IMAQ Vision Image Processing Software

IMAQ hardware and its driver software are designed to work with *IMAQ Vision*. IMAQ Vision software adds high-level machine vision and image processing to LabVIEW, BridgeVIEW, LabWindows/VCI, Component-Works, and ActiveX containers. IMAQ Vision includes an extensive set of MMX-optimized functions for gray scale, color, and binary image display, image processing (statistics, filtering, and geometric transforms), shape matching, blob analysis, gauging, and measurement.

IMAQ Vision is image processing and machine vision software for inspection and quality control in industrial settings. You can use IMAQ Vision to make reliable, high-speed, and critical measurements in semi-conductors, electronic, medical, and pharmaceutical manufacturing processes. You can capture digital images of manufacturing processes in real time and extract critical information such as the location and orientation of parts or the presence or absence of defects.

This software offers various analytical functions to yield the desired information. The functions include routines to measure dimensions, and the form and orientation of objects. The software also includes sophisticated functions from the field of digital signal processing, e.g., one- and two-dimensional FFTs (see Chapter 9).

IMAQ Vision includes a full VI library for LabVIEW and BridgeVIEW. Its tool palette provides image display tools, zoom tools to calculate areas, perimeters, and other parameters, tools to search, shape matching for alignment, measurement, and inspection applications, and a wide range of filters.

12.8   Sample Application in LabVIEW

This section describes a sample application to continuously acquire and represent single images and to calculate and display a histogram. Figure 12-1 shows the source code (diagram in foreground), the front panel (with the histogram in the background), and the acquired image.

The LabVIEW diagram is created as follows. First, the memory space required for the image is created at 8-bit depth and allocated. At the same time, a handle called *image* is created to simplify handling of the image information. This example displays and represents individual images continuously so that the corresponding LabVIEW functions (VIs) have to be run in a loop.   In this

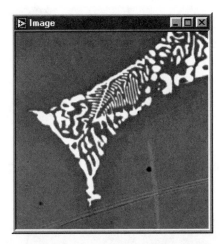

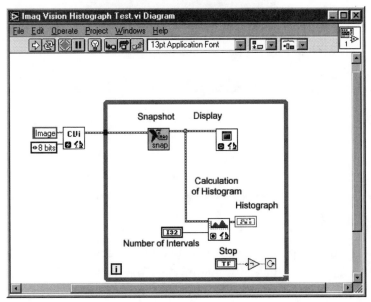

Figure 12-1

Example of how images are acquired and represented in LabVIEW

example, the do...while loop was used; this is denoted as a rectangle around the iterating VIs. The SNAP VI is used to acquire each image. The acquired image is passed on for display and for histogram evaluation. Subsequently, the user presses the STOP key on the front panel to stop the loop.

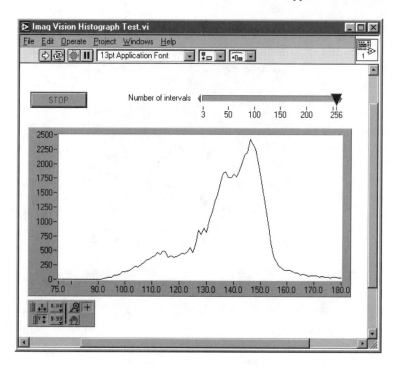

Figure 12-1

(Continued)

12.9   Summary

Industrial image processing has become very popular in view of dropping prices and increasing computing power and rapid developments in semiconductor technology. It can be applied in many different fields of application. NI-IMAQ is designed to meet the requirements of imaging application developers who need to reduce cost and speed up time to market. Transparent memory management, logically named VIs, functions, and parameters make NI-IMAQ easy to use and learn. The functions are built to intuitively work together so you can develop applications faster and use only a few functions. IMAQ Vision software adds high-level machine vision and image processing to LabVIEW. It offers an extensive VI library for LabVIEW and a wide range of image processing and display functions.

13

Quality, Reliability, and Maintainability of LabVIEW Programs

All things in fact begin to change their nature and appearance; one's whole experience of the world is radically different... There is a new vast and deep way of experiencing, seeing, knowing, contacting things.

— Sri Aurobindo
On Yoga II

Computer-controlled measurement, analysis and simulation software, and development systems have been automated to a level that produces significant competitive advantage for companies using such tools — and a significant competitive disadvantage for those without them. On the other hand, companies using such powerful tools are confronted with a catalog of questions they need to answer to select tools suitable for their needs. Among the most important factors to consider when selecting software tools are quality, reliability, and maintainability of software programs. This chapter covers some important background information to facilitate answers to these questions and describes some important capabilities of LabVIEW with regard to quality criteria.

13.1   Introduction

What are software *quality* factors? Software quality factors define attributes that any software should have. These quality factors represent many of your needs and expectations. Two other important factors are reliability and maintainability.

Reliability is a measure of the program's precision — the ability to get the correct answer when given the correct input. Measuring reliability is the responsibility of both the vendor and the user. The vendor should run numerous tests in order to ensure reliability. These tests should include textbook problems for which there are known solutions. However, the tests must also include complex, real-world problems because those are the kinds of models you create. The vendor should provide, upon request, results of these tests as well as the test input files. These tests should not be run only once; they should be run each time a new version is released and on each type of computer platform.

You could also measure reliability using the kinds of models typical of your designs. This kind of testing is often called *acceptance testing*. This means that you will not release the software into production use until it has passed your tests. Depending on the depth of your vendor's test and quality assurance procedures, you may bypass your acceptance testing and put the software into production use immediately.

All vendors say they are committed to quality. To judge for yourself, look at the size of the installed base (especially of users in your industry), clarity and depth of the user documentation, number of the development and support personnel and their knowledge and experience, and overall company track record. And look for a formal quality assurance (QA) program.

Many of the procedures, standards, and regulations defined by quality standardization organizations are similar. For example, many of the specifications found in ISO 9000 are found in other regulations. Each organization defines standards for software design, internal documentation, testing, and corrective action. Software vendors that comply with these regulations submit themselves to internal and external audits to assure compliance.

Maintainability means basically being able to correct and enhance the software, easily and efficiently. This means that the software must be easily testable.

❏ 13.1.1   Additional Quality Criteria

Of course, there are other important factors that tell you a lot about a software tool's quality. For instance, *portability* means being able to use the software on multiple operating systems or hardware platforms. *Flexibility* is a measure of how easy it is to add new functionality. *Reusability* concerns software use in multiple applications. For example, reusability can be thought of as a measure of how much the program is general purpose versus special purpose. And *integrity* is a measure of your confidence that no one has inadvertently or purposely modified or changed the program or the data such that the analysis is adversely affected.

Availability is another important factor — the program must be available on your specific computer type. Likewise, updated versions, including enhancements and error fixes, must also be available on a timely basis on your computer.

Compatibility measures the degree to which different versions of the software work together. For example, if a program is upward compatible, later versions will support the current version. Conversely, downward compatible means that the current version supports previous versions. Both upward and downward compatibility are important because as new versions of the software are available, you do not want to recreate and reanalyze your designs from scratch.

The following sections describe important hardware and software reliability criteria, and how LabVIEW meets these criteria is described in the following sections, including important hardware and software aspects.

13.2   Hardware Reliability

Figure 13-1 shows typical failure rates of components and hardware systems based on the duration of usage. Because of its course, this graph is also referred to as a *tub graph*. According to the Weibull distribution, it is divided into three categories. The first part indicates early failures. In the initial stage of the creation of integrated components (still applied in military and aeronautical applications), an almost total screening of all

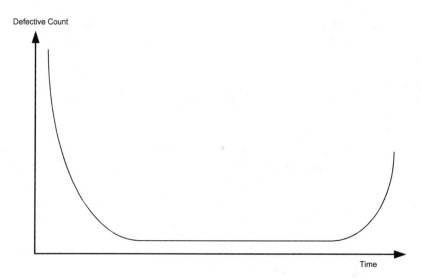

Figure 13-1

Qualitative failure rate curve of components and hardware systems in relation to their use

active components was carried out (including burn-in and stresses of active components before they were being used) in order to detect failure mechanisms described by activating energies and to keep the first part of the tub graph as level as possible. The various crashes of newly manufactured missiles during the past few years clearly show that the curve cannot start at zero.

The second part of the tub curve relates to an almost constant failure rate. (Failure rates are usually denoted in multiples of 1 fit = 1 failure in 10^9 component hours). As a rule, only so-called coincidental failure rates occur during this phase.

The third part of the tub curve shows the extent of failures due to wear and tear. The aging of components (drift of different parameters) has a negative effect on system consistency and may lead from failures to total breakdown of entire systems.

13.3 Software Reliability

In contrast to hardware, software *MTTFF* (Mean Time to First Failure) increases with growing error-free use, which means that a software system tends to become more reliable the longer it is being used.

Looking at a typical software reliability curve (medium failure rate based on duration of use), we get a completely different picture. Only the first part of the curve is comparable to the one emerging as a tub graph, which highlights weaknesses in the design that should have generally been corrected during the vendor's test and validation procedures. In order to avoid "banana developments," software products should only be made available to end users in the curve's second category. (The product matures while being used by the customer.) The level part of this curve represents small design flaws which the developer is able to reduce by installing patches or modifying the design.

Figure 13-2 shows the typical failure rate curve of software components on the basis of duration of use. As a rule, the first curve's category in LabVIEW developments confirms a much more level and shorter course than in customary traditional designs.

Despite all the modern development methods and tools, there is little chance of a perfect development system emerging in future. Failures and breakdowns of software systems cannot be totally avoided. With regard to the curve shown in Figure 13-2, this means that the distance between the curve and the X axis (time) will never reach zero. One of the best methods currently available to minimize that distance is using LabVIEW because this development environment offers outstanding quality characteristics.

Since LabVIEW VIs generally consist of reliable components (Figure 13-3), applications written in G should result in an extremely reliable runtime performance. (A good command of the programming language is essential.)

13.4   Reliability and Performance in LabVIEW

Version 5 of LabVIEW features several capabilities that are of interest in the context of the issues covered in this chapter — quality, reliability, and maintainability of a software development environment.

- **Multithreading** — The benefits of multithreading are numerous, including better reliability and throughput, better microprocessor utilization, and the ability to take advantage of multiprocessor machines. Until now, however, because of the difficulty in writing multithreaded applications in text-based or sequential languages, multithreading technology has been limited to only expert, traditional programmers. All LabVIEW programs are automatically multithreaded without any code modifications. Thus, you benefit from the advantages of multithreading without having to learn new programming techniques.

- **Instrument Wizard** — With the Data Acquisition Wizards (see Chapter 3) introduced in LabVIEW 4.1, you can create sophisticated data acquisition programs with only a few mouse clicks and no prior programming knowledge. You can easily create programs for instrument control with GPIB, VXI, and RS-232 instruments. With the Instrument Wizards, you can see what instruments are connected to your system. The Instrument Wizard automatically installs instrument drivers for those instruments and generates application examples that use these drivers. With only a few clicks of a mouse, you now can be up and running quickly with no programming required. The Instrument Wizard generates the entire application with the LabVIEW block diagram (source code), so you can modify the application if necessary.

- **ActiveX Container** — LabVIEW 5.0 is an ActiveX Container, meaning that you can easily drop any 32-bit custom control (OCX) or ActiveX document onto a LabVIEW front panel to efficiently reuse code that has already been written. Examples of ActiveX controls and documents include a web browser, a Word document, an Excel spreadsheet, a calendar control, ComponentWorks controls, a HiQ document, or any other ActiveX-compatible control available over the Internet or CDs. You can edit these objects directly on the LabVIEW panel or control your methods and properties by using simple graphical programming.

- **LabVIEW Automation and TCP/IP Server** — LabVIEW 5.0 is an ActiveX Automation Server, meaning any ActiveX Automation Client can easily control the LabVIEW application or call LabVIEW programs directly in much the way an application calls a function in a dynamic link library (DLL) (see Chapter 4). With this interface, ActiveX Automation Clients, including C, Visual Basic, Microsoft Excel, LabWindows/CVI, or even another LabVIEW application, can easily call LabVIEW VIs from directly within the program. The LabVIEW Automation Server includes the TCP/IP server interface, which enables LabVIEW client applications to easily control other LabVIEW applications directly over a network.

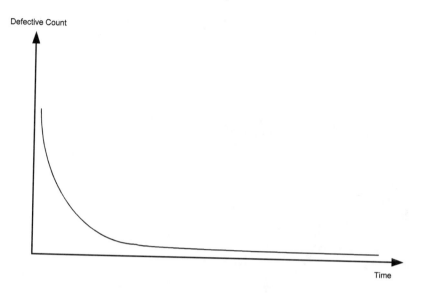

Figure 13-2

Qualitative failure rate curve of software components in relation to their use

- **Distributed Computing Tools** — LabVIEW provides tools that you can use to quickly and easily develop distributed applications, where various sections of code can execute on different machines across a network. These easy-to-use tools make it possible for you to offload processor-intensive routines to other machines for faster execution.

- **Documentation and Translation Tools** — With the documentation tools in LabVIEW, you can automatically generate documentation for your application in HTML and RTF formats — simplifying the generation of user manuals, function reference manuals, or online help systems.

- **Graphical Differencing Tools** — Numerous tools necessary for large application development, including source code control, performance profiling, and extensive debugging capabilities, have been available in LabVIEW for years. Now LabVIEW delivers an important code management tool that lets you compare two LabVIEW programs to determine the differences between them. This tool eases the task of revision control for teams of developers and simplifies large project development.

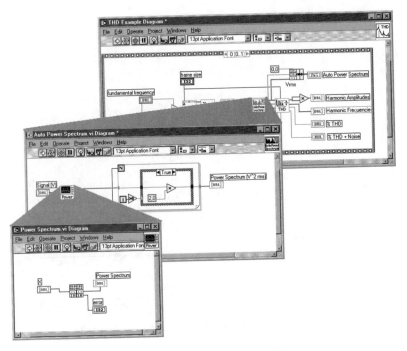

Figure 13-3

Diagram hierarchy in LabVIEW

13.5   LabVIEW Benefits

LabVIEW, unlike other programming languages, allows single modules (virtual instruments) to be fully tested and subsequently used in hierarchic virtual instruments. Its well-defined modularity and hierarchy allow you to reuse programming codes, which has not been possible in conventional systems.

In LabVIEW, the "software revolution" takes place deep inside the LabVIEW system, hidden from the developer. However, LabVIEW lets you integrate add-in modules (i.e., in the form of code interfaces nodes or DLLs) any time when you need specific tools. Notice that you can use existing DLLs as they are or modify them to your requirements.

Implementing new "parts," using LabVIEW's built-in G programming language, you can enhance known code and achieve more security, quality, and reliability than developers working with traditional software. As a LabVIEW developer, you are relieved from the hassle of implementation-specific details and can concentrate fully on your applications. This means that there is greater confidence in the runtime performance of the developed software. Your intuition, creativity, and energy, while designing, are all channeled into the functionality, ergonomics, performance, quality, reliability, and availability of your applications.

Thanks to the inherently positive G characteristics, which massively support rapid prototyping and development, the application developer will derive a high degree of satisfaction within a short period of time. He is thus able to meet his cost and time estimates he submitted to his client, and "Hofstadter's Law" will no longer apply. (It states that no matter how conservative your time estimate for a development may be, it will always take longer than planned.)

The concentration on quality, reliability, and functionality and, above all, on ease of use results in applications that offer the end user the highest degree of flexibility and application security. A well-understood code ensures good maintainability. This has the nice side effect that program developers are more readily inclined to optimize their programs. If you are a LabVIEW G programmer, you will have no problem dealing with your systems and program solutions. You can use various options during different stages of development stages for rapid prototyping.

One of LabVIEW's strengths is troubleshooting management within a graphical flow programming environment. You can use *error clusters* (similar to the structures of conventional programming languages like Pascal and FORTRAN) to easily handle error positions, error rates, and error descriptions.

If you make good use of these troubleshooting options, you can create highly reliable applications. The extra effort involved (error clusters have to be connected to VIs and subVIs in daisy-chain mode) is more than compensated for by the high quality and manageability of your code.

One of the main advantages of LabVIEW is the declarative approach to problem solutions. With a problem described in graphical form, the application is created almost automatically, which may turn out to be even simpler than using the DAQ Wizards (see Chapter 3).

Through its declarative code, LabVIEW facilitates application development in teams. Any module generated by members of the team can easily be included in a specific application. The characteristics of the G programming language have led to a great deal of generalized and specialized VIs in the last few years. Their major benefit is that they can be used "out of the box."

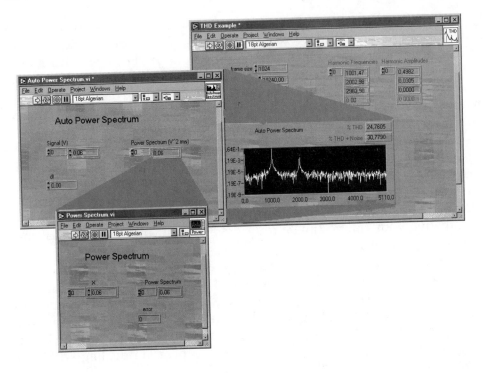

Figure 13-4

Panel hierarchy in a LabVIEW application

LabVIEW handles all tedious tasks such as compiling, linking, environment management, storage management, etc., in the background, leaving you more time for creative work.

Probably the greatest benefit of LabVIEW is its libraries of virtual instruments, mostly written by specialists, which you can use as "templates" to accomplish your tasks at a fraction of the cost of traditional programming. When your needs change, you can modify your virtual instrument in moments. In addition, an extensive choice of specialized toolkits is available, including toolkits, for math, time-frequency analysis, image processing, and SPC.

Overall quality, reliability, and maintainability can be further increased by extensive use of front panel controls and indicators. Mechanical and electromechanical components have a very high failure rate compared to their (VLSI) PC counterparts and should therefore be avoided. LabVIEW supports this strategy in an exemplary way by providing front panel elements which

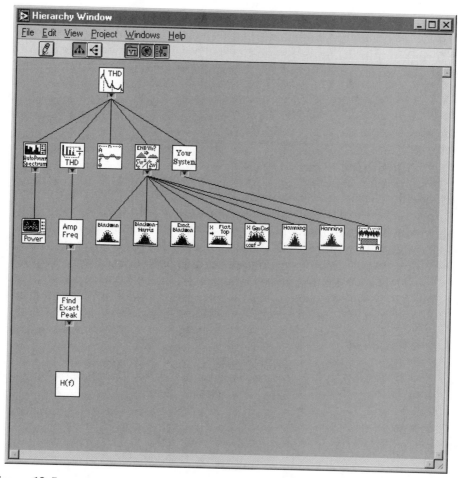

Figure 13-5

Hierarchy window of a LabVIEW application

you can easily modify. The development of proprietary controls is child's play and offers great advantages in developing complete systems.

Many benefits result from virtual instrumentation with regard to purchasing components. Significantly fewer components are required to create a full-function system. This means that LabVIEW-based systems have a much shorter time to market.

An extremely important factor, albeit often overlooked, is security. Although you probably do not write real-time applications to control airplanes, all your applications should be safe. In particular, your system should comply with the relevant official safety regulations. In practice, LabVIEW is considered beneficial in this respect because it tries to make your (programming) life as hassle-free as possible and because LabVIEW applications are streamlined to use a minimum of physical components to build your system.

13.6   Summary

Based on the clearly possible reduction of hardware components of any kind, the use of virtual instruments significantly facilitates systems certification. While time to market and system costs are minimized, the chances of market success are maximized. The use of LabVIEW significantly eases the innovation-induced pressure on developers. While security concerning law abidance increases, the responses anticipated from customers will grow when the first applications have been delivered and follow-up projects are being undertaken.

14

Statistical Process Control in LabVIEW

Hofstadter's Law: It always takes longer than you expect, even when you take into account Hofstadter's Law.

— Douglas Hofstadter
Gödel, Escher, Bach

The manufacturing industry's growing demands for quality make computer-assisted acquisition and processing of measurement values in quality assurance and control (CAQ) more and more important. This arena requires well-performing, readily available tools that can easily be created in the form of virtual instruments, which are known to be highly effective in preventive quality assurance environments (i.e., methods of statistical process control). This chapter describes the basic procedures for statistical process control (SPC) in LabVIEW.

14.1 Introduction

Since the 1940s the industry has been using statistical methods to ensure the quality of both products and services at their place of origin in order to

337

reduce errors, costs, and time to market. The aim of *statistical process control* (*SPC*) is the systematic selection of crucial parameters from the total of all parameters involved in the process in order to control the manufacturing process with decisive parameters and appropriate measures. Not surprisingly, SPC is often also referred to as *statistical quality control* (*SQC*).

An SPC application functions according to the control mechanisms of a classical control cycle, where the process procedure is represented as the control path. The process operator takes on the role of a controller whose task is to monitor the characteristics to be created and, in case of deviations, to make the necessary corrections. Although SPC functions are relatively easy to use, it is advisable to thoroughly train everyone involved (i.e., process operators, the project team, etc.) to ensure a successful outcome of the SPC application. While SPC tools may greatly facilitate the analysis of processes, it is essential to bear in mind that the real interpretation of the results is up to the user alone. It is therefore necessary that well-grounded training takes place at an early stage.

14.2   SPC Methods

Three proven methods are generally used for process analysis: *control charts*, *process capability analysis*, and *Pareto analysis* (also known as *ABC analysis*). To monitor and control a process, its behavior is initially examined and, if necessary, adjusted with the help of standardized control charts. Once it has been established that the process is under control, its long-term behavior is analyzed by examination of the process capability. Basically, the Pareto analysis can be used at any time. It reveals all potentially relevant errors in order of gravity, indicating the sequence of countermeasures to be taken. These three SPC methods are further explained in the following sections.

14.3   Control Charts

Control charts are used to monitor and control processes in order to highlight influences and changes affecting process parameters so that adjustments can be made at an early stage. In the general sense, a control chart is a trend recorder of certain process parameters over time. The simplest form of a control chart provides sequentially obtained measured results deriving from spot checks over time. Border limits in relation to mean value $\overline{x}$ of the measured values x_i are set at upper and lower limits of warning and intervention. If a measured value exceeds the warning limit, process control needs intensifying. If the limit of intervention is exceeded, the process must be adjusted. Regardless of the process model, different formulas of calculation are chosen for intervention limits.

Generally, though, one goes for a distance of $\pm 2\sigma$ for warning limits and $\pm 3\sigma$ for intervention limits, where σ, as is well known, represents the standard deviation.

Control charts are basically classified as having variable (continual) and attributive (qualitative) features. Variable control charts record quantitative features in terms of physical size, such as the diameter of a component or the number of punching holes in tin. Variable quality features are differentiated between $\overline{x} - S$ charts (mean value standard deviation chart) and $\overline{x} - R$ charts (mean value span chart) depending on how a standard deviation has been established.

In attributive classification, production units are counted according to certain qualitative criteria, for instance, the number of defective units. Four additional control charts designated as *np* chart, *p* chart, *c* chart, and *μ* chart, are available for attributive quality features.

Control charts can be updated in real time, which means that sequentially obtained measurements are displayed over time during the measuring process. The LabVIEW toolkit contains all the necessary functions for both offline analysis of process data and online analysis of measured values (Figure 14-1).

14.4   Process Capability Analysis

Taking into account all parameters involved in the process, the process capability describes the long-term behavior of the process in order to assess performance trends over time and provides a measure of the quality of the process. Capturing the mean values of the measured values by way of many single spot checks and applying them as Gaussian distribution will determine the quality of the process. Capability process calculations such as *cp* (capability process, i.e., admissible tolerance versus process distribution ratio) and *cpk* (considers additionally the conformance of the process versus the tolerance limits) indicate both the capability for quality and repetitive

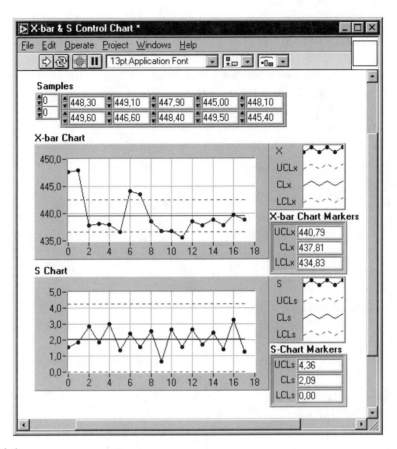

Figure 14-1

Virtual instrument for the $\overline{X}$ – S chart

accuracy of a process. The higher the *cpk* value, the more capable the process is in relation to distribution and compliance. In industrial mass production, a *cpk* value >1.33 will show that all systematic influences are eliminated and is therefore considered the desirable characteristic. Processes that meet these requirements are classified as controlled (Figure 14-2).

14.5   Pareto Analysis

Since, according to the Pareto principle, generally also known as the 20/80 rule or ABC principle, relatively much (80%) can be achieved with few

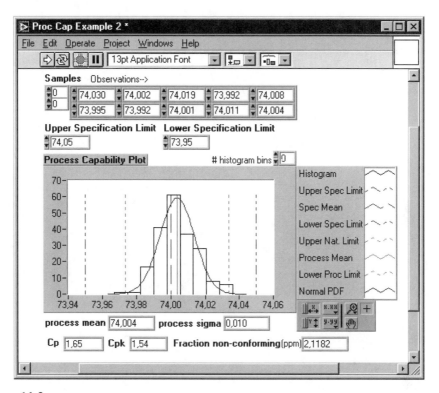

Figure 14-2

Virtual instrument for process capability analysis

errors (20%), it would appear to be sensible to first examine the importance of these errors in order to prioritize any countermeasures to be taken.

The result of a Pareto analysis is a Pareto diagram, a special form of histogram that reveals all potentially crucial errors affecting the product and that ranks these errors by importance and frequency, and thus allows you to set up necessary countermeasures in order of priority. Figure 14-3 depicts a typical Pareto analysis confirming the 20/80 principle.

14.6   SPC Tools in LabVIEW

The SPC Toolkit for LabVIEW contains libraries and functions for SPC applications. You can use LabVIEW (or BridgeVIEW and LabWindows/ VCI) not only to monitor your process, but also to identify problems and

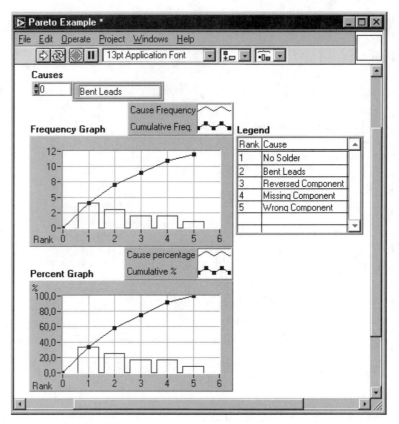

Figure 14-3

Virtual instrument for Pareto analysis

actually improve the quality of the process. You can apply SPC methods to analyze and track process performance. In addition to subVIs or functions that perform the SPC computations, the toolkit contains numerous example programs and custom controls that demonstrate how to incorporate typical SPC methods and displays into LabVIEW.

The toolkit addresses the three areas of SPC — control charts, process statistics, and Pareto analysis. For each of these areas, the toolkit contains analysis functions to compute important information from your data and plotting functions to display analysis results.

- **Control charts** — The control chart libraries include VIs or functions that can compute points to be plotted for different types of attributes and variables charts. You can select from chart types such as X-bar and S, X-bar and R, p, and np. You can apply run rules to detect out-of-control points or process shifts. The Chart Draw VIs and functions create X-Y graphs with center line and upper and lower limits; control charts with zones; control charts with variable limits; and run and tier plots for displaying the spread of subgroups against specification limits.

- **Process statistics** — The SPC Toolkit includes several VIs or functions for analyzing the process capability. It features graphing functions and VIs that can fit a normal distribution curve to a histogram to evaluate whether the process is normally distributed. Additional functions calculate process capability ratios, such as *cp* and *cpk*, as well as the fraction nonconforming to normally distributed processes.

- **Pareto analysis** — You can use Pareto analysis VIs and functions to display the relative importance of problems or assignable causes in your process. You can create Pareto charts to graphically represent the number of occurrences or percentage occurrence of particular causes.

In short, the SPC Toolkit lets you integrate SPC directly into your applications.

14.7 Summary

SPC is frequently used in many quality assurance applications as a standard tool to monitor and control processes. Understandably, such special methods have hitherto predominantly been created in the form of proprietary software solutions. Since the introduction of LabVIEW in 1986, the open concept of virtual instrument creation as an alternative to traditional methods has gained a substantial market share. This chapter introduced virtual instruments for SPC applications. In addition, other methods dealing with preventive quality assurance, such as the fault-tree analysis, can be easily implemented in LabVIEW.

15

LabVIEW and Quality Management

The vast majority of human beings dislike and even dread all notions with which they are not familiar. Hence it comes about that at their first appearance innovators have always been derided as fools and madmen.

— Aldous Huxley
Adonis and the Alphabet

System developers normally have their own ideas as to how they want their quality and test management systems to look and operate. In addition, as testing and quality control become more and more important for building quality products, and the test data collected must be shared and analyzed throughout your organization, connecting to the enterprise is another key issue. Users each have their own internal corporate information systems and analysis tools that must be integrated with their test systems. These are just a few of the reasons why many test and quality control engineers feel pressured to build their own testing environment. When you build your own quality management system, you are taking on an expensive long-term support and maintenance effort. This chapter describes a few basics about quality management. A real-world application shows how this task can be integrated in LabVIEW to deliver a ready-to-run quality management environment with many of the most commonly requested features, such as multilevel user access control, advanced sequencing and branching control logic, and automatic report generation.

15.1   Introduction

A test and quality environment is the control center for any automated test system. The LabVIEW quality management software consists of a variety of software tools and libraries designed to help you develop, schedule, execute, and generate reports from your automated test programs. LabVIEW toolkits provide key capabilities for anyone building or using an automated test system for the production environment — capabilities such as test sequencing and execution control tools, branching logic, automated report generation, user management, etc. In addition, LabVIEW test tools include plug-ins for storing your test results directly to any ODBC database, using our SQL Toolkits.

15.2   Using LabVIEW for Calibration and Qualification

LabVIEW not only proves itself in the implementation of large-scale projects but is also of universal help in the measuring environment. The following examples explain why.

Whether it is a question of calibrating or qualifying a component, there are always more or less extensive measuring sequences required which will subsequently be evaluated by a PC. At least in the sphere of electrical measuring parameters, reference devices as well as many test pieces are equipped with GPIB interfaces so that it seems obvious to automate this in a PC-assisted environment. However, the variety of tested components and appliances — more than 500 types are listed in the universal test bench of Quelle at their comparative test center (Quelle IWU) — consequently shows that measuring procedures are seldom exactly the same. It is therefore advisable to consider if the relation between the time required for manual measurement and the development time of an appropriate program actually warrants automatic measurement.

Through simple graphic programming and the availability of existing VIs — in this case, in particular, measuring device drivers which are usually bundled with the software — LabVIEW makes the program development brief and profitable. Automatic measuring has the advantage that it eliminates human error and can be executed more intensively without extra cost.

The following sections describe several examples.

☐ 15.2.1 Qualifying an AC Voltage Source

One part of the developed testing areas at Quelle contains an AC voltage source for the 80–320 V range. When voltage fluctuations occur, the voltage source automatically re-controls the set-point value (by way of load alternation, for instance) allowing the set point to be predetermined by a 0–10 V direct-voltage input.

When qualifying an AC source, one of the important steps is to find out the characteristic of the control input. As the voltage setting takes place by means of the transformer and motor operator, a certain hysteresis ensues, which means that a specific value does not necessarily lead to an output value. It depends on the position of the regulating transformer within a certain fluctuation range before arriving at the value. To be able to tell which input value results in a certain output voltage, a measuring sequence needs to register as many states as may possibly occur. By use of a PC, a great number of measurements can be executed, so that with many coincidental input values, all possibilities are well covered.

A calibrator (Datron 4808) and a system multimeter (HP 34401) are used as reference devices for set-point reference and for output signal measurement, respectively. Existing LabVIEW drivers are used for both devices.

Writing the program was as easy as copying the panels of the device driver VIs to the new program to provide the user with every possible setting option (Figure 15-1). The control elements for measuring instruments in the program are connected to VIs' input drivers. To address the calibrator, only the output voltage value needs to be inserted. An X-Y diagram displays the characteristic of the distribution of measuring values so that faulty measurements can be detected in an early stage.

The underlying program has the structure shown in Figure 15-2.

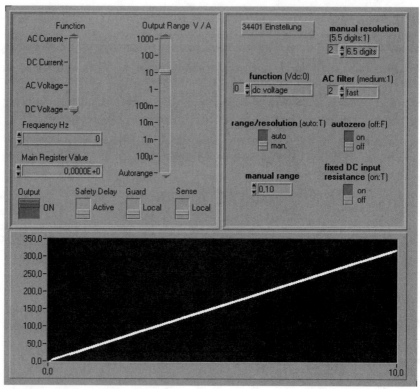

Figure 15-1

Panel to qualify an AC source

After program start, the equipment is initialized according to the chosen settings (frame 0 of the outer sequence). Subsequently, a `for` loop executes all 500 cycles.

The **Random** function generates a random value in the range from 0 to 10 which is inserted in the calibration setting and transmitted in the small sequence (frame 0). After waiting for the regulating transformer to position itself, which takes about 5,000 milliseconds, the system multimeter reads the measuring value. In the last frame, the default value and the measured value are written into a text file. Once the measuring sequence has been processed 500 times and the LabVIEW program terminates, the result file can be opened and evaluated in Excel.

Setting up the program takes about an hour. The entire program run, which does not require the presence of an operator, also takes about one hour. A manual execution with many measuring sequences would probably take several days.

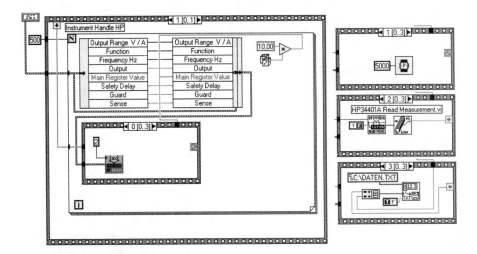

Figure 15-2

Program to qualify an AC source

☐ 15.2.2   Using Excel for Semiautomatic Calibrations

Every measuring device used for testing at Quelle IWU needs to be checked at certain intervals (usually after 1–2 years.) A calibrating protocol generated in Excel will record the result.

For each type of device it is necessary to create a sample protocol based on tolerance defaults (standards, manufacturer information, and user requirements), which contain the areas, measuring points, and the relevant tolerance information needed for calibration. During calibration, the tester adds the values determined to the protocol online. An Excel spreadsheet contains calculated fields to determine deviations automatically, compare these against the tolerance information, and display a good/bad evaluation.

When electrical measuring devices are used, the following steps are necessary: The tester retrieves the default value from the protocol and sets it up in the reference device (normally the Datron 4808 calibrator), scans the display of the test piece, and adds the result to the protocol. The program shown in Figure 15-3 makes this procedure considerably easier.

Basically, this program is a slightly enhanced version of the existing LabVIEW calibration device driver (internally developed by Quelle IWU).

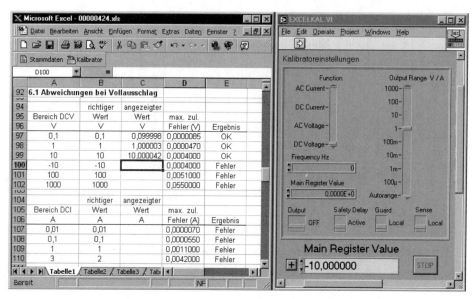

Figure 15-3

LabVIEW and Excel cooperating for calibration

More specifically, an additional input field for the startup situation was added to the calibration settings, and a continuous loop was put around the original driver. The program uses this loop to wait for a change in the settings or in output voltage values. Subsequently, the program transmits all data to the calibrator. This means that it also represents a kind of virtual remote operation.

This job is greatly facilitated by the fact that the remote operation can be achieved not only manually with a mouse and keyboard but also automatically with the help of an Excel macro. The tester moves through the calibration protocol and positions the cell display onto the cell deemed to contain the reading value, and by clicking the mouse he activates the macro. Retrieving the calculated value from the cell on the left of the cursor position (set-point value) and making it a variable activates the calibration application and sends simulated keyboard sequences to the LabVIEW input field. The LabVIEW program accepts these changes and passes them on to the calibrator.

The transmission of keyboard sequences is an unusual yet practical form of data exchange between programs, as the source code in Figure 15-4 shows.

```
Sub Kalibrator()
    ActiveCell.Offset(0, -1).Range("A1").Select
    KWert = ActiveCell.Value
    AppActivate "Excelkal"
    Application.SendKeys ("+{F1}" & CStr(KWert) & "~"), True
    AppActivate "Microsoft Excel"
    ActiveCell.Offset(0, 1).Range("A1").Select
End Sub
```

Figure 15-4

Excel macro used to exchange data via keyboard buffer

The key combination UP+F1 is assigned to the `Main Register Value` input field. Excel activates this input field, overwriting the old value, and terminates the input by adding the Enter (~) command. While LabVIEW transmits the new value to the calibrator, Excel jumps back to the foreground to position the cursor where the tester enters the read value.

This VI is universal and can be reused for many test pieces inasmuch as device-specific measuring sequences can be taken over from available Excel templates. This means that the measuring process can be executed much faster and more easily without the need for a complex piece of special software.

❑ 15.2.3   Calibrating Load Cells

Load cells (also called *weighting cells*) consist of an elastic element to which a wire strain gauge is attached. The wire strain gauge represents a resistance bridge, which is powered by a distribution voltage. This resistance bridge supplies an output voltage that depends on the elasticity. The supply voltage to the output voltage ratio is proportional to the applied power.

During a calibration process, calibrated reference mass pieces are used to stress the load cell and to determine the basic value. Both the supply voltage and the output voltage measurements have to be as accurate as possible to ensure good results. To eliminate physical installation of a reference display device in the calibration lab, the function of such a measuring amplifier is simulated. A calibrator (Datron 4808) provides the exact supply voltage, while a precision multimeter (Datron 1281) is used to measure the output

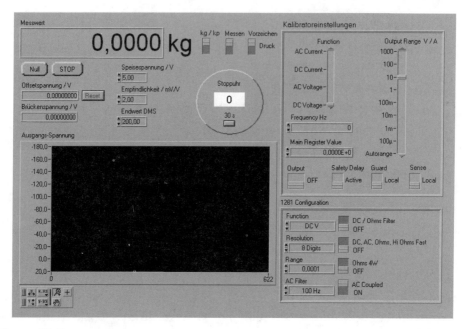

Figure 15-5

Power measurement in LabVIEW

signal, containing only a few millivolts, with extreme accuracy. The control of both reference devices as well as the conversion from voltage ratio into power (or mass) is carried out with the help of LabVIEW.

In this application, too, the program is based on existing measuring device drivers, as Figure 15-5 shows. In addition, the panel contains input fields for calculation parameters such as supply voltage, sensitivity, and rating. The user determines the load cell's zero offset by the push of a button. Notice that the zero offset has to be subtracted from the measuring value to ensure correct reading.

Using a stopwatch during the measuring process can be helpful to indicate the end of the load time by an acoustic signal. A graphical display reveals whether or not the measuring value is properly stabilized. In this case, the value is accepted and added to the protocol.

The LabVIEW diagram in Figure 15-6 depicts the measuring program's main loop. The top left part shows how the voltage value is added to the calibrator settings. Calibration and multimeter configurations are stored in a local loop variable and tested for changes in each cycle. If one of the values changes, the configuration is updated and the new configuration

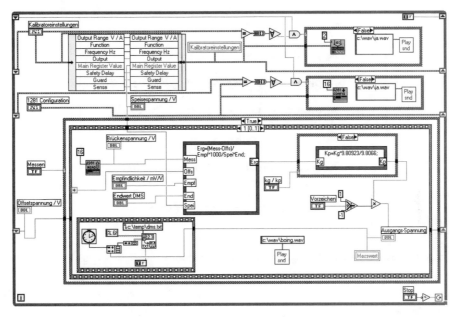

Figure 15-6
Program for power measurement in LabVIEW

is transmitted to the appropriate device (top right), and notified by an acoustic signal.

The lower part of the diagram shows how the multimeter reads the measured value. The multimeter takes the measured value, converts it into a mass or power, depending on the kg/kp switch, and adds the appropriate sign for tension or pressure load. The program displays the results on the screen, adds a time stamp, and writes these contents to a text file that is not evaluated during calibration.

Creating the log requires only one single value at the end of the load time. Again, a macro is used to import the value to Excel. This macro, listed in Figure 15-7, offers an additional unusual way to allow processes to exchange data through the Windows clipboard.

Excel activates LabVIEW and the required field with the UP+F1 (+{F1}) key combination. It simulates the SHIFT+c (~) key combination to copy the value to the clipboard. Excel then uses the same procedure to paste the value into the active cell. Finally, Excel moves the cursor to the cell underneath the active cell.

This small program allows direct reading of the results and graphical display, so the examiner can operate a plausibility check immediately on the values without the need to convert voltages or transmit a protocol.

```
Sub DMS()

    AppActivate "Excel Measurement"

    Application.SendKeys ("+{F1}" & "^c"), True

    AppActivate "Microsoft Excel"

    Application.SendKeys ("^v"), True

    ActiveCell.Offset (1, 0).Range("A1").Select

End Sub
```

Figure 15-7

Excel macro to exchange data through the Windows clipboard

☐ 15.2.4   Calibrating a Multimeter

If measuring devices with their own PC interface are calibrated, it would seem obvious to execute the entire calibration procedure automatically. However because of the number of different devices needed, the development costs of a calibration routine may be quite important. The program illustrated in Figure 15-8 was created to implement a low-cost, automatic calibration procedure for the Hewlett-Packard system multimeter HP34401A, where savings were achieved by using VIs available in LabVIEW.

The calibration routine is based on the simple concept of putting all operating elements for measuring devices used into one array. Each array element represents a test step. The steps involved are initializing the multimeter, transmitting the calibrator settings, reading the measured values, and writing these results to a text file. The tester enters the device settings step by step for each particular measuring point and saves them as default values directly in the program file.

As the user proceeds, a text field displays instructions about certain steps involved, for example, to change cable connections. The program tests the code's statement field for text and, if true, waits until the user presses the appropriate key to continue. Some measurement steps require a longer wait time than others before reading takes place. With this particularity in mind, an additional field for *extra time* was created for the user to enter a wait time

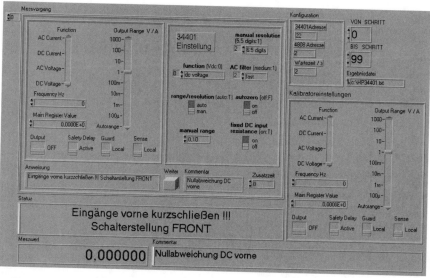

Figure 15-8

Calibrating an HP multimeter

value. The result, together with the text entered in the **Comment** field, is then saved to disk. Figure 15-9 shows the compactness of the underlying diagram.

The *Measuring Process* array contains a cluster which also incorporates the clusters for device settings and additional fields. A `for` loop then deals with the individual steps. All other program parts are included in a `case` structure to allow the user to run individual steps, if necessary. One runtime criterion is to test whether the current step is within the defined **From – To** range. The statement text contained in the `case` structure is displayed in the **Status** field. **Index Value** is an attribute node used to display the current step on the screen. To wait for a user input, the cluster cycles the `while` loop and is released only when the user presses the **Continue** button.

The actual measuring process takes place in the *Sequence* part. First, the multimeter is configured for the required measuring environment. The calibrator receives the calibrator settings and issues the measuring parameter. When the wait time expires, the multimeter receives a read command and returns the measured value to the program. The result is written to a file, together with comments. The calibrator is then returned to its basis position, then the sequence terminates and initiates the next loop cycle. After the completion of the calibration routine, the text file can be further processed directly in Excel and the calibration protocol can be set up.

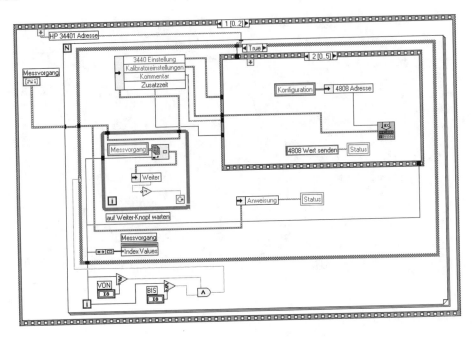

Figure 15-9
Program to calibrate an HP multimeter

The whole procedure takes about 15 minutes, during which the tester can deal with other tasks in the interim between measuring range changes. To manually run the required measuring sequence, the time required would be about one hour. This means that the development cost of approximately four hours is compensated for after a few calibrations.

15.3   Summary

Modern automated test and quality control tools are built around state-of-the-art test development environments, for example, the LabVIEW programming environment. This environment combines the execution speed of a compiled language, which is of key importance for production test systems, and the ease of interactive development for shorter time to market of new products. The examples illustrated in this chapter show that LabVIEW facilitates the development of effective tools to achieve important cost and time savings.

16

LabVIEW in Medical Applications

The seat of the soul is where the inner world and the outer world meet.
Where they overlap, it is in every point of the overlap.

— Novalis
Fragments

Continuous monitoring of physiological parameters has become an indispensable instrument in medical applications. Constant control of hemodynamic and respiratory parameters, supplying important information on the patient's condition, support or facilitate decision processes during the course of a therapy. However, conventional monitoring has some known drawbacks. This chapter introduces biosignals briefly and describes how biosignals can be measured and processed in LabVIEW. A real-world application illustrates practical aspects of this topic.

16.1   Introduction

Most conventional signal monitoring systems have their specific benefits and drawbacks. First, each medical instrument has a specific form of representation. The set of measuring devices used are normally not installed in the same room, so data cannot be monitored "at a single source." In addition, the time axes of these devices are normally not synchronized, which means that the parameters acquired by each one cannot be directly compared. Second, these are normally stand-alone instruments, so there is no way to provide central data storage. This means for the physician or the nursing staff that they have to manually enter data acquired from a device into a data sheet. Subsequent retrospective evaluation of these forms is normally difficult and time consuming. Third, these instruments usually do not support signal analysis capabilities.

Simultaneous data acquisition of biological signals and their online visualization and analysis on a screen allows users to view and understand the interplay of all parameters and the clinical condition of a patient. In addition, acquiring all parameters in arbitrary intervals and storing them on data media opens ways to retrospectively evaluate a patient's collective data. Moreover, signal analysis allows early detection of changes in vital patient information.

16.2   Biosignals

Of course, to develop a digital data acquisition and analysis system, it is necessary to know the properties of biosignals to be processed. The following sections briefly describe five biosignals.

❏ 16.2.1   Bioelectric Signals

Electrical signals recorded in and upon the body carry information about the functioning of (parts of) organs. The signal sources are within the organs, and the signals are carried through the tissues between the sensor and the organ. The recorded signals may be used to derive the properties of the signal source or to develop a diagnostic tool to distinguish abnormal from normal signals.

Each muscle is composed of muscle fibers, functionally organized in so-called *motor units*. Nearly all muscles contain motor units of different properties ranging from slow to fast for mechanical and electrical properties. Fast motor units fatigue rather rapidly. Fatigue expresses itself in the electrical signal of the motor unit and in the mechanical performance. The changes in the signals can be used to prevent muscular overload.

The load distribution of a nerve cell can be described in a simplified way as a *triplet* (two sources of different amplitudes and one broad sink; see Figure 16-1). In contrast to this, the distribution of the current density along the heart muscle cell is composed of two different sources and sinks, where one pair each can be allocated to excitation (depolarization) and excitation discharge (repolarization), as shown in Figure 16-1.

❑ 16.2.2   Bioimpedance Signals

The electric properties of biological tissues are described by the relative dielectric constant and the specific resistance (or specific conductivity). Both parameters are functions of measurement frequency *and* temperature. They differ depending on the tissue, and mainly morphological structure properties, water content, and macromolecule concentrations have an important effect. The electric properties of blood depend also on the quantity of corpuscular components and flow speed. Measurement is normally carried out in a four-electrode arrangement, where two electrodes serve to apply a sine-shaped current ($f = 20 - 200\ Hz$ and $I = 1 - 5\ mA$) into the tissue. The two measuring electrodes that are connected to the impedance measuring instrument detect the voltage dropping over the tissue.

❑ 16.2.3   Biomechanical Signals

Biomechanical signals have their origin in the mechanical functions of the biological system, for example, pressure and flow measurements. The arterial blood pressure can be measured directly (invasive) or indirectly (noninvasive). Noninvasive pressure measurement is particularly significant in clinical routine diagnostics. It supplies quantitative values for maximum (systolic) and minimum (diastolic) arterial pressures. In contrast to this, an online acquisition system can directly determine the pressure curve only by invasive methods. The *mean arterial pressure* (*MAP*) is calculated by the following equation over the heart cycle duration:

$$MAP = \frac{1}{t_H} \int_{t_H} P_{a(t)}\, dt \tag{1}$$

where $P_{a(t)}$ is the arterial blood pressure, and t_H is the heart cycle time.

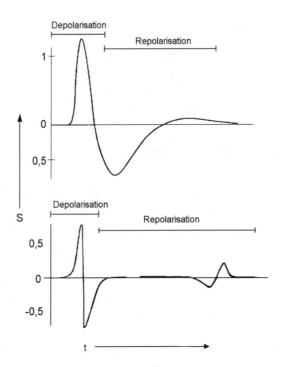

Figure 16-1

Distribution of current density (S) in the cell membrane of a nerve cell and a heart muscle cell

MAP can be calculated alternatively by the following formula:

$$MAP = P_d + C_{korr}(P_S - P_d) \tag{2}$$

where P_s is the systolic pressure, P_d is the diastolic pressure, and C_{corr} is the correction factor (0.38 for normotonic cases and 0.42 for hypertonic cases).

The frequency domain has to reach to the 10th harmonic of the basic frequency to ensure correct representation of the pressure curve. The basic frequency in an adult is approximately 1.2 Hz (rest) and 3 Hz (stress). The transmission properties of a catheter filled with liquid and connected to a pressure transformer correspond to an anti-aliasing filter with resonance rise at f_R, i.e., an RC element with serial inductivity,

$$f_R = \frac{d}{4}\sqrt{\frac{E}{\pi ds}} \tag{3}$$

where f_R is the resonance frequency in Hz, d is the catheter inner diameter in mm, E is the volume elasticity coefficient of the entire system in Pa/ml, l is the catheter length in mm, and s is the liquid density in the catheter.

Pressure transformers used in medical applications are based mainly on the following transformation methods, where the actual pressure recording is over an elastic membrane:

- A wire strain gauge, applied directly on the elastic membrane or over a mechanical transmission system. Measurement is taken with half or full bridge and direct or alternate voltage.

- Capacitive transformers.

- Inductive transformers.

Other biomechanical signals include brain pressure (measured by epidural, subdural/subarachnoidal, intraventricular, or intraparenchymal methods), heart sound (frequency domain of 15–150 Hz, heart noises up to 800 Hz), blood flow (ultrasonic dual method), and breathing signals (e.g., tidal volume).

❑ 16.2.4   Electrochemical Biosignals

An example for these biosignals is the determination of oxygen partial pressure in brain tissue. To measure tissue pO_2, a highly flexible pO_2 micro-catheter, probe-type Clark sensor (see Figure 16-2) is normally used. It is introduced through a bore hole in the skull to the brain tissue at a depth of approximately 3 cm, as shown in Figure 16-3.

The probe has a diameter of 0.55 mm. The measuring principle is based on the polarographic method, i.e., oxygen is diffused through the sensor membrane and reduced at the cathode so that it changes the polarization current that exists between anode and cathode.

This change in polarization current is proportional to the oxygen partial pressure. The measured pO_2 value depends on the temperature, so that the measuring data has to be corrected concurrently when the brain tissue temperature is measured. The carbon dioxide partial pressure is determined by the electrochemical method to measure the pH. A CO_2-permeable membrane separates the measuring medium from a reaction chamber, which contains a bicarbonate solution. As soon as CO_2 molecules enter the reaction chamber, the pH changes according to the Henderson-Hasselbach equation. This change is measured by means of a pH electrode.

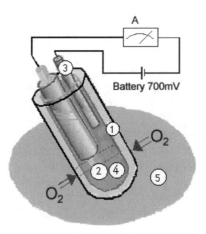

Figure 16-2

Arrangement to measure the O_2 partial pressure by the polarographic method (Clark electrode):
(1) PE hose, diffusion membrane; (2) polarographic gold cathode; (3) polarographic anode;
(4) inner space with electrolyte filling; (5) tissue

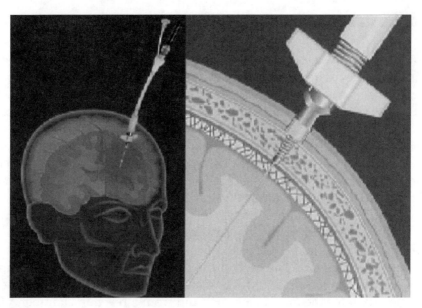

Figure 16-3

Placing the Clark probe

❑ 16.2.5   Biooptical Signals

An example for biooptical signals is the determination of cerebro-vascular oxygen saturation ($SJvO_2$). $SJvO_2$ is measured by means of a fiberoptic catheter. A photometric measuring instrument (Oxymetrix of Abbott, Wiesbaden, Germany) is used to place this catheter in the jugular vein red cells. The color of the red cells in blood changes in line with the degree of their oxygen saturation. If light falls on these red cells, the quantity of reflected light changes in relation to the oxygen content in blood.

The oxygen saturation is calculated from the following formula:

$$\text{Oxygen saturation} \quad SO_2 = \frac{HbO_2}{Hb + HbO_2} \qquad (4)$$

where HbO_2 is the oxygen-saturated hemoglobin, and Hb is the unsaturated hemoglobin.

16.3   Characterizing Biosignals

To obtain optimum signal quality, the recording chain has to be adapted to the characteristics of the respective biosignal and to interferences included in the signal. These requirements concern primarily the amplification and the bandwidth of the recording chain, which can be derived from the frequency spectrum of the desired signal. The amplification depends on the maximum amplitude of the biosignal and interferences contained in it. Therefore, there should be a high ratio between the desired signal's amplitude and noise at the output of the recording chain. The frequency spectrum of a biosignal is normally determined by the type of derivation. For instance, measurements at the body surface tend to have smaller amplitudes. Table 16-1 summarizes the characteristics of biomedical signals.

In general, we define two classes of biosignals: *continuous* signals and *discrete* signals. Continuous signals are described by a function, *s(t)*, and contain information on the signal's time curve. Discrete signals are described by a sequence, *s(m)*, which supplies information on a discrete point on the time axis. Biosignals are mostly continuous signals. To digitally

Table 16-1   *Biomedical signals*

Biosignal	Acquisition	Frequency Range	Dynamic Range
Action potential	Microelectrode	100 Hz–2 kHz	10 µV–100 mV
Electroneurogram (ENG)	Needle electrode	100 Hz–1 kHz	5 µV–10 mV
Electroencephalo-gram (EEG)	Surface electrode	0.5–100 Hz	2–100 µV
Evoked potential (EP)	Surface electrode		0.1–10 µV
Visual evoked potential (VEP)		1–300 Hz	1–20 µV
Somatosensoric evoked potential (SEP)		2 Hz–3 kHz	
Acoustic evoked potential (AEP)		100 Hz–3 kHz	0.5–10 µV
Electromyography (EMG)	Needle electrode	500 Hz–10 kHz	1–10 mV
Electrocardiogram (ECG)	Surface electrode	0.05–100 Hz	1–10 mV
High frequency ECG	Surface electrode	100 Hz–1 kHz	100 µV–2 mV

process a signal in a computer, the analog continuous signal $s(t)$ has to be transformed into a digital discrete signal, $s(m)$

$$s(m) = s(t)\big|_{t=mT_s} \tag{5}$$

where T_s is the sampling interval.

To prevent the formation of additional frequency components due to undersampling during the transformation of the analog signal, the sampling frequency must be at least five times the highest frequency f_0 that exists in the signal. In addition, the bandwidth of the analog signal should be limited with a steep, active, anti-aliasing filter.

Both the decomposition of the biosignal and the dynamic range of the digital signal are defined by the size of a quantization step and the number of possible quantization steps. Amplitude changes in the analog signal below a certain discrete step are lost during digitization and cannot be recovered. Therefore, the biosignal should be digitized with a decomposition of at least ≥10 bits.

16.4 Developing a Digital Acquisition System

Before you buy data acquisition hardware and start programming the control software, you should work out a detailed design of your data acquisition system, considering the following points:

- Type and number of physiological signals

- Sampling rate

- Data storage interval and display interval

- Format for data storage (binary or ASCII format)

- Measurement instruments to be integrated

☐ 16.4.1 Measurement Instrument Assessment

In the next step, you document the following points for each measurement instrument you want to integrate:

- Available outputs (XY writer, RS-232 interface, etc.)

- Interval to output parameters

- Transmission protocol in case of serial transmission (baud rate, parity, stop and data bits, handshake, etc.)

- Internal signal processing (analog filtering, A/D and D/A conversions)

- Dynamic range of analog signals

- Galvanic separation of interfaces

- Device offset

- Conversion factors to calculate the physical units from the volt signal

☐ 16.4.2 Available Hardware

Once you have assembled all data about your equipment and the biosignals to be acquired, you can select hardware for computer-assisted data acquisition.

In addition to their graphical programming environment — LabVIEW — National Instruments offers a wide range of sophisticated hardware for almost all measurement problems. For example, to acquire and digitize analog signals, both simple A/D converter boards and easy-to-use plug-and-play multifunction boards are available. Cascading multiplexers can be installed to acquire up to 128 differential channels concurrently. Stand-alone devices, such as bed-mounted hemodynamic monitors or respirators, are normally equipped with an RS-232 interface. You can use the serial AT232/4 expansion card to read data from up to four instruments concurrently.

☐ 16.4.3 Software Requirements

Once you have established all your hardware requirements, you need to define the properties of the control software. This assessment involves issues like: "Should data be represented in trend curves, or will digital indicators be sufficient?" The most important software characteristics are:

- Graphical trend representation

- Graph updating interval

- Data storage format (e.g., binary, ASCII)

- Permanent configuration or multimodal design

- Software, including setup routines

- Digital filtering and/or down-sampling

- Switches and editors for manual data input

- Parameter extraction

- Real-time x axis, including expansion and contraction options

- Online signal analysis (cross-correlation, power spectra, etc.)

Consider that the software you plan for your system will most likely be used or operated by several other people, so you should ensure that it has a user-friendly interface.

16.5 Sample Application in LabVIEW

This section describes a real-world application developed at the Clinical Center of Mannheim, Germany. Specifically, this is a multimodal monitoring system for patients with serious skull or brain trauma. This system expands the established monitoring systems installed in the Clinic's intensive care department.

☐ 16.5.1 Requirements

The system was implemented with the following requirements in mind:

- Maximum patient safety

- Maximum flexibility by integration of various medical instruments from various vendors in any combination

- Easy-to-use graphical user interface (GUI)

- Online trend representation of yielded data from a span of up to 72 hours, including x-axis options

- Storage of data in spreadsheet format to make the data available in popular software applications, e.g., Microsoft Excel

- Open software architecture to allow easy system modification

- International hardware and software compatibility

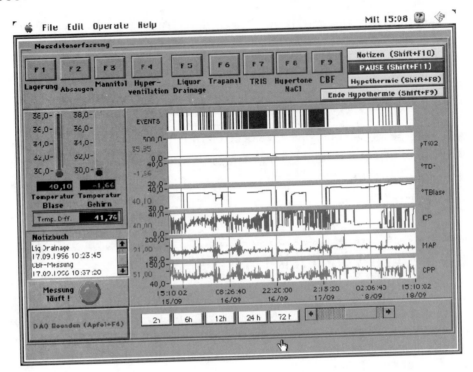

Figure 16-4

Front panel of enhanced monitoring system

Figure 16-4 illustrates the front panel of such a system.

The parameters listed in Table 16-2 must be acquired in intervals of minutes and stored, including time and date stamp.

In addition, the following inputs, including time stamps, should be input manually and in an easy way:

- Important events, e.g., relocating patients, administering mannite

- Medication

- Blood gas analysis values

- Coagulation parameters

- Lab values

- General patient data

Table 16-2 *Parameters, measuring instruments, and interface used in the sample system*

Parameter	Measuring Instrument	Interface
Heart frequency	1281 Monitor	RS-232
Arterial pressure, systolic	1281 Monitor	RS-232
Arterial pressure, diastolic	1281 Monitor	RS-232
Arterial mean pressure	1281 Monitor	RS-232
Intracranial pressure	Camino	RS-232 via 1281
Bladder temperature	1281 Monitor	RS-232
Brain temperature	LICOX Computer	RS-232
Cerebral perfusion pressure	1281 Monitor	RS-232
Oxygen partial pressure in brain	LICOX Computer	RS-232
Mixed venous oxygen saturation	Oxymetrix	RS-232
FiO_2	SERVO 300	RS-232
Inspiratory tidal volume	SERVO 300	RS-232
Peak breathing pressure	SERVO 300	RS-232
PEEP level	SERVO 300	RS-232
Mean breathing pressure	SERVO 300	RS-232

☐ 16.5.2 Implementation

In this application, minute polling of the instruments and data storage has priority over the sampling speed, so we opted to transmit data over the RS-232 interface. Using the serial AT232/4 expansion card from National Instruments, concurrent communication between a personal computer and

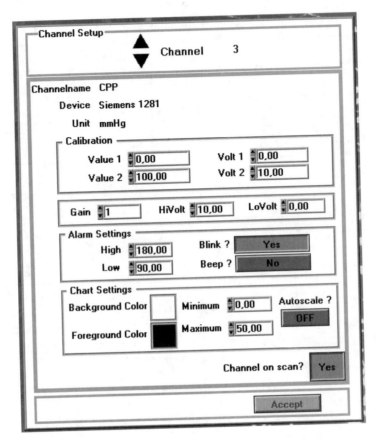

Figure 16-5

Setup VI to configure the data acquisition system

the four instruments (Siemens 1281 monitor, Siemens Servo 300 respirator, Abbott Oxymetrix 3 SO_2/CO computer, and GMS Licox pO_2 computer) is supported.

The Abbott instrument is the only unit of the instruments listed above that can be operated over a conventional zero modem cable. Both the Siemens instruments and the Licox computer require special manufacturer cables.

Servo 300 and Oxymetrix offer an extensive command set for serial communication that allow polling of almost all parameters, from serial number to the actual measured data.

The Siemens 1281 monitor supplies up to eight parameter blocks, each identified by ID numbers. The powerful string functions of LabVIEW make it

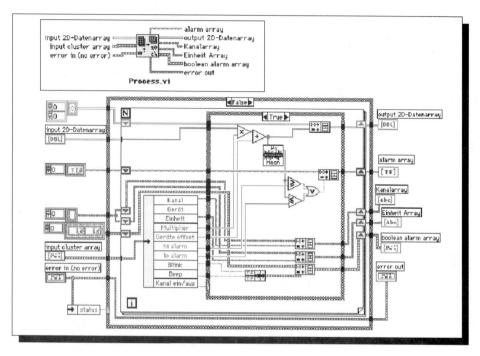

Figure 16-6

SubVI within the Setup VI to extract the channels selected during Setup

very easy to extract the desired parameters, to convert them into numerical values, and to represent the results on the computer screen. Programming parallel while loops makes it possible to acquire all data concurrently and to represent it on a monitor. The integration of various measuring instruments into the system was implemented by one of LabVIEW's setup subVIs. This means that channels can be enabled and disabled, amplification and conversion factors can be input, and alarm thresholds can be set. All setup data can be stored and retrieved on demand. The data acquisition system is configured in the setup program by setting appropriate parameters.

To enable manual input of parameters, events, or comments, various editors were programmed as subVIs so that the users can activate them by the push of a button. Data can then be time-stamped and saved in ASCII files. Input and saved lab values or medication information can also be displayed and printed in graphical form.

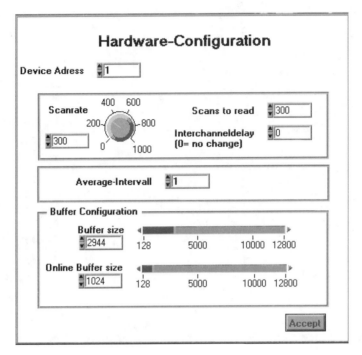

Figure 16-7

SubVI hardware configuration VI to create hardware parameters

16.6 Summary

This chapter described important background information on biosignal acquisition and ways to use LabVIEW in biomedical applications. The real-world system described in this chapter to illustrate the potentials of LabVIEW in this field is a multimodal monitoring system that enhances the existing conventional monitoring system. This system demonstrates that LabVIEW represents a powerful development tool to quickly turn advanced concepts into program code for biomedical applications.

17

BioBench

It is probably true quite generally that in the history of human thinking the most fruitful developments frequently take place at those points where two different lines of thought meet. These lines may have their roots in quite different parts of human culture in different times or different cultural environments or different religious traditions: hence, if they actually meet, that is, if they are at least so much related to each other that a real interaction can take place, then one may hope that new and interesting developments may follow.

— Werner Heisenberg
Physics and Philosophy

*I*n the previous chapter, we discussed a number of physiological parameters in a medical environment and a real-world application illustrating the basic concepts. In this chapter, we introduce BioBench, a ready-to-run, general-purpose software tool, designed for physiological data acquisition and analysis.

17.1   Introduction

Virtual instrumentation is a powerful paradigm providing flexibility to engineers, researchers, and scientists. Flexibility enables solutions to be built with ease and less invested time, empowering users to finish applications more quickly and with greater efficiency. However, the most efficient solution to a problem is that which is already complete. National Instruments has recognized this need for turnkey software applications and has developed solutions for applications when programming skills are underdeveloped or time is of critical importance. The first turnkey solution from National Instruments was VirtualBench, a suite of instruments common to the engineering or research laboratory. The suite includes a digital multimeter, function generator, arbitrary waveform generator, oscilloscope, and data logger. With these software tools and a National Instruments data acquisition card, an engineer can perform a variety of tasks using one instrument, the PC, rather than a number of instruments. This concept has provided solutions to user applications in hours rather than weeks or even months of programming. Virtual instrumentation and turnkey solutions continue to change the way engineers and scientists accomplish tasks and realize their goals.

The virtual instrumentation concept has been applied to the areas of biomedicine and physiology as well. Although there are many medical researchers who have the expertise to program a custom application, many have neither the training nor the time to program their own solutions. In fact, many scientists would rather just focus on the data itself and its analysis rather than on the methods by which data was entered into the computer. In this arena, National Instruments recognized another prime opportunity for a turnkey application — thus the inception of BioBench, a ready-to-run software tool designed for physiological data acquisition and analysis. For the medical researcher acquiring physiological data but without the programming expertise or time to analyze that data, BioBench is an ideal solution.

17.2   BioBench Capabilities

BioBench is easy to use and intuitive in its operation. Similar in many respects to the chart recorder, BioBench provides a real-time display of up to 8 signal channels at one time while having the capability to continuously store 16 channels of data to disk. You begin a BioBench session by selecting your particular instrument from a vendor database. This action automatically configures all channels on your instrument for data collection. Pressing the **GO** button begins the acquisition of data from the specified instrument. During acquisition, you can enable or disable logging data to disk, create user-defined markers for specific data areas, and record application-specific notes in a journal. Real-time trending may also be enabled to display the data history for any particular signal. Using BioBench, you can also configure alarms to activate when signal values are out of range. If a stimulus is needed, BioBench has stimulus control through a front-panel user control as well as from a user-defined pattern stimulus. For the customization of acquisition timing, BioBench can begin recording data after a specified time period, software trigger value, or front-panel manual control.

Once data acquisition is finished, BioBench employs powerful analysis routines to meet your research needs. You can view up to eight signals, each from a different file for direct comparison. Each of these eight signals can be plotted on top of the others with the overlaid analysis view. You can also view the data with a 2, 4, or XY graph analysis view. After highlighting the data of interest, you can calculate the integral, derivative, standard deviation, and other mathematical functions. More in-depth analysis can be accomplished with histogram, XY graph, peak detection, and power spectrum utilities. Using these general analysis tools, you can accomplish much of your analysis without having to do any programming. You can also employ a number of different filters and transforms on their signals for featured analysis results. All files within BioBench are exportable to a spreadsheet format (.txt) for reading into a spreadsheet program. For customized analysis, you can read the BioBench binary data files into LabVIEW and write your own set of analysis routines.

17.3   Compatibility and Cost Effectiveness

Using BioBench with a National Instruments data acquisition device is the best choice for physiological data acquisition and analysis. With the data acquisition device, you can acquire any type of signal pertinent to your research. Because the device is an open configuration, you can connect any existing instruments for recording data directly into the PC. This open architecture allows you to continue to use your instruments rather than buying new hardware. This compatibility also makes BioBench a very cost-effective solution.

Because there is no need to buy additional instrument hardware, you save both time and money. You save time because no getting-up-to- speed time is required. You save money because you need not reinvest in instruments you already have. Continue to use your existing instrument hardware while implementing a flexible, cost-effective solution with virtual instrumentation.

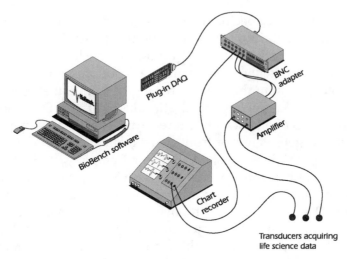

Figure 17-1

Transducers acquiring physiological data

17.4   BioBench Applications

BioBench is a turnkey application software designed for the life sciences. With the appropriate hardware, BioBench can be used to monitor any type of physiological signal such as EEG, ECG, EMG, EOG, ERG, pressure, volume, flow, temperature, and force. A typical laboratory arrangement that uses BioBench to acquire physiological signals is shown in Figure 17-1.

One of the most important and innovative features of BioBench is that it allows you to take advantage of your existing transducers and amplifiers while implementing a PC-based solution for acquisition, storage, and analysis. Most of the time, instrumentation solutions require the use of both hardware and software from the same vendor. In BioBench, a vendor database is provided, to which you can add specific vendor information if it is not already included. This feature allows you to calibrate your measurements based on the specific instruments that are being used in your environment.

Acquisition with BioBench is compatible with any existing instrument that outputs a linear analog voltage. As Figure 17-1 shows, it is a matter of connecting your instrument to any National Instruments BNC connector and to any of our data acquisition (DAQ) boards. Figure 17-2 shows the results of an experiment in which ECG, heart sounds, and carotid pulse

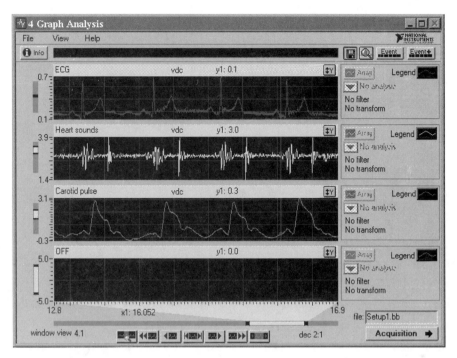

Figure 17-2

ECG, heart sounds, and carotid pulse

were monitored using a standard six-lead configuration, phonocardiogram, and a pressure-wave transducer, respectively. These devices were connected to a chart recorder (Physiograph Four, E&M Inc., from the 1970s) that both filtered and amplified the signal. The chart recorder analog output was connected to a DAQ-1200 PCMCIA card via a connector block. The signals were acquired, saved, and analyzed using a laptop computer running BioBench. This experimental setup demonstrates the versatility of BioBench in integrating modern technology with equipment developed before the PC revolution.

Another experiment performed with a similar configuration involved monitoring cardiac output. This diagnostic is important in detecting potentially life-threatening abnormalities in the performance of the heart. Thermodilution techniques can be used to indirectly measure cardiac output. The temperature-time history of the patient's blood is monitored with an intravascularly positioned thermistor after injection of a cold stimulus into the circulatory system. The signals that were measured are shown in Figure 17-3; these include the ECG, blood pressure, and the change in temperature detected by the thermistor.

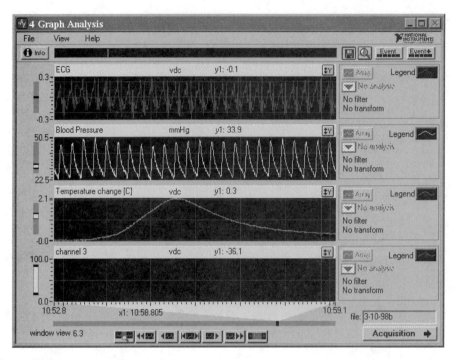

Figure 17-3

BioBench image with three channels: ECG, blood pressure, and temperature

Using the same physiograph, connector block, and data acquisition card, investigators performed other experiments involving the implementation of different types of transducers. For example, a drop-counter transducer was used to measure urine output during drug delivery in order to study the effects of different drugs on the renal system. Another example includes the use of an ultrasound flow meter as another means of monitoring cardiac output.

17.5   Summary

BioBench is an easy-to-use and powerful instrument for signal acquisition and analysis in both medical research and academic settings. It can be used as a stand-alone tool, or it can be combined with LabVIEW for user-specific data analysis. BioBench's flexibility and versatility allow the user to combine new technologies with existing equipment, achieving a faster, less expensive, and powerful automated solution.

18

Control and Simulation in G — An Integrated Environment for Dynamic Systems

> *Anyone could make new combinations with mathematical entities... to create consists precisely in not to make useless combinations and in making those which are useful, and which are only a small minority: invention is discernment, choice... Among chosen combinations the most fertile will often be those formed of elements drawn from domains which are far apart.*
>
> — Henry Poincaré
> *in J. Hadarmard: An Essay on the Psychology of Invention in the Mathematical Field*

*T*he development of control systems involves simulation as well as real-world implementation. The control and simulation software GSIM for LabVIEW with its ability to model, analyze, and simulate dynamic systems provides an integrated environment for such applications. In this chapter we discuss GSIM based on the graphical and data-driven paradigm LabVIEW and practical applications.

18.1   Introduction

The following pages discuss one theme that throws a stronger light on the simulation and control aspects of graphical languages. The question is whether such languages or programming environments are suitable simulation tools. There is no need to discuss these capabilities in specific areas, such as signal processing, because this is a field of application in which graphical languages have been well represented since its introduction.

After reading thus far, it might seem that a graphical language is highly suitable for the fields of work mentioned. There is a range of arguments which would strengthen this position. Thus, the conversion of algorithms in such a program system is not only efficient, but frequently even much more obvious than the classic sequentially oriented description. As a rule, the runtime response is very good, and very often the platform independence is guaranteed. Finally, and this is an item on the credit side of a graphical language such as LabVIEW's G which is hardly to be underestimated and entirely in contrast to the vast majority of other known mathematics and simulation programs, the coupling with the real world is by no means an insurmountable bottleneck.

It would really be desirable if all the algorithmic knowledge accumulated to the present day could be fitted in paradigms such as the graphical representation. Even if all other advantages were to be left out of account, the legibility and internal structure of algorithms are revealed to the learner much more simply and even transparently.

For dynamic system control and simulation, many tools are available to help engineers and scientists to efficiently accomplish their research and development tasks. People have often asked:

- Is there a genuine engineering programming tool that is more intuitive and straightforward?

- Is it possible to implement the results of simulation immediately to a physical, e.g., experimental, process?

- Is it possible to integrate theoretical prototype testing, simulation, and control system design in *one* integrated programming environment?

The answer is YES. The control and simulation software for LabVIEW — GSIM for short — is completely based on the graphical and data-driven paradigm of LabVIEW. Control and simulation tasks are significantly simplified by the direct online user interactions and connections to real processes through a variety of hardware devices with built-in drivers.

Control and simulation software for LabVIEW (GSIM) is a collection of routines for modeling, analyzing, and simulating dynamic systems. The systems can be both linear and nonlinear, where all operations are based on continuous or discrete time. GSIM supports the combinations of these two time models. The GSIM package can handle real-world control problems, simulation tasks, and a mixture of both. The latter can be of great importance during the development of a new control system which is partly simulated and partly realized.

The systems are completely defined as graphical objects (LabVIEW diagrams). Special elements such as transfer functions, relays, PID (proportional integrator and differential) controllers, or signal generators, to name a few, are realized in form of subVIs with well-defined connectors. Almost all necessary parameters can be altered by the end user during run time, which permits a great deal of flexibility (e.g., adaptive controllers).

The user interface of LabVIEW is an essential part of the package. All elements of the LabVIEW user interface can be utilized. Moreover, all existing LabVIEW VIs (Full Development System and all Toolkits) can be used in conjunction with GSIM. For example, one can add the functionality of G Math to define input signals by formulas. Another example: Results of GSIM can be analyzed by special transformations such as wavelets.

All GSIM models are hierarchical. Though there is usually no need to construct deeper structures, more complicated models or control systems can be built up by hierarchical structures. All predefined GSIM VIs are written in G and can be opened and investigated at any time. These VIs are reentrant, which means that any number of the same VI can be used in a solution (independently of their positions in the hierarchy).

The control and simulation software is under further development in at least two directions. The first direction is determined by more sophisticated simulation routines. On the other hand, GSIM can not only simulate or analyze systems but it can also control real-world processes. With a typical data acquisition board the system can handle more than 1,000 control loops a second. One of the most promising ideas is the so-called embedded LabVIEW. This version uses a special real-time kernel. Embedded LabVIEW will run on the intelligent hardware. Data acquisition functions will be provided by compact PCI cards that reside on the board.

18.2   Need for Control and Simulation Software

Control and simulation for dynamic systems are becoming increasingly important due to cumulative economic, safety, and environmental demands being placed upon companies as they strive to remain competitive in the world market. As a result, control and simulation for dynamic systems have undergone significant changes since the 1970s, when the availability of inexpensive digital technology began a radical change in instrumentation technology. Pressures associated with increased competition, rapidly changing economic conditions, more stringent environmental regulating, and the need for more flexible yet more complex processes have given engineers and scientists an expanded role in the design and operation of almost all industrial sectors.

In the foreseeable future, there will be an expanded use of micro-processor-based instrumentation and networks of digital computers. More sophisticated control strategies — including feed-forward, supervisory, multivariable, and adaptive control features as well as sophisticated digital logic — are easily justified to maintain plant operation closer to the economic optimum. On the other hand, conventional analog control systems continue to be used in many existing plants.

An effective tool for developing a strategy should reflect this rather divers milieu of theory and applications. It should incorporate process dynamics, computer simulation, feedback control, a discussion of measurement, and control hardware (both analog and digital).

There are a number of specially developed simulation languages which have also partly become widespread in science and technology. Such systems should have the following properties in the ideal case.

- Considerable closeness to the problem, i.e., it should be easy to translate the tasks into the programming language

- Good visualization possibilities

- Simple treatment of variant calculations and parameter domains

- Range of data structures that are significant for simulation calculations

- Well-constructed mathematics and signal theory

- Possibility of linkage to the real world

National Instrument's LabVIEW was primarily developed to improve applications concerning general problems in processing measured values. Until recently, it was not actually clear whether LabVIEW is also capable of handling more demanding mathematical or simulation problems. Although there was a series of examples which converted more complex mathematical routines on the basis of LabVIEW's language G, the majority of these examples were designed purely for demonstration purposes. However, a few of these programs can be termed the first examples of a new generation of algorithmic descriptions with a graphically determined background. This gave rise to the question whether LabVIEW's G has the expressive power to build a sufficiently large selection of important functions and procedures. It will be necessary in a second stage to consider how far such a strategy is efficient.

18.3   GSIM and LabVIEW

Graphical objects such as LabVIEW diagrams completely define systems. SubVIs with well-defined connectors provide special elements such as transfer functions, relays, PID controllers, or signal generators. Because the end user can alter most necessary parameters during run time, you have a lot of flexibility when designing systems.

With GSIM, you can use all elements of the LabVIEW user interface and all existing LabVIEW VIs from the full development system and toolkits. With complete compatibility, you can add the functionality of G Math to define input signals in formulas. You can use special transformations to analyze GSIM results (for example, wavelets when you use the LabVIEW Wavelet and Filter Bank Design Toolkit).

One usually can complete a GSIM project in four steps. Depending on the problem, you might be able to combine some steps.

1. Define and program the control system or simulation problem.

2. Choose parameters and initial values.

3. Execute the solution.

4.   Process the results (for example, visualization or further investigation).

The following list enumerates the main features of GSIM.

- Simulates and offers real-time control of linear and nonlinear systems

- Connects directly to data acquisition (DAQ) boards

- Displays solutions in graphical form

- Supports many predefined special control routines and elements (for example, PID, relay, and filter)

- Allows direct user actions on the front panel during run time

- Handles both continuous and discrete problems

- Supports LabVIEW and offers complete G compatibility (that is, you can combine GSIM with other LabVIEW VIs and toolkits)

- Includes many VIs completely written in G

- Offers an open system where a user can add new elements and control structures

- Animates results

- Executes compiled code quickly

Furthermore, GSIM also includes the following special features.

- Three different continuous integrators: Euler, Adams, and Runge-Kutta

- Three different discrete integrators: Euler backward, Euler forward, and trapezoidal

- Zero-pole, transfer-function, and state-space representations of systems

- Graphical analysis tools such as Bode plot, Nyquist plot, and root-locus plot

- Design of linear state feedback for arbitrary pole placement

- Elements and control structures in VI form with user-controlled, runtime-adjustable parameters that are important for applications such as autotuning PID parameters

- GSIM Manager VI and GSIM Synchronizer VI watch real-time behavior and provide warnings

- Large systems can be decomposed based on the subVI technique

18.4   General Simulation Structure

The main idea of GSIM is that the majority of control and simulation problems can be reformulated in graphical form. In fact, this graphical form is the most appropriate representation of these tasks. Figure 18-1 shows a differential equation with a given initial condition. The time-dependent function y (which stands for a physical, technical, biological, or such quantity) is unknown, whereas the function f and the initial value are known. The formula is a very compressed description of this problem. On the other hand, there is a one-to-one relationship between this formula and the graphical depiction of the initial value problem. This relation is based on a feedback structure, as shown in Figure 18-1. The box with the name "Integration" outputs y if the input is the derivative of y. The second box calculates f as a function of y and time t. The feedback structure can be interpreted as an equation, namely, the derivative of y and f have to be the same. This is exactly what the formula describes. The initial condition is part of the left box, "Integration."

There are at least three advantages to a graphical representation.

1.   Many control and simulation problems are much more complex than the one depicted. Very often, one has to take into account nonlinearities, numerous variables, and special cases — the list is very long. In all these cases, the graphical form is the best choice. Moreover, graphical representations can describe some problems that cannot be described by formulae.

2.   The graphical concept is extremely intuitive. One can add new features, produce immediate results by simply adding a new branch (wire), and check and test the entire model or parts of it without influencing or destroying the existing system.

Principle of Simulation

$$dy(t)/dt = f(t,y(t))$$
$$y(0) \quad = y_0$$

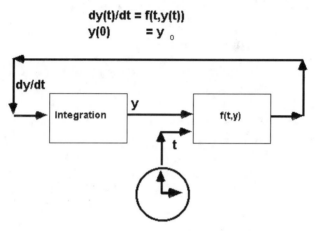

Figure 18-1

Differential equations (simulations) are realized in the form of feedback structures.

Principle of Control

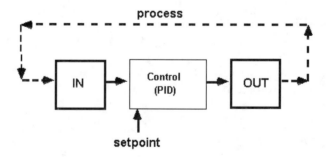

Figure 18-2

Control systems are organized in a straightforward manner (no feedback on the diagram; the process provides the feedback)

3. Many control and simulation problems are described graphically, so that there is no need to reformulate the system.

Pure control systems have a graphical representation, as shown in Figure 18-2. Usually, real-world control systems receive feedback through hardware inputs, so that there is no need to build a feedback structure. The control block can have any degree of complexity and is constructed by an appropriate combination of GSIM routines. The IN and OUT boxes are part of GSIM. A mixture of both concepts shown in Figure 18-1 and Figure 18-2 is possible.

18.5   Typical GSIM Applications

Control and simulation applications fall into many domains. The following list explains some of the possible applications and provides examples of those applications.

- **Analysis and optimization of linear time-invariant systems** — Calculates the overall transfer function of two systems in series, parallel, or feedback connection. Computes the poles, zeros, and other useful parameters such as damping ratios, natural frequency, and settling time. Converts one type of representation to another. Designs linear state feedback for arbitrary pole placement. Analyzes a control system using graphical methods such as the root-locus plot, the Bode plot, or the Nyquist plot.

- **Real-time control of deployed systems** — Controls inverted pendulums, PID, heating with relays or more sophisticated strategies, cascaded water tanks, and chemical reactors. These applications are based on DAQ boards.

- **Development of real-time control systems** — Accomplishes the same tasks as those listed for real-time control of deployed systems. Developers can optimize PID parameters, relay thresholds, and limited integrators. With GSIM, you can alter values during run time or partly implement and partly simulate a new system under development.

- **Simulation and simulation with data acquisition** — Simulates situations in fields such as engineering, physics, chemistry, biology, astronomy, and math. Many simulations are based on input data (for example, the current position of a space craft or the current flow into a tank), and you can predict the behavior of the system in the near future (for example, the position of that space craft in one hour or the water level of that tank in one day with given inflow).

■ **Education** — Simulates linear and nonlinear systems of any desired complexity. For example, you can simulate mechanics, analog circuits, biology (predator-prey model), celestial bodies, physics, inverted pendulum, study of numerical algorithms, and animations.

18.6   A First Example

Figure 18-3 depicts the front panel of the House Temperature Control Example.vi. The upper-left part of the front panel consists of the input parameters. The values of *dt* (step rate of the control or simulation), end time representing simulation time, not real time, and continuous integrator have to be fixed at the beginning of the calculation. All other parameters can be altered during run time — the Model/DAQ switch, which switches between simulated and actual process; the set point; the controller switch; and for all parameters determining these control strategies in detail. The behavior of the control system is shown by a chart. The boolean "timing" indicates a violation of the real-time behavior of the realized control system and provides an important warning, i.e., the LED turns on when the system is running slower than real time.

Figure 18-4 represents the main diagram (there are some subVIs) pertaining to the House Temperature Control Example.vi. All control and simulation tasks in this package are based on a while loop. Feedback structures are realized by the use of local variables. In the example, there is only one feedback variable, namely, the temperature. The complete control or simulation task is controlled by the GSIM Manager.vi (see the lower part of the while loop). The VI on the left side of the while loop initializes the step rate, the end time, and the chosen integrator. All other parameters are part of the while loop and can be altered during run time of the control or simulation task. The set point, in particular, can be modified by the user or as a result of a direct measurement. The example shows only the first case.

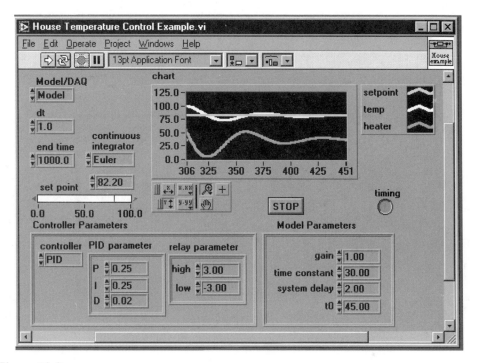

Figure 18-3

Front panel of the House Temperature Control Example.vi. All parameters are accessible during run time.

The graphical presentation of the calculated results is realized in the upper-right part of the `while` loop. All LabVIEW graph-handling elements can be used. In this example, a chart consisting of the values for set point, temperature, and heater was chosen.

The strength of LabVIEW and control and simulation software is based on the following features.

- Instrument control, e.g., libraries for VISA, GPIB, VXI, PXI, and serial connections

- Data acquisition and control, e.g., numerous plug-in boards, PLCs, data loggers

- Data analysis, e.g., Advanced Analysis library, G Math

- Connectivity, e.g., TCP/IP, DLL, UDP, and AppleEvents

- Numerous specific toolkits

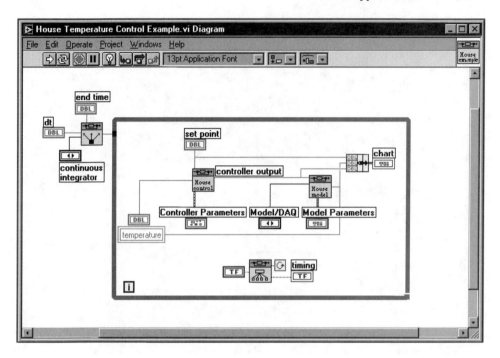

Figure 18-4

Diagram of the House Temperature Control Example.vi

18.7   A Second Example

Figure 18-5 shows both the main diagram and the user interface pertaining to the realization of a general transfer function *H(s)* with given initial conditions. All control and simulation tasks in GSIM are based on a while loop. Feedback structures are realized by the use of so-called local variables. In the example there is no feedback variable. The complete control or simulation task is controlled by the GSIM Manager.vi (cf. lower part of the while loop). The VI on the left side of the while loop initializes the step rate, the end time, and the chosen integrator. All other parameters are part of the while loop and can be altered during run time of the control or simulation task. The *H(s)* icon has some input parameters, namely, numerator and denominator belonging to the transfer function, the initial conditions, and the current value of the signal.

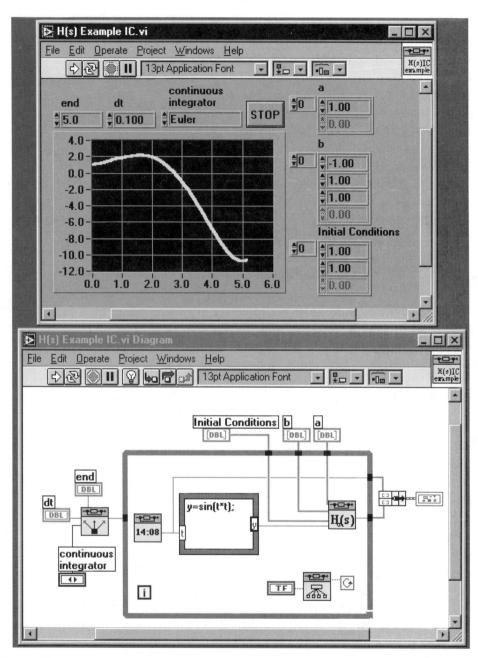

Figure 18-5
User interface and program (diagram) of a general transfer function with initial conditions

The graphical presentation of the calculated results is illustrated in the upper-right part of the diagram. All LabVIEW graph-handling elements can be used.

The position of control parameters on the diagram is of fundamental importance. All controls that are part of the `while` loop can directly and immediately determine the behavior of the GSIM structures. A control parameter that is not part of the `while` loop, on the other hand, delivers a valid value only at the very beginning of the program. Depending on the given problem, the user can choose the appropriate method.

18.8   Case Study 1: Simulation Analysis of Relay Feedback Technique

The Relay Feedback autotuning method proposed by K. J. Astrom and T. Hagglund in 1984 is quite popular in the area of PID autotuning because it is simple to understand, easy to use, and applicable to a wide class of common industrial processes. If you want to try this technique and gain some experience with it, you can simulate it in an integrated environment; include the following tasks.

- Obtain a process model simulator (transfer function, ODE calculation).

- Continuously monitor the simulation process, perform the identification, i.e., calculate the oscillation period, ultimate gain, etc.

- Calculate the new PID parameters.

All the above tasks can be accomplished in the Control and Simulation package, as illustrated in Figure 18-6. In the user interface, you can easily define the process model and experimental conditions; moreover, you can change these parameters *on line* to observe how the system behaves in different situations; you can also observe the oscillation curve through the plot, which can be rescaled, zoomed in, and zoomed out; the results of identification and tuning can also be shown in the interface; you are able to define any other interfaces with which you are more comfortable.

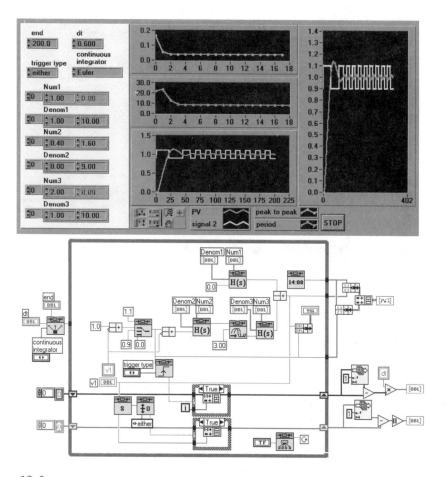

Figure 18-6

User interface and program (diagram) of the Relay Feedback example.vi

The real programming is the diagram, where you are doing the graphical programming, which means instead of writing commands and sentences, you just select existing VIs from the Control and Simulation package and LabVIEW, and then wire them. This manner is very natural to engineers and scientists because it just reminds them of the experience of connecting a circuit or deploying the toys.

The debugging process is very easy to handle, too. In the Run mode, you can choose the single step to observe how data is transferred into individual VIs, and you can put some probes on any part of the data path to watch the specific variable.

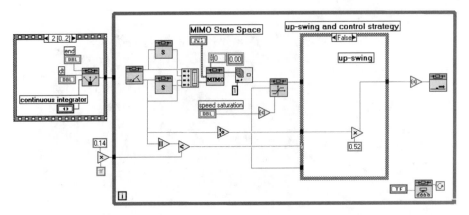

Figure 18-7

Realization of the inverted pendulum under GSIM

18.9   Case Study 2: Inverted Pendulum

Implementing control solutions from scratch requires knowledge not only in automatic control but also in sensor technology and in computer engineering, such as hardware interfacing and real-time programming. LabVIEW and GSIM simplify this process significantly. The inverted pendulum has been widely used as an example in the understanding of more complicated control strategies. The development of controllers for this highly unstable system is a key that opens the door to a much broader class of real-world applications.

The inverted pendulum consists of a rod on a pivot which rotates on top of a cart. The cart moves horizontally on a rail while the pendulum's angle of rotation is perpendicular to the direction of the motion of the cart. The objective is to keep the rod in a relatively stable upward direction.

An appropriate control scheme to handle this problem is based on state-feedback control. The state vector consists of four different values. Only two of them can directly be measured (position of the cart, angle of the rod); the other two components must be derived from the other ones.

Figure 18-7 shows the implementation of a real inverted pendulum experiment, where the so-called upswing is part of the solution. The sampling rate is 0.08 seconds. GSIM can not only control this real pendulum but also simulate the entire system. This capability enables the developer to tune the parameters in advance.

18.10   GSIM Programs — A Step-by-Step Approach

All GSIM programs have a similar structure. Though a user can realize solutions of any desired complexity, some general rules have to be followed. This concept can be explained in a very straightforward manner based on simple steps. The package contains some predefined frames that can be used as starting points for projects under development. The structure of these frames is demonstrated in a step-by-step approach.

Step 1:

Start with a new VI. Name it and save it. Add a while loop to the diagram and choose the Manager.vi and the Initializer.vi which are in the SHARE.llb. Place both VIs as shown in Figure 18-8. On the front panel, create controls for end time, *dt*, and the continuous integrator to GSIM Initializer.vi. Connect the dummy output of GSIM Initializer.vi to the while loop. This guarantees the correct execution sequence. Connect the upper output of Manager.vi — the name is stopped? — to the conditional terminal of the while loop.

   The GSIM Manager.vi controls all data flows and handles error conditions. Usually, a control or simulation task stops when the end time is reached or if an error occurred. Additionally, one can connect a stop button to GSIM Manager.vi. The Manager.vi accepts a boolean true as an explicit stop command.

Step 2:

Add a numerical control representing the feedback to the front panel and name it. Go to the diagram and create a local variable with this name. Convert the local variable to an indicator. Create a second control with the name "initial value". Open the LINEAR.llb library and choose GSIM Integrator.vi. Connect all controls, the local variable, and GSIM Integrator.vi as shown in Figure 18-9. The numerical control *y* and its local variable form the two parts of a feedback structure. This VI is your first self-made GSIM program. You can run it under various conditions. You can watch the behavior of the numerical control and alter the initial conditions.

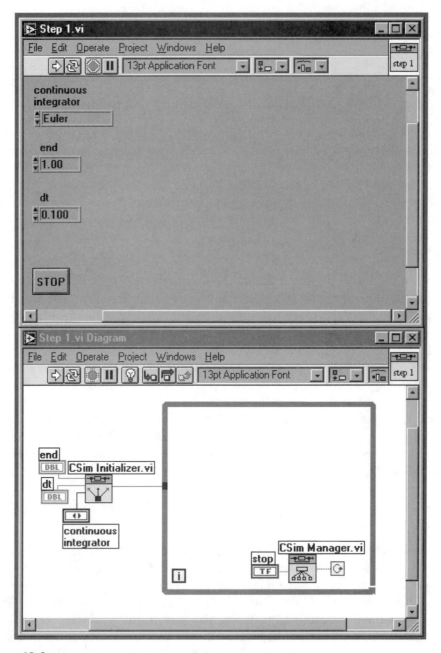

Figure 18-8

The first step generates a very general GSIM frame

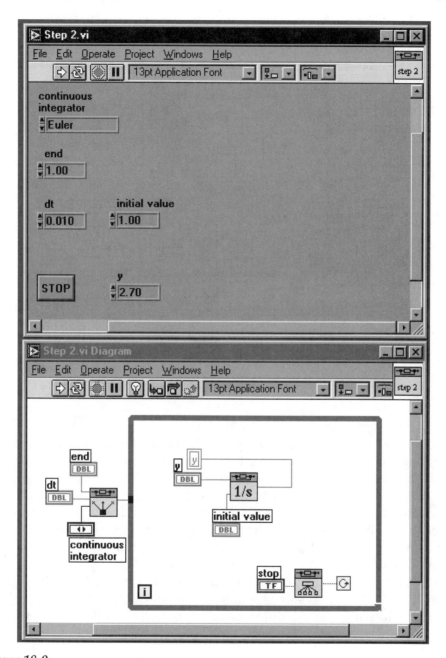

Figure 18-9

The integrator and a control for the initial conditions are connected as shown

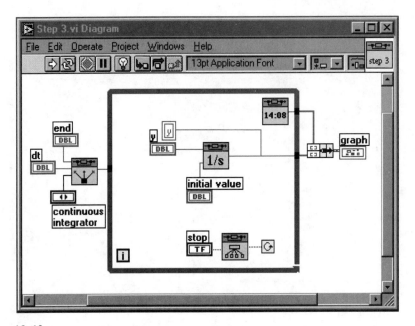

Figure 18-10

The graphical user interface is programmed

Step 3:

Add the user interface as shown in Figure 18-10. You can find GSIM Simulation Clock.vi in SOURCE.llb. Hide the numerical control y. Though this is not absolutely necessary, it results in an essential increase of speed. Fix the following values: *end time = 1, dt = 0.01, continuous integrator = Euler, initial condition = 1.0.* The calculated graph depicts the *exp(t)* function in the interval (0,1). It is extremely important to understand this solution. The central point of this fairly simple GSIM program is the integrator. The input can be interpreted as the derivative *dy(t)/dt* of an unknown function *y(t)*; the output is the function *y(t)* itself. Because of the feedback structure, both *dy(t)/dt* and *y(t)* have to be the same. In other words, the differential equation *dy(t)/dt = y(t)* is solved where the initial condition is given (in our case, the value of initial condition is equal to 1). This differential equation has the unique solution *y(t) = exp(t)*.

Step 4:

Modify the developed GSIM program as shown in Figure 18-11. Figure 18-1 explains the underlying theory. This VI solves all first-order, ordinary differential equations with given initial conditions. It is highly recommended that you alter all parameters and the contents of the formula node. Watch the calculated results. Furthermore, you should run at least some tests in the debugging mode (choose *end time = 1* and *dt = 0.1*).

Step 5:

Study the GSIM program shown in Figure 18-5. This VI is part of the example section delivered with this package (H(s) Example.vi). The program realizes a general transfer function where the input is given by a formula.

18.11 Another Example: PID Controllers in GSIM

GSIM PID.vi is part of LINEAR.11b and represents a general PID controller in GSIM notation. Figure 18-12 shows that the underlying GSIM program is simply a combination of a proportional factor, an integrator, and a differentiator. All necessary parameters are controls of this VI. GSIM can perform approximately 5,000 PID control loops a second (Pentium, 166 MHz). A PID Benchmark is part of the example library delivered with the control and simulation software for LabVIEW.

18.12 Some Restrictions and Hints

GSIM gives you many degrees of freedom during the development of new control and simulation projects. Nevertheless, some rules must be followed by the developer. It is strongly recommended that a developer adheres to the following guidelines.

Guideline 1:

Almost all GSIM VIs are reentrant. Do not change the execution options of any of these VIs. Doing so can result in an unexpected behavior of the solution. Moreover, do not place any of the GSIM VIs in a loop. If you need, for instance, an integrator for an array of variables, choose the array version of GSIM Integrator, which is GSIM Integrator (array).vi.

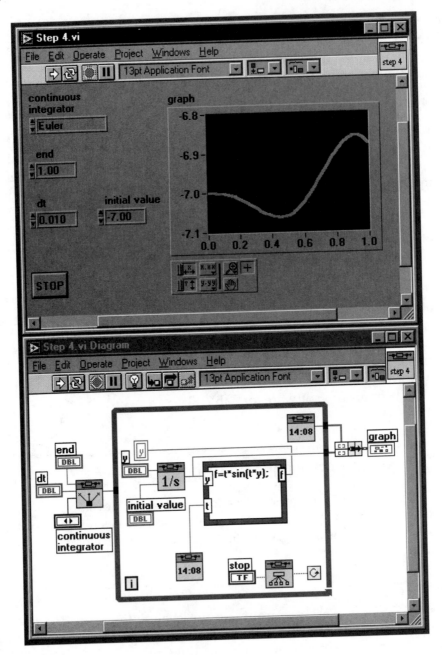

Figure 18-11

This GSIM program solves all differential equations of first order with exactly one unknown function

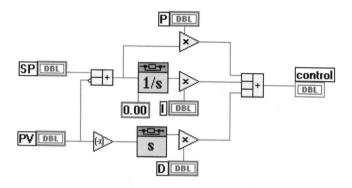

Figure 18-12

A PID controller in GSIM notation

Guideline 2:

GSIM `Initializer.vi` and GSIM `Manager.vi` *must be* part of all GSIM programs. Moreover, watch for the above-mentioned connections.

Guideline 3:

Do not run two or more GSIM programs concurrently. You can build up programs of any complexity based on an arbitrarily deep hierarchy, but two or more concurrently running VIs will interfere with each other.

Guideline 4:

Hide all controls and indicators that represent feedback structures. Doing so improves the runtime behavior of the solution significantly.

Guideline 5:

Evaluate all results very carefully. No mathematical method can guarantee the correctness in all cases. Try different integrators and choose the value *dt* carefully. This is a tradeoff between speed and correctness, and it depends strongly on the problem under development. In general, control routines can be realized by the Euler method; the Adams and the Runge-Kutta methods are more sophisticated and can be the preferred choice in case of simulation tasks, but a general rule does not exist.

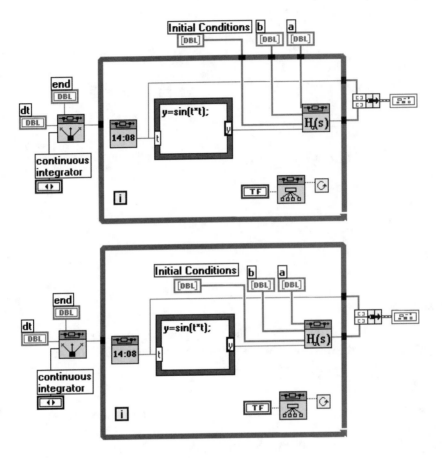

Figure 18-3

Two slightly different programs. The positions of the controls a, b, and initial conditions determine the runtime behavior of a GSIM program.

Guideline 6:

The position of control parameters on the diagram is of fundamental importance. All controls that are part of the while loop can directly and immediately determine the behavior of GSIM VIs. A control parameter that is not part of the while loop, on the other hand, delivers a valid value only at the very beginning of the program. Depending on the given problem, you should choose the appropriate method. Figure 18-13 explains both situations.

18.13   Further Developments

Control and simulation software for LabVIEW is under further development in at least two directions. The first direction is determined by more sophisticated simulation routines. These include new continuous integrators such as those for strict differential equations, use of discrete systems with independent sampling rates, introduction of algebraic loops, and tools for system identification, to name a few.

On the other hand, GSIM can not only simulate or analyze systems but it can also control real-world processes. With a typical data acquisition board, the system can handle more than 1,000 control loops a second. This is far beyond the number needed for most control tasks. But there is a second problem. Currently, the underlying operating systems cannot guarantee the real-time behavior of such control tasks. One of the most promising ideas is the so-called embedded LabVIEW, i.e., a special version of LabVIEW and a number of LabVIEW managers. It is essentially a dedicated execution engine, tuned for determination and real-time performance. This LabVIEW version uses a special real-time kernel. Users can interact with embedded LabVIEW through a master running on the host. In this sense, a fully developed control or simulation task can be downloaded. In other words, the graph is running on the embedded system. There is no need to translate the original simulation code into other languages or special DSP code; this feature will shorten the development cycle significantly. From the point of view of an end user, the entire process is completely transparent. Embedded LabVIEW will run on the intelligent hardware. Data acquisition functions will be provided by compact PCI cards that reside on the board.

19

Network-centric Test and Measurement System

Actualities seem to float in a wider sea of possibilities from out of which they were chosen; and somewhere, indeterminism says, such possibilities exist, and form part of the truth.

— William James
Diaries

Concurrent measurement, control, and testing of one or several devices represents a complex task which cannot normally be handled by an instrumentation platform based on single-processor systems. This chapter describes how a network-centric test and measurement system can be used to handle complex testing and measuring tasks.

19.1 Introduction

To ensure continuous flow of processes, it is necessary to consider various processes involved, such as measurement, control and test processes, office applications, database queries, etc., as individual entities. This view is normally achieved by distributing the processes over various computer nodes in a network (e.g., within a workstation, a LAN or a WAN). It has been shown that this is the only way to ensure the required CPU capacity for time-critical tasks.

For this reason, the Quelle Institute for Merchandise Testing and Environment has developed a *network-centric test and measurement system* (*NCTMS*), capable of handling a number of different tests. Almost all technologies (hardware and software) converge into this system. This project represents an attempt to implement a cost-efficient and scalable platform for current and future test processes. A major part of the system, described in the following sections, has already become reality.

NCTMS is a low-cost, scalable, reliable instrumentation system based on LabVIEW and Windows NT. It offers high-performance distributed automation, instrument control, and connectivity features. NCTMS has been designed to run various test processes for one or several test pieces in parallel.

19.2   Fields of Application

The potential fields of use for NCTMS extend to product characterization and qualification, design characterization, type tests, quality tests, production tests, etc., for commercial, industrial, and aviation and space products.

The following list summarizes the potential uses of NCTMS:

- Heat tests according to VDE (e.g., 0700) (up to several hundred measuring points)

- Improper use

- Handling tests (e.g., by using robots)

- Storage programmable control

- Control systems (PID, fuzzy, adaptive control)

- Network simulation tests

- Synchronization/servo-motor tests

- Illumination tests

- AC laser tests

- Relay and switch tests (single testing)
- TRICS, SCR, and passive components tests
- Supply voltage tests
- USV functional tests
- Aviation and space equipment tests
- Air-conditioning system tests
- Transformer tests
- Protection equipment tests
- Residual voltage tests
- Starting current tests
- Flicker measurements
- Measurement of current reactions
- and many others

19.3   Hardware

NCTMS is an extremely cost-efficient measurement, control, and test system that eliminates the limitations found in measurement and control systems based on single-processor PCs. NCTMS consists of a 19-inch rack with two or more motherboards (single/dual Pentium II, Windows NT 4.0), a high-resolution (17–24 inch) color monitor, a video mouse keyboard switch, a (4–16 port) Fast Ethernet hub/switch, redundant (2–3) hot-plug power supplies, and hard disks (optionally RAID5). Each motherboard can be equipped with different plug-in boards (NI-DAQ, NI-GPIB, NI-MXI, transputer, NI-CAN, RS-422, RS-485, etc.), serial-port instrumentation hardware, or GPIB- and VXI-based instruments. Various SPC and field bus systems can be optionally implemented.

Figure 19-1 shows a stand-alone NCTMS test and measurement (T&M) system capable of handling a wide range of different measurement, control, and test processes.

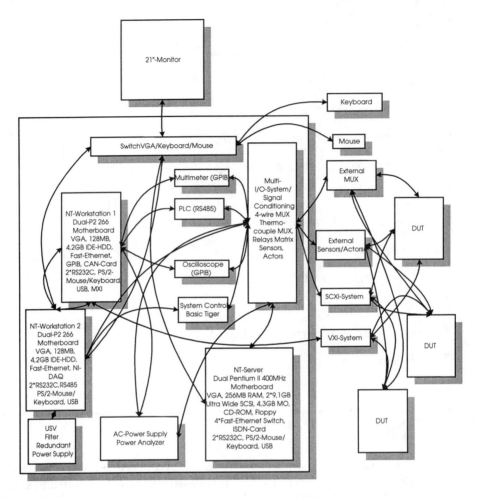

Figure 19-1

Schematic representation of typical NCTMS hardware

This system offers many features that increase productivity and reduce system cost dramatically. One single 19-inch rack accommodates two or more independent T&M systems, improving the system's EMC properties and reducing system cost. Redundant hot-plug and additional uninterruptible power supplies ensure that the system can be used as a long-time data logging and control system. Almost all redundancies that do not contribute to the system security and stability are eliminated. There is only one monitor and one keyboard with a built-in glide pad. A switch (multiplexer) connects these components to the active workstation.

An embedded controller (low-cost "basic tiger") controls the boot sequences of all system modules (including PC components); it is also responsible for system security and reliability (e.g., system temperature monitoring).

An NT Server in each NCTMS T&M system offers various services, for example, archive, communication, and print services. The server has a multiple Fast Ethernet switch that provides for extremely high communication bandwidths between all connected PC nodes. The latency times (time delays) are reduced to a minimum, thanks to short switching times and a point-to-point bandwidth of 100 MHz. Real-time capabilities (hardware and software) are integrated in the system in many different ways (e.g., by separating screen input and output processes from program processes and by using special hardware such as DAQ cards or external devices).

There is a wide range of options for data acquisition, control, and test hardware. Each motherboard (standard high-volume ATX motherboard) can be equipped with different plug-in cards, depending on the number of available slots. NCTMS features a self-test utility, and it is protected against overload, overtemperature, overcurrent, and ventilation failures, so that it offers a high-quality and reliable platform for top-level measurement and test applications.

NCTMS features a built-in AC power source based on the *PWM* (*pulse width modulation*) technology and a so-called *PFC* (*power factor correction*) module, which provides a power factor, *pf*, of more than 0.98. This is an invaluable feature in view of many applicable legal provisions (e.g., EMC guidelines) that have to be observed in practice.

NCTMS can simulate a number of different network conditions (power failure, power interruption, harmonics, etc.) and analyze the impact of such conditions on the test piece. A 3000-V-A AC source supplies a voltage of 0–300 V at a frequency between 15 Hz and 2 kHz. In addition, the AC power source generates very clean graphs (sine, rectangle, triangle, arbitrary) with a typical distortion factor of less than 0.5%.

A *DDS* (*direct digital synthesis*) waveform generator ensures this low distortion factor and serves to create the firmware or LabVIEW-based graphs. For example, it is possible to vary amplitude and distortion factor between 0% and 100%. The generator can simulate power interferences, such as half-wave failures, transient spikes and glitches, phase-angle specific bounces, frequency and voltage modulations, etc. More than 30 graphs are integrated in the firmware. LabVIEW VIs are used to download and execute as many graphs as one may need.

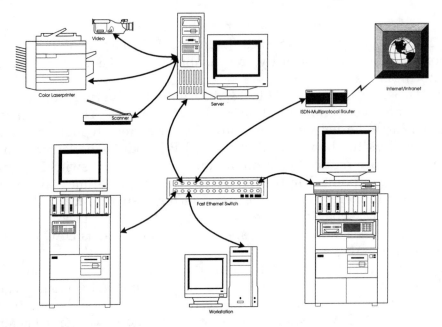

Figure 19-2

NCTMS LAN/WAN connectivity

The AC power supply and measurement system of NCTMS has a built-in 16-bit data acquisition system, capable of acquiring and evaluating real effective voltages and currents, peak currents, power, frequency, crest factor, power factor, starting currents (phase-angle triggered!). By adding an optional efficiency analyzer, NCTMS is capable of conducting tests according to IEC 1003-2 (harmonics) and IEC 1000-3-3 (flicker, voltage fluctuations).

Figure 19-2 shows two NCTMS stand-alone systems in a LAN (local area network) environment with Internet/intranet connectivity and access to peripheral components.

Table 19-1 lists important NCTMS specifications.

19.4 Software

NCTMS can be built as a system based on PCs with two or several nodes. A central server (Windows NT Server) acts as domain controller and MS Backoffice platform. System administrators can easily manage applications,

Table 19-1 *NCTMS basic specifications*

Power/Phase	1 Phase	3000 VA
Voltage	Range	150 V/300 V/Auto
	Accuracy	0.2% F.S. (45 Hz–2 kHz)
	Resolution	0.1 V
	Distortion factor	0.5% (>45 Hz–500 Hz; 1% (15–45 Hz, >500–1 kHz), 2% (1–2 kHz)
	Power adjustment	0.1%
	Load adjustment	0.1%
	Temperature coefficient	0.02% per degree C
Maximum current/phase	Effective value	30 A/15 A (150 V/300 V ranges)
	Peak value	90 A/45 A (15–100 Hz); 75 A/38 A (>100 Hz – 1 kHz); 60 A/30 A (>1–2 kHz)
Frequency	Range	15 Hz – 2 kHz
	Accuracy	0.15%
	Resolution	0.01 Hz (15–99.99 Hz); 0.1 Hz (100–999.9 Hz); 0.2 Hz (1–2 kHz)
Input voltage	Voltage range	190–254 V
	Frequency range	47–63 Hz
	Power consumption (max.)	23 A
	Power factor (full load)	0.98

Table 19-1 (Continued)

Power/Phase	1 Phase	3000 VA
Measuring value acquisition — current/phase	Range	0–140 A
	Accuracy (effective value)	0.1% F.S. + 0.4%
	Accuracy (peak)	0.2% F.S. + 0.4%
	Resolution	0.01 A
Measuring value acquisition — voltage/phase	Range	0–150 V/0–300 V
	Accuracy	0.1% F.S. + 0.25%
	Resolution	0.1 V
Measuring value acquisition — Frequency	Range	15–2000 Hz
	Accuracy	0.01% + 2 C
	Resolution	0.01 Hz
Measuring value acquisition — power/phase	Accuracy	1% F.S.
	Resolution	0.01 W
Temperature range	Operation	0 – +40 degrees C
	Storage	+40 – +85 degrees C
Miscellaneous	Efficiency factor (typical)	80%
	Protection	OPP, OLP, OTP, FAN, FAIL

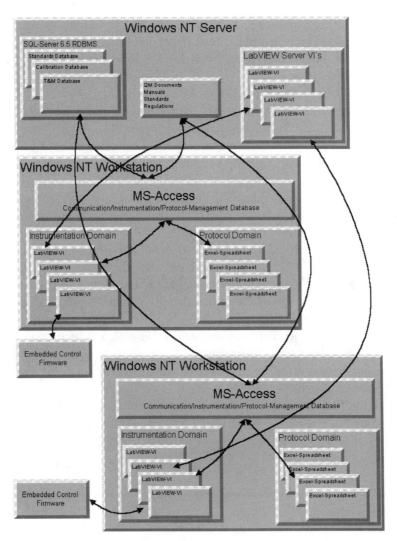

Figure 19-3

Schematic representation of typical NCTMS software

computers, and users by means of the MS System Management Server. MS Exchange Server is used to allow the individual workstations access to the Internet/intranet. Optional servers (Internet Information Server, proxy servers, etc.) control external client access. The domain controller runs various TCP/IP-based LabVIEW server VIs, offering time synchronization options and VI scheduling functions. LabVIEW VIs read data from a *GPS* (*Global Positioning System*) time module and a recyclable calibrated compressed-air sensor over the serial interface, and output this information to all workstations in the network.

The MS-SQL Server 6.5 implements a bundle of databases, which close the gaps in the areas of calibration management, document management, standardization and rules management, quality management, project management, instrumentation management, and others. Client applications (Access 97 on T&M workstations) obtain SQL Server information to control LabVIEW VIs.

There are a number of LabVIEW server and client VIs to control the communication between the central domain controller, the internal server nodes, and the internal workstation nodes.

LabVIEW VIs acquire data from various measuring value acquisition and evaluation systems (DAQ, GPIB, serial, parallel, Transputer, USB, etc.).

The Access 97 T&M management database controls the flow of information between SQL Server data and LabVIEW instrumentation and control software. MS Access starts LabVIEW VIs. The T&M VI reads stored data (Access-based test procedure control and DUT data), executes its actual test program, saves data to a temporary file, and passes control back to the server-based Access database. Older communication options such as OLE and DDE are supported.

Eventually, the user can decide whether or not a test log has to be created. If so, Access will open an MS Excel spreadsheet to generate a test report. Figure 19-3 illustrates a typical NCTMS software configuration.

19.5   Summary

NCTMS is an ideal technology for low-cost, parallel, high-performance measurement with control and test applications for a wide choice of test pieces or systems. Standard components, reliability, security, connectivity, scalability, and a number of LabVIEW VIs are the most important features of NCTMS that make it a superior T&M solution.

Bibliography

Introduction and Basics
(Chapters 1–2)

Burnett, M. M., Goldberg, A., Lewis, T. [1995], *Visual Object Oriented Programming*, Manning Publications.

Griemert, R., Erhart, W. [1994], LabVIEW Examples — The Art of Graphical Programming, National Instruments, Munich, Germany.

Heinrichs, G., Rongen, H., Jamal, R. [1994], Using LabVIEW for the Design and Control of Digital Signal Processing Systems, Simulation of the Ultra Slow Extraction at COSY, *Nuclear Instruments & Methods in Physics Research*, v. 352, Nos. 1/2, p. 352.

Jamal, R. [1993], Graphical Object-Oriented Programming with LabVIEW, *Proceedings of ICALEPCS 93* (International Conference on Accelerator and Large Experimental Physics Control Systems), Berlin, Germany, Oct. 1993.

Jamal, R. [1996], Virtual Instruments — Yesterday, Today, Tomorrow, *Elektronik*, v. 26.

Jamal, R.   [1996], Virtual Instruments in Practice (VIP '96), User Symposium, National Instruments Germany GmbH, Munich, Germany.

Jamal, R., Erhart, W.   [1995], LabVIEW — Fourth Generation Programming Language, *F&M Feinwerktechnik, Mikrotechnik, Messtechnik*, Carl Hanser Verlag, Munich, Germany, Jan./Feb. 1995.

Jamal, R., Heinze, R.   [1998], Virtual Instruments in Practice (VIP '98), User Symposium, VDE-Verlag, Germany.

Jamal, R., Jaschinski, H.   [1997], Virtual Instruments in Practice (VIP '97), User Symposium, Praxiswissen Elektronik Industrie, Huethig Verlag, Heidelberg/Munich, Germany.

Jamal, R., Krauss, P.   [1998], *LabVIEW Basics*, Prentice Hall, Germany.

Jamal, R., Wenzel, L.   [1995], The Applicability of the Visual Programming Language LabVIEW to Large Real-World Applications, *11th IEEE Symposium on Visual Languages* (IEEE VL '95), Darmstadt, Germany, Sept. 5–8, 1995.

Kodosky, J.   [1993], Objects and Messages in the LabVIEW Graphical Programming System, *Proceedings of SEAM 93*, Aug. 2, 1993.

Kodosky, J., et al.   [1991], Visual Programming Using Structured Data Flow, *Proceedings of the 1991 IEEE Workshop on Visual Languages*, Kobe, Japan, Oct. 8–11, 1991.

Pichlik, H.   [1992], Universal Test-Site Development System, *elrad*, Oct. 1992.

Pichlik, H.   [1993], Instrument on Image, *elrad*, Apr. 1993.

Pichlik, H.   [1996], Universal Testing of Household Equipment using LabVIEW, *Proceedings — NI-Week*, Austin, Texas, July 1996.

Pichlik, H.   [1997], Networkcentric Test and Measurement System, *Proceedings — NI-Week*, Austin, Texas, Aug. 1997.

Wells, L.   [1994], *LabVIEW — Student Edition User's Guide*, Prentice Hall.

Wells, L., Travis J.   [1996], *LabVIEW for Everyone*, Prentice Hall.

DAQ (Chapter 3)

Illig, H., Jamal, R. [1997], DAQ Instruments, *Elektronik Industrie*, 5/97.

Illig, H., Jamal, R. [1997], Instrument on Card, *Elektronik*, 9/97.

Jamal, R., Erhart, W. [1994], *Using LabWindows for Measuring Data Acquisition*, Franzis Verlag.

Jamal, R., Illig, H. [1994], Multifunction Cards — Important Criteria, *Elektronik-Plus*, 5/94, pp. 47–54.

Jamal, R., McConnell, E. [1994], New Achievements in Counter/Timer Data Acquisition, *Proceedings of MessComp '94*, pp. 492–498.

Jamal, R., McConnell, E., Erhart, W. [1995], Equivalent-Time Sampling for PC-based Multifunction Data Acquisition Boards, *Proceedings of MessComp*, Sept. 1995, pp. 227–231.

McConnell, E., Jamal, R. [1996], Performance Considerations in PCI-based Data Acquisition and Control Applications, *Measuring Technology and Measuring Signal Processing*, Expert Verlag, pp. 32–37.

National Instruments [1993], Measuring Temperature with Thermocouples, *Application Note 43*, National Instruments GmbH, Munich, Germany.

National Instruments [1993], Is Your Data Inaccurate Because of Instrument Amplifier Settling Time?, *Application Note 45*, National Instruments GmbH, Munich, Germany.

National Instruments [1993], Signal Conditioning Fundamentals for PC-Based DAQ Systems, *Application Note 48*, National Instruments GmbH, Munich, Germany.

Pichlik, H. [1991], High-Speed High-Resolution Analog/Digital Converter Circuit, United States Patent 4,999,624, Mar. 12, 1991.

Pichlik, H. [1996], Switching Arrangement for Analog/Digital Conversion, European Patent (EP 0349 793 B1), Jan. 17, 1996.

Potter, D. [1994], Reading Specifications of Data Acquisition Boards and Systems, *Sensors Expo West*, Feb. 1994.

Schwetlick, H. [1997], *PC Measuring Technology: Basic and Applications in Computer-Assisted Measuring Technology*, Vieweg Verlag.

GPIB (Chapter 3)

IEEE [1997], IEEE Standard Codes, Formats, Protocols and Common Commands, *IEEE Std 488.2*, Feb. 1997.

Jamal, R. [1991], IEEE 488.2 — A Standard for GPIB, *Elektronik Industrie*, 11/91, pp. 98–100.

Jamal, R. [1993], TNT 4882 — Get 8 Mbytes/s with an IEEE 488.2 IC, *Elektronik Industrie*, 11/93, pp. 72–75.

Jamal, R., Erhart, W. [1994], *Using LabWindows for Measuring Data Acquisition*, Franzis Verlag.

Jamal, R., Carter, V. [1995], Increasing the Performance of MXI- and GPIB-Controlled VXI Systems, *Open Bus Systems* (OBS '95), Swiss Institute of Technology ETH Zurich, Switzerland, Oct. 1995, pp. 357–361.

Schwetlick, H. [1997], *PC Measuring Technology: Basic and Applications in Computer-Assisted Measuring Technology*, Vieweg Verlag.

SCPI Consortium [1991], *SCPI Specification*, May 1991.

Virtual Instrumentation Software Architecture (VISA) (Chapter 3)

Hogen, M. [1995], An Introduction to Win Frameworks, *VXI Journal*, Winter 1995/96.

Jamal, R. [1996], *Further Development of the Graphical Programming Language LabVIEW*, VDE Verlag, Apr. 1996, pp. 30–32.

Jamal, R. [1996], Free Access: VISA — A New Input/Output Software Standard for VXI Systems, *Elektronik Praxis*, Vogel Verlag, 8/96, pp. 90–94.

Jamal, R., Erhart, W. [1994], *Using LabWindows for Measuring Data Acquisition*, Franzis Verlag.

Jaschinski, H. [1996], VISA — A Software Architecture for Virtual Measuring Technology, *Elektronik Industrie*, Huethig Verlag, May 1996, pp. 70–74.

Mitchell, B. [1995], Understanding VISA, *EE Evaluation Engineering*, Feb. 1995.

VXI*plug&play* Systems Alliance [1995], VPP-4x: Virtual Instrument Software Architecture Specifications, Feb. 21, 1995.

Control Engineering (Chapter 4)

Jamal, R., Erhart, W. [1993], Implementing PID Control by Software, *VFI*, Jan. 1993.

Pichlik, H., Heinloth, K. [1995], Self-Training Control, *Design & Elektronik*, Oct. 1995.

LabVIEW in Automation Technology (Chapter 5)

Dahmen, N., Hofmeier, R., Jamal, R. [1996], Fuzzy for On-line Process Control, *Elektronik*, 14/96, pp. 68–72.

Erhart, W., Jamal, R. [1996], A Bridge to the Process, *Electrical Engineering for Automation*, 9/96, pp. 80–82.

Jamal, R. [1996], Further Development of the Graphical Programming Language LabVIEW, *etz*, 4/96, pp. 30–32.

Neumann, D. et al. [1995], A Graphical User Interface for Bioreactor Control, *atp Automatisierungstechnische Praxis*, v. 37, No. 3, pp. 16–24.

Controller Area Network (CAN) (Chapter 5)

Jamal, R. [1996], Virtual Instruments — Yesterday, Today, Tomorrow, *Elektronik*, v. 26.

Jamal, R., Roenpage, T. [1997], A Graphical Programming Environment for CAN Applications, *Elektronik Industrie*, 4/97.

Lawrenz, W. [1994], *CAN — Theory and Application*, Huethig Verlag.

Fuzzy Logic (Chapter 6)

Altrock von, C. [1993], *Fuzzy Logic*, Vol. I (Technology), Oldenbourg Verlag.

Dahmen, N., Hofmeier, R., Jamal, R. [1996], Fuzzy for On-line Process Control, *Elektronik*, v. 14, pp. 68–72.

Dahmen, N., Hofmeier, R., Jamal, R. [1996], Draft and Implementation of Fuzzy Controllers in LabVIEW for On-line Process Control, *Proceedings of Real-Time '96*.

Dahmen, N., Toszkowski, G., Jamal, R., Limroth, J. [1997], Knowledge-based Concepts for On-line Process Control Using Graphical Programming Techniques, *Measuring Technology and Measuring Signal Processing*, Vol. 2, K. W. Bonfig (Ed.), Expert Verlag.

Kahlert, J., Frank, H. [1993], *Fuzzy Logic and Fuzzy Control*, Vieweg Verlag.

National Instruments [1997], LabVIEW Fuzzy Logic Toolkit Reference Manual.

Fast Fourier Transform (FFT) (Chapter 9)

Brigham, E. O. [1988], *The Fast Fourier Transform and Its Applications*, Prentice Hall Signal Processing Series.

Jamal, R. [1996], Using Fast Fourier Transform — FFT — for Signal Analysis in the Frequency Range, *Application Note*, National Instruments GmbH, Munich, Germany, Jan. 1996.

Jamal, R., Erhart, W. [1995], Virtual Instruments for Dynamic Signal Analysis, *Design & Elektronik*, 5/95.

Marko, H. [1977], *Methods of System Theory: The Spectral Transform and its Applications*, Springer Verlag.

Oppenheim, A., Schafer, R. [1989], *Discrete-Time Signal Processing*, Prentice Hall.

Rabiner, L., Gold, B. [1975], *Theory and Application of Digital Signal Processing*, Prentice Hall.

Stockhausen, N., Jamal, R. [1995], Using Digital Signal Analysis for Acoustic Locating of Leakage in Waste Water Pipes, *Proceedings of MessComp*, 9/95.

Time-Frequency Analysis (Chapter 10)

Chen, D., Qian, S., Wenzel, L. [1997], Signal Analysis Made Easier, *Elektronik*, v. 11.

Jamal, R. [1996], Filters, in *Measurement, Instrumentation and Sensors Handbook*, J. G. Webster (Ed.), CRC & IEEE Press.

Jamal, R., Cerna, M., Hanks, J. [1996], Designing Filters Using the Digital Filter Design Toolkit, *Application Note 97*, National Instruments Corporation.

Jamal, R., Griemert, R. [1993], Practical Application of Gabor Transform in Measuring and Testing Technology, *MessComp Proceedings*.

Jamal, R., Griemert, R. [1994], *Using Gabor Transform*, VFI 1.

Nef, C. [1996], Combined Time-Frequency Analysis of Signals, Part I, *Megalink*, v. 8.

Nef, C. [1996], Combined Time-Frequency Analysis of Signals, Part II, *Megalink*, v. 10.

Qian, S., Chen, D. [1996], *Joint Time-Frequency Analysis*, Prentice Hall.

Image Processing (Chapter 12)

Jamal, R., Erhart, W. [1993], *Using LabWindows for Measuring Data Acquisition*; How to acquire, analyze and present measuring data successfully, Franzis Verlag.

Jamal, R., Erhart, W. [1996], LabVIEW — Fourth Generation Programming Language, R. Jamal: Virtual Instruments — Yesterday, Today, Tomorrow, *Elektronik*, v. 26.

National Instruments [1997], Seminar papers on image processing, National Instruments GmbH, Munich, Germany.

Sinkovic, G., Jamal, R. [1997], Image Processing in the Age of Virtual Instruments, *F&M Feinwerktechnik, Mikrotechnik, Messtechnik*, 7/97, Carl Hanser Verlag, Munich, Germany.

State of NRW [1995], Status and Trend in Image Processing, BridgeVIEW Study of the Federal State of North Rhine Westphalia, Germany.

Quality (Chapters 13 and 15)

Pichlik, H. [1990], Reliability, *elrad*, Aug. 1990.

Pichlik, H., Bertuch, M. [1991], Temperature Specifications and Reliability of VLSI Chips, *c't*, Apr. 1991.

Pichlik, H., Schmidt, R. [1994], ISO Shock, *elrad*, Apr. 1994.

Statistical Process Control (SPC) (Chapter 14)

Hering, E., Triemel, J., Bnak, H. P. [1993], *Quality Assurance for Engineers*, VDI Verlag.

Jamal, R. [1995], Virtual Instruments for SPC Applications, *Elektronik Industrie*, Mar. 1995.

Montgomery, D. C. [1991], *Introduction to Statistical Quality Control*, J. Wiley & Sons, 2nd Edition.

National Instruments [1994], *LabVIEW SPC Toolkit Reference Manual*.

Medical Applications (Chapter 16)

Abbott Laboratories [1990], *Oximetrix 3SO$_2$/CO Computer RS-232 I/O Port User's Guide*, North Chicago, IL 60064, USA.

Cohen, A. [1995], Biomedical Signals: Origin and Dynamic Characteristics; Frequency Domain Analysis, in *The Biomedical Engineering Handbook*, J. D. Bronzino (Ed.), Trinity College Hartford, Connecticut, CRC Press, IEEE Press, pp. 806–825.

Herrmann. Dr. M. Q., Jamal, R.   [1994], Computer-aided Measurement and Instrumentation in Intensive Care Medicine, *Proceedings of the 8th International Conference on Biomedical Engineering*, Singapore, Dec. 10, 1994.

Johnson, G. W. [1994], *LabVIEW Graphical Programming, Practical Applications in Instrumentation and Control*, McGraw-Hill, Inc.

Mainardi, L. T., Bianchi, A. M., Cerutti, S.   [1995], Digital Biomedical Signal Acquisition and Processing, in *The Biomedical Engineering Handbook*, J. D. Bronzino (Ed.), Trinity College Hartford, Connecticut, CRC Press, IEEE Press, pp. 828–850.

Neumann, M. R.   [1995], Biomedical Sensors, in *The Biomedical Engineering Handbook*, J. D. Bronzino (Ed.), Trinity College, Hartford, Connecticut, CRC Press, IEEE Press, pp. 728–779.

Siemens AG   [1991], Description of Digital Serial Interface (RS-232) for SIRECUST Monitors 960/961/1280/1281, Jan. 1991.

Siemens-Elema AB   [1995], *Servo Ventilator 300, Computer Interface Reference Manual*, Solna, Sweden.

Silny, J., Rau, G.   [1992], Bioelectrical Signals, in *Biomedical Technology*, Part 1: Diagnostics and Image-Producing Methods, H. Hutten (Ed.), Springer Verlag, Berlin/Heidelberg/New York, pp. 1–62.

Glossary

A

ActiveX Controls (formerly known as OLE Controls)
A special form of component Automation Object. ActiveX Controls are similar to Visual Basic custom controls (VBXs), but their architecture is based on Object Linking and Embedding (OLE). Unlike VBXs, ActiveX Controls can be freely plugged into any OLE-enabled development tool, application, or web browser.

A/D
Analog-to-digital conversion; refers to the operation electronic circuitry does to take a real-world analog signal and convert it to a digital form (as a series of bits) that the computer can understand.

ADC
Analog-to-Digital Converter; an electronic device, often an integrated circuit, that converts an analog voltage to a digital number.

Aliasing
A false lower frequency component that appears in sampled data acquired at too low a sampling rate.

ALU
Arithmetic Logic Unit; the element or elements in a processing system that perform the mathematical functions such as addition, subtraction, multiplication, division, inversion, AND, OR, NAND, and NOR.

Amplitude Flatness
A measure of how close to constant the gain of a circuit remains over a range of frequencies.

Analog Trigger
A trigger that occurs at a user-selected point on an incoming analog signal. Triggering can be set to occur at a specific level, on either an increasing or a decreasing signal (i.e., a positive or negative slope).

ANSI
American National Standards Institute.

ASIC
Application-Specific Integrated Circuit; a proprietary semiconductor component designed and manufactured to perform a set of specific functions for a specific customer.

Asynchronous
(1) Hardware — A property of an event that occurs at an arbitrary time, without synchronization to a reference clock. (2) Software — A property of a function that begins an operation and returns prior to the completion or termination of the operation.

ATE
Automated Test Equipment; a term typically applied to computer-based systems for testing semiconductor components or circuit card assemblies.

B

Background Acquisition
The collecting of data acquired by a DAQ system while concurrently another program or processing routine is running without apparent interruption.

Backplane (VXI)
An assembly, typically a printed circuit board (PCB), with 96-pin connectors and signal paths that bus the connector pins. VXIbus systems have two sets of bused connectors, called the J1 and J2 backplanes, or have three sets of bused connectors, called the J1, J2, and J3 backplanes.

Barrel Shifter
An element in a high-performance processing system that logically shifts the bits of a data word to multiple locations in one instruction cycle.

Base Address
A memory address that serves as the starting address for programmable registers. All other addresses are located by adding to the base address.

Baud Rate
Serial communications data transmission rate expressed in bits per second (bps).

Bipolar
A signal range that includes both positive and negative values (e.g., –5 V to +5 V).

Bit
One binary digit, either 0 or 1.

Block Diagram
Pictorial description or representation of a program or algorithm. In LabVIEW, the block diagram, which consists of executable icons (called nodes) and wires that carry data between the nodes, is the source code for the VI. The block diagram resides in the block diagram window of the VI.

Block Mode
A high-speed data transfer in which the address of the data is sent and followed by a specified number of back-to-back data words.

Break-Before-Make
A type of switching contact that is completely disengaged from one terminal before it connects with another terminal.

Breakdown Voltage
The voltage high enough to cause breakdown of optical isolation, semiconductors, or dielectric materials; see also Working Voltage.

BridgeVIEW
A tool for industrial automation technology based on the LabVIEW concept and the G programming language.

Burst Mode
A high-speed data transfer in which the address of the data is sent and followed by back-to-back data words while a physical signal is asserted.

Bus
The group of conductors that interconnect individual circuitry in a computer. Typically, a bus is the expansion vehicle to which I/O or other devices are connected. Examples of PC buses are the AT bus, NuBus, Micro Channel, and EISA bus.

Bus Master
A type of a plug-in board or controller with the ability to read and write devices on the computer bus.

Byte
Eight related bits of data, an eight-bit binary number. Also used to denote the amount of memory required to store one byte of data.

C

CAN
Controller Area Network; a serial bus finding increasing use as a device-level network for industrial automation. CAN was developed by Bosch and Intel to address the needs of in-vehicle automotive communications.

Cache
High-speed processor memory that buffers commonly used instructions or data to increase processing throughput.

Chromatograph
An instrument used in chemical analysis of gases and liquids.

CIN
Code Interface Node; allows integration of C code in LabVIEW.

CMRR
Common-Mode Rejection Ratio; a measure of an instrument's ability to reject interference from a common-mode signal, usually expressed in decibels (dB).

Code Generator
A software program, controlled from an intuitive user interface that creates syntactically correct high-level source code in languages such as C or BASIC.

CodeLink
A tool embedded in LabVIEW for automatic generation of LabVIEW block diagrams from dynamic link libraries (DLLs) created in LabWindows/CVI.

Cold-Junction Compensation
The means to compensate for the ambient temperature in a thermocouple measurement circuit.

Command (VXI)
Any communication from a Commander to a Message-Based Servant that consists of a write to the Servant's Data Low register, possibly preceded by a write to the Data High or Data High and Data Extended registers.

Commander (VXI)
A message-based device that is also a bus master and can control one or more Servants.

Common-Mode Range
The input range over which a circuit can handle a common-mode signal.

Common-Mode Signal
The mathematical average voltage, relative to the computer's ground, of the signals from a differential input.

Compiler
A software utility that converts a source program in a high-level programming language, such as BASIC, C, or Pascal, into an object or compiled program in machine language. Compiled programs run 10 to 1,000 times faster than interpreted programs; see also Interpreter.

Component Software
An application that contains one or more component objects that can freely interact with other component software. Examples include OLE-enabled applications such as Microsoft Visual Basic and OLE Controls for virtual instrumentation in ComponentWorks.

Configuration Registers (VXI)
The A16 registers of a device that are required for the system configuration process.

Control Flow
A model for programming in which an instruction counter sequentially executes instructions in memory; program control flows from one instruction to another. Programs in languages such as FORTRAN and BASIC follow the control flow model.

Conversion Time
The time required, in an analog input or output system, from the moment a channel is interrogated (such as with a read instruction) to the moment that accurate data is available.

Counter/Timer
A circuit that counts external pulses or clock pulses (timing).

Coupling
The manner in which a signal is connected from one location to another.

Crosstalk
An unwanted signal on one channel due to an input on a different channel.

Current Drive Capability
The amount of current a digital or analog output channel is capable of sourcing or sinking while still operating within voltage range specifications.

Current Sinking
The ability of a DAQ board to dissipate current for analog or digital output signals.

Current Sourcing
The ability of a DAQ board to supply current for analog or digital output signals.

D

D/A
Digital-to-analog; the opposite operation of A/D.

DAC
Digital-to-Analog Converter; an electronic device, often an integrated circuit, that converts a digital number into a corresponding analog voltage or current.

DAQ
Data Acquisition; (1) Collecting and measuring electrical signals from sensors, transducers, and test probes or fixtures and inputting them to a computer for processing; (2) Collecting and measuring the same kinds of electrical signals with A/D or digital input/output (DIO) boards plugged into a PC, and possibly generating control signals with D/A or DIO boards in the same PC.

Data Flow
A model for programming in which instructions or operators execute only when all inputs are available. In this model, data flows into and out of operators.

dB
Decibel; the unit for expressing a logarithmic measure of the ratio of two signal levels: $dB = 20 \log_{10} (V1/V2)$, for signals in volts.

DC
Direct Current.

DCS
Distributed Control System; a large-scale process control system characterized by a distributed network of processors and I/O subsystems that encompass control, user interfacing, data collection, and system management. DCSs are commonly used in large industrial facilities, such as a petroleum refinery or a paper mill.

DDE
Dynamic Data Exchange; a standard software protocol in Microsoft Windows for interprocess communication. DDE is used when applications such as LabVIEW send messages to request and share data with other applications such as Microsoft Excel.

Delta-Sigma Modulating ADC
A high-accuracy circuit that samples at a higher rate and lower resolution than is needed and (by means of feedback loops) pushes the quantization noise above the frequency range of interest. This out-of-band noise is typically removed by digital filters.

Derivative Control
A control action with an output that is proportional to the rate of change of the error signal. Derivative control anticipates the magnitude difference between the process variable and the set point.

Device (VXI)
A component of a VXIbus system, normally one VXIbus board. However, multiple-slot devices and multiple-device modules can operate on a VXIbus system as a single device. Some examples of devices are computers, multimeters, multiplexers, oscillators, operator interfaces, and counters.

Differential Input
An analog input consisting of two terminals, both of which are isolated from computer ground, whose difference is measured.

DIN
(Deutsches Institut für Normung) German Institute for Standardization.

DIO
Digital input/output.

Discrete Gabor Transform
A new algorithm for transforming data from the discrete time domain to the joint time-frequency domain.

DLL
Dynamic Link Library; a software module in Microsoft Windows containing executable code and data that can be called or used by Windows applications or other DLLs. Functions and data in a DLL are loaded and linked at run time when they are referenced by a Windows application or other DLLs.

DMA
Direct Memory Access; a method by which data can be transferred to/from computer memory from/to a device or memory on the bus while the processor does something else. DMA is the fastest method of transferring data to/from computer memory.

DNL
Differential Nonlinearity; a measure in LSB of the worst-case deviation of code widths from their ideal value of 1 LSB.

Driver
Software that controls specific hardware devices, such as DAQ boards, GPIB interface boards, programmable logic controllers (PLCs), remote terminal units (RTUs), and other I/O devices.

DSP
Digital signal processing.

Dual-Access Memory
Memory that can be sequentially accessed by more than one controller or processor, but not simultaneously accessed. Also known as shared memory.

Dual-Ported Memory
Memory that can be simultaneously accessed by more than one controller or processor.

Dynamic Configuration (VXI)
A method of automatically assigning logical addresses to VXIbus devices at system startup or other configuration times. Each slot can contain one or more devices. Different devices within a slot can share address decoding hardware.

Dynamic Range
The ratio of the largest signal level a circuit can handle to the smallest signal level it can handle (usually taken to be the noise level), normally expressed in dB.

E

EEPROM
Electrically Erasable Programmable Read-Only Memory; ROM that can be erased with an electrical signal and reprogrammed.

EISA
Extended Industry Standard Architecture

Encoder
A device that converts linear or rotary displacement into digital or pulse signals. The most popular type of encoder is the optical encoder, which uses a rotating disk with alternating opaque areas, a light source, and a photodetector.

EPROM
Erasable Programmable Read-Only Memory; ROM that can be erased (usually by ultraviolet light exposure) and reprogrammed.

Events (VXI)
Signals or interrupts generated by a device to notify another device of an asynchronous event. The contents of events are device dependent.

Extended Devices (VXI)
A device that has VXIbus configuration registers and a subclass register. This category is intended for definition of additional device types.

Extended Longword Serial (VXI)
A form of Word Serial communication for 48-bit data transfers between Commanders and Servants.

External Trigger
A voltage pulse from an external source that triggers an event such as A/D conversion.

F

Fast Handshake (VXI)
A high-speed mode of operation that uses the same communication registers as the Word Serial Protocol. Data can be transferred without the need to poll after each transfer.

Fetch-and-Deposit
A data transfer in which the data bytes are transferred from the source to the controller, and then from the controller to the target.

FFT
Fast Fourier Transform.

Fieldbus
An all-digital communication network used to connect process instrumentation and control systems; it will ultimately replace the existing 4–20 mA analog standard.

Fieldbus Foundation
The organization located in Austin, Texas, that is developing a standard digital communication network (fieldbus) for process control applications. The network developed by the Foundation is referred to as the Foundation Fieldbus.

FIFO
First-In First-Out memory buffer; the first data stored is the first data sent to the acceptor.

Fixed Point
A format for processing or storing numbers as digital integers.

Flash ADC
An Analog-to-Digital Converter whose output code is determined in a single step by a bank of comparators and encoding logic.

Floating Point
A format for processing or storing numbers in scientific exponential notation (digits multiplied by a power of 10).

Flyby
A type of high-performance data transfer in which the data bytes pass directly from the source to the target without being transferred to the controller.

Foreground
In a PC system, the activity subject to direct operator intervention. Other (background) activities continue.

488-VXIbus Interface Device (VXI)
A message-based device that communicates between the IEEE 488 bus and VXIbus instruments.

Function
A set of software instructions executed by a single line of code that may have input and/or output parameters and returns a value when executed. Examples of functions are $y = COS (x)$; $status = AO\_config(board, channel, range)$.

Fuzzy Logic
A rule-based control system. Used for process control or for expert decision-making, such as pattern recognition or diagnosis, fuzzy logic control designs are well suited for complex or highly nonlinear control systems. They are easier to implement than traditional linear control systems.

Fuzzy Logic Control Design Software for G
An application software to design fuzzy logic control systems for LabVIEW or BridgeVIEW. In addition, Fuzzy Logic Design for G can be combined with NI-DAQ, PID Control Toolkit, and the Statistical Process Control (SPC) Toolkit for advanced control applications.

G

G
Graphical programming language embedded in LabVIEW.

Gabor Spectrogram
A new algorithm for joint time-frequency analysis that uses the discrete Gabor transform. The Gabor Spectrogram produces a three-dimensional plot of signal power versus frequency and time. Used in sonar, acoustics, and vibration analysis.

Gain
The factor by which a signal is amplified, sometimes expressed in dB.

Gain Accuracy
A measure of deviation of the gain of an amplifier from the ideal gain.

GPIB
General-Purpose Interface Bus, synonymous with HP-IB; the standard bus used for controlling electronic instruments with a computer. Also called the IEEE 488 bus because it is defined by ANSI/IEEE Standards 488-1978, 488.1-1987, and 488.2-1987.

GUI
Graphical User Interface; an intuitive, easy-to-use method of interacting with a computer program by means of graphical screen displays. GUIs can resemble the front panels of instruments or other objects associated with a computer program.

H

Half-Flash ADC
An Analog-to-Digital Converter that determines its output code by digitally combining the results of two sequentially performed, lower-resolution flash conversions.

Half-Power Bandwidth
The frequency range over which a circuit maintains a level of at least −3 dB with respect to the maximum level.

Handler
A device driver that is installed as part of the operating system of the computer.

Hardware
The physical components of a computer system, such as the circuit boards, plug-in boards, chassis, enclosures, peripherals, cables, and so on.

Hierarchical
A method of organizing computer programs with a series of levels, each with further subdivisions, as in a pyramid or tree structure.

HMI
Human-Machine Interface; see MMI.

HS488
High-Speed 488; an enhancement to the IEEE 488.1 protocol, combining significantly higher speed with the benefits of the current GPIB standards, while maintaining compatibility with conventional GPIB equipment.

Hybrid Device (VXI)
A VMEbus-compatible device that has application-specific subsets of VXIbus protocols.

I

IAC
Interapplication Communication; protocol by which applications can pass messages. Messages can be either blocks of data and information packets, or instructions and requests for application(s) to perform actions. An application can send messages to itself, to other applications on the same machine, or to applications located anywhere on the network.

IEC
International Electrotechnical Commission.

IEEE
Institute of Electrical and Electronics Engineers.

IEEE 488
The shortened notation for ANSI/IEEE Standards 488-1978, 488.1-1987, and 488.2-1987. See also GPIB.

Image Processing
Processes and methods to retrieve and manipulate information from two-dimensional signals (images) on a computer system.

IMD
Intermodulation Distortion; the ratio, in dB, of the total rms signal level of harmonic sum and difference distortion products to the overall rms signal level. The test signal is two sine waves added together according to the relevant standards.

Industrial Device Networks
Standardized digital communications networks used in industrial automation applications; they often replace vendor-proprietary networks so that devices from different vendors can communicate in control systems.

INL
Integral Nonlinearity; a measure in LSB of the worst-case deviation from the ideal A/D or D/A transfer characteristic of the analog I/O circuitry.

Input Bias Current
The current that flows into the inputs of a circuit.

Input Impedance
The measured resistance and capacitance between the input terminals of a circuit.

Input Offset Current
The difference in the input bias currents of the two inputs of an instrumentation amplifier.

Instrument Driver
A set of high-level software functions that controls a specific GPIB, VXI, or RS-232 programmable instrument or a specific plug-in DAQ board. Instrument drivers are available in several forms, ranging from a function callable from a programming language to a virtual instrument (VI) in LabVIEW.

Instrumentation Amplifier
A circuit whose output voltage with respect to ground is proportional to the difference between the voltages at its two inputs.

Integral Control
A control action that eliminates the offset inherent in proportional control.

Integrating ADC
An Analog-to-Digital Converter whose output code represents the average value of the input voltage over a given time interval.

Internet
A collection of interconnected networks spanning the globe, formed of commercial public networks, not-for-profit public networks, private or corporate networks, and governmental or publicly funded networks.

Interpreter
A software utility that executes source code from a high-level language such as BASIC, C, or Pascal by reading one line at a time and executing the specified operation. See also Compiler.

Interrupt
A computer signal indicating that the CPU should suspend its current task to service a designated activity.

Intranet
A corporate network based on the Internet.

I/O
Input/Output; the transfer of data to/from a computer system and involving communications channels, operator interface devices, or data acquisition and control interfaces.

IPC
Interprocess Communication; protocol by which processes can pass messages. Messages can be either blocks of data and information packets, or instructions and requests for a process or processes to perform actions. A process can send messages to itself, other processes on the same machine, or processes located anywhere on the network.

IRQ
Interrupt Request.

ISA
Industry Standard Architecture

Isolation Voltage
The voltage that an isolated circuit can normally withstand, usually specified from input to input or from any input to the amplifier output, or to the computer bus.

J

JTFA
Joint Time-Frequency Analysis; a technique for spectral analysis of rapidly changing waveforms.

L

Linearity
The adherence of device response to the equation $R = KS$, where R = response, S = stimulus, and K = a constant.

Listener
A device on the GPIB that receives information from a Talker on the bus.

Logical Address (VXI)
An 8-bit number that uniquely identifies each VXIbus device in a system. It defines the A16 register addresses of a device and indicates Commander and Servant relationships.

Longword Serial
A form of Word Serial communication for 32-bit data transfers between Commanders and Servants.

LSB
Least significant bit.

M

Mainframe (VXI)
The chassis of a VXI system that mechanically contains VXI modules inserted into the backplane, ensuring that connectors fit properly and that adjacent modules do not contact each other. It also provides cooling airflow and ensures that modules do not disengage from the backplane due to vibration or shock.

Memory Device (VXI)
A memory storage device that implements the defined VXIbus registers and communication protocols.

Memory Window
Continuous blocks of memory that can be accessed quickly by changing addresses on the local processor.

Message-Based Device (VXI)
An intelligent device that implements the defined VXIbus registers and communication protocols.

MFLOPS
Million FLoating-point Operations per Second; the unit for expressing the computational power of a processor.

MIPS
Million Instructions Per Second; the unit for expressing the speed of processor machine-code instructions.

MMI
Man-Machine Interface, also Human-Machine Interface; the means by which an operator interacts with an industrial automation system; often a GUI.

Module (VXI)
Typically, a board assembly and its associated mechanical parts, front panel, optional shields, and so on. A module contains everything required to occupy one or more slots in a mainframe.

Multifunction Card
A card used to both acquire and output analog values and to create digital signals.

Multitasking
A property of an operating system where several processes can be run simultaneously.

Mux
Multiplexer; a switching device with multiple inputs that sequentially connects each of its inputs to its output, typically at high speeds, in order to measure several signals with a single analog input channel.

MXIbus
Multisystem eXtension Interface bus; a multidrop, parallel bus architecture designed for high-speed communications between devices. It can extend the VXIbus across multiple mainframes and directly and transparently couple the VXIbus to industry-standard computers.

N

Noise
An undesirable electrical signal. Noise comes from external sources such as AC power lines, motors, generators, transformers, fluorescent lights, soldering irons, CRT displays, computers, electrical storms, welders, radio transmitters, and internal sources such as semiconductors, resistors, and capacitors.

Nyquist Sampling Theorem
A law of sampling theory stating that if a continuous bandwidth-limited signal contains no frequency components higher than half the frequency at which it is sampled, then the original signal can be recovered without distortion.

O

Object Technology
A broad term that refers to the use of objects to analyze, to model or design, or to implement some aspect of a computer system. In terms of actual application implementation (as opposed to object-oriented analysis or design), objects are self-contained software modes that encapsulate both data and processing logic, and they can be accessed only through well-defined interfaces.

OLE

Object Linking and Embedding; a set of system services that provides a means for applications to interact and interoperate. Based on the underlying Component Object Model, OLE is object-enabling system software. Through OLE Automation, an application can dynamically identify and use the services of other applications to build powerful solutions using packaged software. OLE also makes it possible to create compound documents consisting of multiple sources of information from different applications.

OLE Controls

See ActiveX Controls.

OPC

OLE for Process Control.

OpenDoc

A compound document architecture created by the joining of several technologies supplied by Apple (the base OpenDoc architecture, the Bento file system, and the Open Scripting Architecture) and IBM (the System Object Model).

Operating System

Base-level software that controls a computer, runs programs, interacts with users, and communicates with installed hardware or peripheral devices.

Operational Register (VXI)

Any device register not required for the system configuration process.

Optical Isolation

The technique of using an optoelectric transmitter and receiver to transfer data without electrical continuity in order to eliminate high-potential differences and transients.

Output Settling Time

The amount of time required for the analog output voltage to reach its final value within specified limits.

Output Slew Rate

The maximum rate of change of analog output voltage from one level to another.

Overhead

The amount of computer processing resources, such as time or memory, required to accomplish a task.

P

Paging
A technique used for extending the address range of a device to point into a larger address space.

PC Card
A credit-card-sized expansion card that fits in a PCMCIA slot — often referred to as a PCMCIA card.

PCI
Peripheral Component Interconnect; a high-performance expansion bus architecture originally developed by Intel to replace ISA and EISA. It is achieving widespread acceptance as a standard for PCs and workstations; it offers a theoretical maximum transfer rate of 132 Mbytes/s.

PCMCIA
Personal Computer Memory Card International Architecture; an expansion bus architecture that has found widespread acceptance as a de facto standard in notebook-size computers. It originated as a specification, written by the Personal Computer Memory Card International Association, for add-on memory cards.

Photoelectric Sensor
An electrical device that responds to a change in the intensity of the light falling upon it.

PID Control
A three-term control mechanism combining proportional, integral, and derivative control actions. Also see proportional control, integral control, and derivative control.

Pipeline
A high-performance processor structure in which the completion of an instruction is broken into its elements so that several elements can be processed simultaneously from different instructions.

PLC
Programmable Logic Controller; a highly reliable, special-purpose computer used in industrial monitoring and control applications. PLCs typically have proprietary programming and networking protocols, and special-purpose digital and analog I/O ports.

Plug&Play ISA
A specification prepared by Microsoft, Intel, and other PC-related companies that will result in PCs with plug-in boards that can be fully configured in software, without jumpers or switches on the boards.

Port
A communications connection on a computer or a remote controller.

Post-triggering
The technique used on a DAQ board to acquire a programmed number of samples after trigger conditions are met.

Potentiometer
An electrical device that can have its resistance manually adjusted; used for manual adjustment of electrical circuits and as a transducer for linear or rotary position.

Pre-triggering
The technique used on a DAQ board to keep a continuous buffer filled with data so that when the trigger conditions are met, the sample includes the data leading up to the trigger condition.

Programmed I/O
The standard method a CPU uses to access an I/O device; each byte of data is read or written by the CPU.

Propagation Delay
The amount of time required for a signal to pass through a circuit.

Proportional Control
A control action with an output that is to be proportional to the deviation of the controlled variable from a desired set point.

Protocol
The exact sequence of bits, characters, and control codes used to transfer data between computers and peripherals through a communications channel, such as the GPIB.

Proximity Sensor
A device that detects the presence of an object without physical contact. Most proximity sensors provide a digital on/off relay or digital output signal.

Q

Quantization
In digital signal processing, the process by which the amplitude of a signal sample is transformed to take only a limited number of values.

Quantization Error
The inherent uncertainty in digitizing an analog value due to the finite resolution of the conversion process.

R

Real Time
A property of an event or system in which data is processed as it is acquired instead of being accumulated and processed at a later time.

Register-Based Device (VXI)
A Servant-only device that has VXIbus configuration registers. Register-based devices are typically controlled by message-based devices via device-dependent register reads and writes.

Relative Accuracy
A measure in LSB of the accuracy of an Analog-to-Digital Converter (ADC). It includes all nonlinearity and quantization errors. It does not include offset and gain errors of the circuitry feeding the ADC.

Resolution
The smallest signal increment that can be detected by a measurement system. Resolution can be expressed in bits, in proportions, or in percent of full scale — e.g., 12-bit resolution, one part in 4,096 resolution, or 0.0244 percent of full scale.

Resource Locking
A technique whereby a device is signaled not to use its local memory when the memory is in use from the bus.

Resource Manager (VXI)
A message-based Commander, located at logical address 0, that provides configuration management services such as address map configuration, Commander and Servant mappings, and self-test and diagnostics management.

Responses (VXI)
Signals or interrupts generated by a device to notify another device of an asynchronous event. Responses contain the information in the Response register of a sender.

Ribbon Cable
A flat cable in which the conductors are side by side.

RTD
Resistance Temperature Detector; a metallic probe that measures temperature based upon its coefficient of resistivity.

RTSI Bus
Real-Time System Integration bus; the National Instruments timing bus that connects DAQ boards directly, by means of connectors on top of the boards, for precise synchronization of functions.

RTU
Remote Terminal Unit; an industrial data-collection device similar to a programmable logic controller (PLC), designed for location at a remote site, that communicates data to a host system by using telemetry (such as radio, dial-up telephone, or leased lines).

S

SCADA
Supervisory Control and Data Acquisition; a common PC function in process control applications, where programmable logic controllers perform control functions but are monitored and supervised by a PC.

SCPI
Standard Commands for Programmable Instruments; an extension of the IEEE 488.2 standard that defines a standard programming command set and syntax for device-specific operations.

SCXI
Signal Conditioning eXtensions for Instrumentation; the National Instruments product line for conditioning low-level signals within an external chassis near sensors so that only high-level signals are sent to DAQ boards in the noisy PC environment.

SE
Single-Ended; a term used to describe an analog input that is measured with respect to a common ground.

Self-Calibrating
A property of a DAQ board that has an extremely stable on-board reference and calibrates its own A/D and D/A circuits without manual adjustments by the user.

Sensor
A device that responds to a physical stimulus (heat, light, sound, pressure, motion, flow, and so on), and produces a corresponding electrical signal.

Servant (VXI)
A device controlled by a Commander; there are message-based and register-based Servants.

S/H
Sample-and-Hold; a circuit that acquires and stores an analog voltage on a capacitor for a short period of time.

Shared Memory
See Dual-Access Memory.

Shared Memory Protocol (VXI)
A communication protocol for message-based devices that uses an area of memory accessible to both. The protocol specifies connection and operation sequences to be followed by both devices.

Signal (VXI)
Any communication between message-based devices consisting of a write to a signal register.

Slot (VXI)
A position where a module can be inserted into a VXIbus backplane. Each slot provides the 96-pin J connectors to interface with the board P connectors. A slot can have one, two, or three connectors.

Slot 0 Device (VXI)
A VXIbus device that provides basic resources to VXI Slots 1 through 12. For B- and C-size systems, the resources provided are CLK10 and MODID. For a D-size system, the Slot 0 device also provides CLK100.

SNR
Signal-to-Noise Ratio; the ratio of the overall rms signal level to the rms noise level, expressed in dB.

Software Trigger
A programmed event that initiates an event such as data acquisition.

SPC
Statistical Process Control; a statistical analysis methodology in which characteristics of a process are measured or counted and then tracked. SPC is commonly used to evaluate, track, and improve the performance of a product-producing process.

SPDT
Single-Pole Double Throw; a property of a switch in which one terminal can be connected to one of two other terminals.

SQL
Structured Query Language; a nonprocedural computer language used to interact with databases.

SS
Simultaneous Sampling; a property of a system in which each input or output channel is digitized or updated at the same instant.

Stand-alone Program
A compiled program that runs with the operating system but without any other software programs or environments.

Strain Gauge
A sensor whose resistance is a function of the applied force.

Subroutine
A set of software instructions executed by a single line of code that may have input and/or output parameters. Examples of subroutines are `Call Mean (values, mean_value)`; `Call SendMessage (port, message, count)`.

Successive-Approximation ADC
An Analog-to-Digital Converter (ADC) that sequentially compares a series of binary-weighted values with an analog input to produce an output digital word in n steps, where n is the bit resolution of the ADC.

Synchronous
(1) Hardware — a property of an event that is synchronized to a reference clock. (2) Software — a property of a function that begins an operation and returns only when the operation is complete.

Syntax
The set of rules to which statements must conform in a particular programming language.

System (VXI)
One or more mainframes that are connected, all sharing a common Resource Manager. Each device in a system has a unique logical address.

System Noise
A measure of the amount of noise seen by an analog circuit or an Analog-to-Digital Converter when the analog inputs are grounded.

T

Talker
A device on the GPIB that sends information to a Listener on the bus.

TCP/IP
A set of standard protocols for communicating across a single network or inter-connected set of networks. There are three basic suites of protocols: The Internet Protocol (IP) for the low-level service of taking data and packaging of components, the Transmission Control Protocol (TCP) for high-reliability data transmissions, and the User Datagram Protocol (UDP) for low-overhead transmissions.

Test Executive
An application for automated sequencing of test programs. A test executive presents an operator interface for the testing process, determines pass/fail status, and logs test data.

T/H
Track-and-Hold; a circuit that tracks an analog voltage and holds the value on command.

THD
Total Harmonic Distortion; the ratio of the total rms signal due to harmonic distortion to the overall rms signal, in dB or percent.

Thermistor
A semiconductor sensor that exhibits a repeatable change in electrical resistance as a function of temperature. Most thermistors exhibit a negative temperature coefficient.

Thermocouple
A temperature sensor created by joining two dissimilar metals. The junction produces a small voltage as a function of the temperature.

Throughput Rate
The data, measured in bytes/s, for a given continuous operation, calculated to include software overhead.

Transducer
See Sensor.

Transfer Rate
The rate, measured in bytes/s, at which data is moved from source to destination after software initialization and setup operations; the maximum rate at which the hardware can operate.

U

UART
Universal Asynchronous Receiver/Transmitter.

Unipolar
A signal range that is always positive (for example, 0 to +10 V).

V

VDI
(Verein Deutscher Ingenieure) Association of German Engineers.

VI
Virtual Instrument; (1) A combination of hardware and/or software elements, typically used with a PC, that has the functionality of a classic stand-alone instrument; (2) a LabVIEW software module (VI) that consists of a front panel user interface and a block diagram program.

VISA
Virtual Instrumentation Software Architecture; a new driver software architecture developed by National Instruments to unify instrumentation software VXI, GPIB, and serial port. It has been accepted as a standard for VXI by the VXI*plug&play* Systems Alliance.

Visual Basic Custom Control (VBX)
A specific form of binary packaged object that can be created by different companies and integrated into applications written with Visual Basic.

VMM
Virtual multimeter.

VXI
VME (Versa-Modular Eurocard) eXtensions for Instrumentation; A very high-performance system for instrumentation based on an open industry standard.

VXIbus 488 Instrument (VXI)
A message-based device with the communication capability to use the 488-VXIbus interface.

VXIbus Subsystem (VXI)
A subsystem consisting of a central timing module, referred to as Slot 0, with up to 12 additional adjacent VXIbus modules. The VXIbus subsystem bus defines the lines on the P2 and P3 connectors.

VXI*plug&play* Systems Alliance (VXI)
A group of VXI developers dedicated to making VXI devices as easy to use as possible, primarily by simplifying software development.

W

WFS
Windows File Server; a server application allowing transputers to run software under MS Windows.

Word Serial (VXI)
The simplest required communication protocol used by message-based devices in the VXIbus system. It uses the A16 communication registers to transfer data with a simple polling handshake method.

Working Voltage
The highest voltage that should be applied to a product in normal use, normally well under the breakdown voltage for safety margin. See also Breakdown Voltage.

Z

Zero-Overhead Looping
The ability of a high-performance processor to repeat instructions without requiring time to branch to the beginning of the instructions.

Zero-Wait-State Memory
Memory fast enough that the processor does not have to wait.

Index

Timely, Technical Information for LabVIEW® Users!

Subscribe to LabVIEW Technical Resource! Receive a new issue each quarter full of technical solutions to get the most out of LabVIEW. Each issue includes a Resource Disk packed with VIs, Utilities and Source Code. Tools, Tips, Techniques, Tutorials, and more.

Written *by* LabVIEW programmers *for* LabVIEW programmers.

To order a subscription or Library of Back Issues on CD-ROM, please fill out the attached order form.

•tel 214.706.0LTR(0587)   •fax 214.706.0506   •email ltr@ltrpub.com

"The examples on the resource disk contain real solutions to real-world problems...no LabVIEW programmer should be without LTR!"

Jeffrey Travis - Creator of AppletVIEW® and co-author of "LabVIEW For Everyone"

"Please renew my subscription, which paid for itself many times last year!"

David Moschella - Pronto Product Development Corporation

"Saved me hours if not days of development time!"

Brad Hedstrom - Advanced Measurements

visit our website **WWW.LTRPUB.COM**

ORDER FORM

THE ONLY LABVIEW SUBSCRIPTION WITH VI SOFTWARE INCLUDED

TEL: 214-706-0587 FAX: 214-706-0506

WHAT IS LTR?

LabVIEW Technical Resource (LTR) is a quarterly journal for LabVIEW users and developers available by subscription from LTR Publishing, Inc. Each LTR issue presents powerful LabVIEW tips and techniques and includes a resource disk packed with VI source code, utilities, and documentation. Technical articles on LabVIEW programming methodology, in-depth tutorials, and time-saving tips and techniques address everyday programming issues in LabVIEW.

In its sixth year of publication, LTR has subscribers in over 45 countries and is well-known as a leading independent source of LabVIEW-specific information.

Purchase the LabVIEW Technical Resource CD-ROM Library of Back Issues and browse this searchable CD-ROM for easy access to over 100 articles and VIs from LTR Vol I-IV.

To subscribe to the LabVIEW Technical Resource or to order the CD-ROM Library of Back Issues, fax this form to LTR Publishing at **(214) 706-0506** or visit the LTR web page at **www.ltrpub.com** to download a free sample issue.

CONTACT INFORMATION

Name _____ Company _____

Address _____

City _____ State _____ Zip/Post Code _____

Tel (required) _____ FAX _____ E-mail _____

ORDER INFORMATION

QTY	MAC/PC	PRODUCT	U.S.	INTL.	EXTENDED PRICE
		1 year subscription (4 issues / 4 diskettes)	$95	$120	
		2 year subscription (8 issues / 8 diskettes)	$175	$215	
		CD-ROM library of back issues (15 issues / over 100 VIs)	$295	$320	
		Back issues – [Article Index available at www.ltrpub.com]	$25	$30	
		Download sample issue at **www.ltrpub.com**			

SUBTOTAL	
TX TAX @ 8.25%	
TOTAL	

PAYMENT INFORMATION

✔	PAYMENT METHOD	
	Check enclosed (U.S. BANK ONLY* – Make check payable to LTR Publishing) (Texas residents please add 8.25% sales tax)	
	Bill company / P.O. number required (U.S. Only)	
	Visa / MC Card Number	Exp.
	Signature	
	* Wire information available for international orders	

Fill out the form above and Fax it to: 214-706-0506 with your credit card information and signature,
OR fill out the form above and send order form with U.S. check to:
LTR Publishing, Inc., 6060 N. Central Expressway, Dallas, Texas 75206.

You may also include your own Federal Express or Airborne #. If you are ordering
a product for delivery within Texas, please include Texas Sales Tax at 8.25%

W W W . L T R P U B . C O M

LICENSE AGREEMENT AND LIMITED WARRANTY

READ THE FOLLOWING TERMS AND CONDITIONS CAREFULLY BEFORE OPENING THIS SOFTWARE MEDIA PACKAGE. THIS LEGAL DOCUMENT IS AN AGREEMENT BETWEEN YOU AND PRENTICE-HALL, INC. (THE "COMPANY"). BY OPENING THIS SEALED SOFTWARE MEDIA PACKAGE, YOU ARE AGREEING TO BE BOUND BY THESE TERMS AND CONDITIONS. IF YOU DO NOT AGREE WITH THESE TERMS AND CONDITIONS, DO NOT OPEN THE SOFTWARE MEDIA PACKAGE. PROMPTLY RETURN THE UNOPENED PACKAGE AND ALL ACCOMPANYING ITEMS TO THE PLACE YOU OBTAINED THEM FOR A FULL REFUND OF ANY SUMS YOU HAVE PAID.

1. **GRANT OF LICENSE:** In consideration of your payment of the license fee, which is part of the price you paid for this product, and your agreement to abide by the terms and conditions of this Agreement, the Company grants to you a nonexclusive right to use and display the copy of the enclosed software program (hereinafter the "SOFTWARE") on a single computer (i.e., with a single CPU) at a single location so long as you comply with the terms of this Agreement. The Company reserves all rights not expressly granted to you under this Agreement.

2. **OWNERSHIP OF SOFTWARE:** You own only the magnetic or physical media (the enclosed CD-ROM) on which the SOFTWARE is recorded or fixed, but the Company retains all the rights, title, and ownership to the SOFTWARE recorded on the original CD-ROM copy(ies) and all subsequent copies of the SOFTWARE, regardless of the form or media on which the original or other copies may exist. This license is not a sale of the original SOFTWARE or any copy to you.

3. **COPY RESTRICTIONS:** This SOFTWARE and the accompanying printed materials and user manual (the "Documentation") are the subject of copyright. You may not copy the Documentation or the SOFTWARE, except that you may make a single copy of the SOFTWARE for backup or archival purposes only. You may be held legally responsible for any copying or copyright infringement which is caused or encouraged by your failure to abide by the terms of this restriction.

4. **USE RESTRICTIONS:** You may not network the SOFTWARE or otherwise use it on more than one computer or computer terminal at the same time. You may physically transfer the SOFTWARE from one computer to another provided that the SOFTWARE is used on only one computer at a time. You may not distribute copies of the SOFTWARE or Documentation to others. You may not reverse engineer, disassemble, decompile, modify, adapt, translate, or create derivative works based on the SOFTWARE or the Documentation without the prior written consent of the Company.

5. **TRANSFER RESTRICTIONS:** The enclosed SOFTWARE is licensed only to you and may not be transferred to any one else without the prior written consent of the Company. Any unauthorized transfer of the SOFTWARE shall result in the immediate termination of this Agreement.

6. **TERMINATION:** This license is effective until terminated. This license will terminate automatically without notice from the Company and become null and void if you fail to comply with any provisions or limitations of this license. Upon termination, you shall destroy the Documentation and all copies of the SOFTWARE. All provisions of this Agreement as to warranties, limitation of liability, remedies or damages, and our ownership rights shall survive termination.

7. **MISCELLANEOUS:** This Agreement shall be construed in accordance with the laws of the United States of America and the State of New York and shall benefit the Company, its affiliates, and assignees.

8. **LIMITED WARRANTY AND DISCLAIMER OF WARRANTY:** The Company warrants that the SOFTWARE, when properly used in accordance with the Documentation, will operate in substantial conformity with the description of the SOFTWARE set forth in the Documentation. The Company does not warrant that the SOFTWARE will meet your requirements or that the operation of the SOFTWARE will be uninterrupted or error-free. The Company warrants that the

media on which the SOFTWARE is delivered shall be free from defects in materials and workmanship under normal use for a period of thirty (30) days from the date of your purchase. Your only remedy and the Company's only obligation under these limited warranties is, at the Company's option, return of the warranted item for a refund of any amounts paid by you or replacement of the item. Any replacement of SOFTWARE or media under the warranties shall not extend the original warranty period. The limited warranty set forth above shall not apply to any SOFTWARE which the Company determines in good faith has been subject to misuse, neglect, improper installation, repair, alteration, or damage by you. EXCEPT FOR THE EXPRESSED WARRANTIES SET FORTH ABOVE, THE COMPANY DISCLAIMS ALL WARRANTIES, EXPRESS OR IMPLIED, INCLUDING WITHOUT LIMITATION, THE IMPLIED WARRANTIES OF MERCHANTABILITY AND FITNESS FOR A PARTICULAR PURPOSE. EXCEPT FOR THE EXPRESS WARRANTY SET FORTH ABOVE, THE COMPANY DOES NOT WARRANT, GUARANTEE, OR MAKE ANY REPRESENTATION REGARDING THE USE OR THE RESULTS OF THE USE OF THE SOFTWARE IN TERMS OF ITS CORRECTNESS, ACCURACY, RELIABILITY, CURRENTNESS, OR OTHERWISE.

IN NO EVENT, SHALL THE COMPANY OR ITS EMPLOYEES, AGENTS, SUPPLIERS, OR CONTRACTORS BE LIABLE FOR ANY INCIDENTAL, INDIRECT, SPECIAL, OR CONSEQUENTIAL DAMAGES ARISING OUT OF OR IN CONNECTION WITH THE LICENSE GRANTED UNDER THIS AGREEMENT, OR FOR LOSS OF USE, LOSS OF DATA, LOSS OF INCOME OR PROFIT, OR OTHER LOSSES, SUSTAINED AS A RESULT OF INJURY TO ANY PERSON, OR LOSS OF OR DAMAGE TO PROPERTY, OR CLAIMS OF THIRD PARTIES, EVEN IF THE COMPANY OR AN AUTHORIZED REPRESENTATIVE OF THE COMPANY HAS BEEN ADVISED OF THE POSSIBILITY OF SUCH DAMAGES. IN NO EVENT SHALL LIABILITY OF THE COMPANY FOR DAMAGES WITH RESPECT TO THE SOFTWARE EXCEED THE AMOUNTS ACTUALLY PAID BY YOU, IF ANY, FOR THE SOFTWARE.

SOME JURISDICTIONS DO NOT ALLOW THE LIMITATION OF IMPLIED WARRANTIES OR LIABILITY FOR INCIDENTAL, INDIRECT, SPECIAL, OR CONSEQUENTIAL DAMAGES, SO THE ABOVE LIMITATIONS MAY NOT ALWAYS APPLY. THE WARRANTIES IN THIS AGREEMENT GIVE YOU SPECIFIC LEGAL RIGHTS AND YOU MAY ALSO HAVE OTHER RIGHTS WHICH VARY IN ACCORDANCE WITH LOCAL LAW.

ACKNOWLEDGMENT

YOU ACKNOWLEDGE THAT YOU HAVE READ THIS AGREEMENT, UNDERSTAND IT, AND AGREE TO BE BOUND BY ITS TERMS AND CONDITIONS. YOU ALSO AGREE THAT THIS AGREEMENT IS THE COMPLETE AND EXCLUSIVE STATEMENT OF THE AGREEMENT BETWEEN YOU AND THE COMPANY AND SUPERSEDES ALL PROPOSALS OR PRIOR AGREEMENTS, ORAL, OR WRITTEN, AND ANY OTHER COMMUNICATIONS BETWEEN YOU AND THE COMPANY OR ANY REPRESENTATIVE OF THE COMPANY RELATING TO THE SUBJECT MATTER OF THIS AGREEMENT.

Should you have any questions concerning this Agreement or if you wish to contact the Company for any reason, please contact in writing at the address below.

Robin Short
Prentice Hall PTR
One Lake Street
Upper Saddle River, New Jersey 07458

About the CD-ROM

The CD-ROM included with this book contains an evaluation version of LabVIEW for Windows 98/95/NT and PowerPC (Mac OS). If you do not have the full version of LabVIEW already installed on your computer, you can use the evaluation version.

How to Install the Software

Install the LabVIEW evaluation software on your computer. To install this software, run the setup.exe program from the LabVIEW folder on the CD-ROM. Follow the instructions on the screen. If you already have the full version of LabVIEW installed, you do not need to install the evaluation version.

System Requirements

Mac OS (PowerPC only)   12 MB RAM, 16 MB recommended
                               Mac OS 7.1.2 or later
                               75 MB disk space for full installation
                               CD ROM Drive

Windows NT   16 MB RAM, 32 MB recommended
                               486 DX, Pentium recommended
                               Windows NT 4.0
                               75 MB disk space for full installation
                               CD ROM Drive

Windows 95   16 MB RAM, 32 MB recommended
                               486 DX, Pentium recommended
                               75 MB disk space for full installation
                               CD ROM Drive

IN ALL THE ABOVE CASES, TO START THE INSTALLATION YOU MUST HAVE **100 MB** FREE ON YOUR HARD DRIVE EVEN THOUGH AFTER INSTALLATION IT MAY NOT REQUIRE 100 MB.

Restrictions on the LabVIEW Evaluation Version

To run the evaluation version, launch the Labview.exe program from the folder in which you installed LabVIEW. Select the Exit to LabVIEW button in the lower right corner. This opens a window which gives you general information about LabVIEW and National Instruments. After reading this information, click on the OK button. This will then open the LabVIEW window. You can access all the features of the full version in the evaluation version with the following restrictions:

1. After you have installed the evaluation version, it will expire in 30 days. If you want to use the evaluation version after this time, you must reinstall the software.

2. A VI will execute for 5 minutes at any one time. After 5 minutes, you will have to run the VI again. In a single session, the total execution time is limited to 60 minutes. You can edit the block diagrams and front panels after this time-out. However, in order to run any VI, you will have to start LabVIEW again.

3. The evaluation version does not come with tools to build your own Code Interface Nodes (CINs) or with tools to build DLLs that can call back into LabVIEW.

Technical Support

Prentice Hall does not offer technical support for this software. However, if there is a problem with the media, you may obtain a replacement copy by e-mailing us with your problem at:

disc_exchange@prenhall.com